London Mathematical Society Student Texts 19

# Steps in Commutative Algebra

R. Y. SHARP
Department of Pure Mathematics,
University of Sheffield

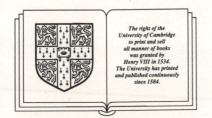

The right of the
University of Cambridge
to print and sell
all manner of books
was granted by
Henry VIII in 1534.
The University has printed
and published continuously
since 1584.

CAMBRIDGE UNIVERSITY PRESS

Cambridge

New York   Port Chester

Melbourne   Sydney

# LONDON MATHEMATICAL SOCIETY STUDENT TEXTS

Managing editor: Professor E.B. Davies, Department of Mathematics,
King's College, Strand, London WC2R 2LS

To my parents

Published by the Press Syndicate of the University of Cambridge
The Pitt Building, Trumpington Street, Cambridge CB2 1RP
32 East 57th Street, New York, NY 10022, USA
10 Stamford Road, Oakleigh, Melbourne 3166, Australia

© Cambridge University Press 1990

First published 1990

Printed in Great Britain at the University Press, Cambridge

*Library of Congress cataloging in publication data: available*

*British Library cataloguing in publication data available*

ISBN 0 521 39338 8 hardback
ISBN 0 521 39732 4 paperback

# Contents

# Preface

Why write another introductory book on commutative algebra? As there are so many good books already available on the subject, that seems to be a very pertinent question.

This book has been written to try to persuade more young people to study commutative algebra by providing 'stepping stones' to help them into the subject. Many of the existing books on commutative algebra, such as M. F. Atiyah's and I. G. Macdonald's [1] and H. Matsumura's [8], require a level of experience and sophistication on the part of the reader which is rather beyond what is achieved nowadays in a mathematics undergraduate degree course at some British universities. This is sad, for students often find some undergraduate topics in ring theory, such as unique factorization in Euclidean domains, attractive, but this undergraduate study does leave something of a gap which needs to be bridged before the student can approach the established books on commutative algebra with confidence. This is an attempt to help to bridge that gap.

For definiteness, I have assumed that the reader's knowledge of commutative ring theory is limited to the contents of the book 'Rings and factorization' [15] by my colleague David Sharpe. Thus the typical reader I have had in mind while writing this book would be either a final year undergraduate or first year postgraduate student at a British university whose appetite for commutative ring theory has been whetted by a course like that provided by [15], but whose experience (apart from some basic linear algebra and vector space theory) does not reach much beyond that. It should be emphasized that, for a reader who has these prerequisites at his or her fingertips, this book is largely self-contained.

Experienced workers in commutative algebra will probably find that the book makes slow progress; but then, the book has not been written for them! For example, as [15] does not work with ideals, this topic is introduced from scratch, and not until Chapter 2; modules are not studied until Chapter 6; there is a digression in Chapter 10 to discuss finitely generated modules over a principal ideal domain, in the hope that this will help to strengthen readers' experience in the techniques introduced earlier in the book; the ideas of Chapter 10 are applied in Chapter 11 to the study of canonical forms for square matrices over fields; and the theory of transcendence degrees of field extensions is developed in Chapter 12, for use in connection with the dimension theory of finitely generated commutative algebras over fields. Otherwise, the topics included are central ones for commutative algebra.

The hope is that a reader who completes this book will feel inspired and encouraged to turn to a more advanced book on commutative algebra, such as H. Matsumura's [8]. It must be emphasized that the present book will not in itself provide complete preparation, because it does not include any introduction to the homological algebra of the functors Ext and Tor, and a good understanding of these is highly desirable for the serious student of commutative algebra. The student will have to turn elsewhere for these, and even for the theory of tensor products. The latter have been avoided in this book because the risk of putting youthful readers off with unnecessary technicalities at an early stage (after all, one can do a lot of commutative algebra without tensor products) seemed to outweigh the advantages that would be gained by having them available.

The reader's attention is drawn to possible further avenues of study, such as tensor products, homological algebra, applications of the Nullstellensatz to algebraic geometry, or even, in Chapter 12, to the Galois theory and ruler and compass constructions which often feature in books on field theory, in items called 'Further Steps' towards the ends of some of the later chapters. No attempt has been made in this book to provide historical background comments, as I do not feel able to add anything to the comments of this type which are already available in the books listed in the Bibliography.

A feature of the book is the large number of exercises included, not only at the ends of chapters, but also throughout the text. These range from the routine checking of easy properties to some quite tricky problems; in some cases, a linked series of problems form almost a 'mini-project', in which a line of development related, but peripheral, to the work of the chapter, is explored. Some of the exercises are needed for the main development later in the book, and these exercises are marked with a symbol '♯': those so marked which, in my opinion, seemed among the harder ones have been provided with hints. (Exercises which are used later in the book but only in other exercises have not been marked with a '♯'; also, there are some unmarked exercises which are not particularly easy but have not been provided with hints.) It is hoped that this policy will help the reader to make steady progress through the book without becoming seriously held up on details which are important for the subsequent work. Indeed, I have tried to provide very full and complete arguments for all the proofs presented in the book, in the hope that this will enable the reader to develop his or her own expertise: there are plenty of substantial exercises available for consolidation of that expertise.

The material included (none of which is new or due to me) has been selected to try to give prominence to topics which I found exciting when I was a postgraduate student. In this connection, I would like to record here my gratitude to three British mathematicians who considerably influenced my own development, namely Ian Macdonald, who first excited my interest in commutative algebra (in an Oxford lecture course which was a forerunner to the book [1]), Douglas Northcott, from whose writings I have benefited greatly over the years, and David Rees, whose infectious enthusiasm for local ring theory has often been a source of inspiration to me.

The presentation of the material in the book reflects my experiences in teaching postgraduate students at the University of Sheffield, both through MSc lecture courses and through reading programmes for beginning PhD students, over the past 15 years. Most of the presentation has grown out of MSc lectures gradually refined over many years. Preliminary versions of most of the chapters have been tried out on classes of postgraduate students at Sheffield University during the sessions 1988-89 and 1989-90, and I am grateful to those students for acting as 'guinea pigs', so to speak. I would particularly like to thank Ian Staniforth and Paul Tierney, whose eagle eyes spotted numerous misprints in preliminary versions.

I would also like to thank David Tranah, the Senior Mathematics Editor of Cambridge University Press, for his continual encouragement over many years, without which this book might never have been completed; Chris Martin of Sheffield University Computing Services and my colleague Mike Piff for their patient advice over many months which has helped to make the world of computers less daunting for me than it was at the outset of this project; and my wife, Alice Sharp, not only for many things which have nothing to do with mathematics, but also for casting her professional mathematical copy-editor's eye over preliminary versions of this book and providing very helpful advice on the layout of the material.

Rodney Y. Sharp
Sheffield
April 1990

# Chapter 1

# Commutative rings and subrings

This book is designed for students who have followed an elementary undergraduate course on commutative ring theory, such as that covered in D. W. Sharpe's little book [15], and who wish to learn more about the subject. The aim of the book is to assist the reader to attain a level of competence in the introductory aspects of commutative algebra sufficient to enable him or her to begin with confidence the study of a more advanced book on the subject, such as H. Matsumura's [8].

We begin by introducing some of the notation that will be used throughout this book.

**1.1** NOTATION. The symbol $\mathbb{Z}$ will always denote the set of integers; in addition, $\mathbb{N}$ (respectively $\mathbb{N}_0$) will always denote the set of positive (respectively non-negative) integers. The set of rational (respectively real, complex) numbers will be denoted by the symbol $\mathbb{Q}$ (respectively $\mathbb{R}$, $\mathbb{C}$).

The symbol $\subseteq$ will stand for 'is a subset of'; the symbol $\subset$ will be reserved to denote strict inclusion. Thus, for sets $A, B$, the expression $A \subset B$ means that $A \subseteq B$ and $A \neq B$.

The symbol '$\square$' will be used to denote the end, or absence, of a proof.

We shall reserve the symbols

$$X, Y, X_1, \ldots, X_n$$

to denote indeterminates.

We shall denote the number of elements in a finite set $\Omega$ by $|\Omega|$.

A comment should perhaps be made about the distinction between a family and a set. We shall often use round parentheses ( ), as in $(a_i)_{i \in I}$,

1

to denote a family indexed by the set $I$; here $a_i$ should be thought of as situated in the 'position' labelled by $i$; and the family $(a_i)_{i \in I}$ is considered to be equal to $(b_i)_{i \in I}$ if and only if $a_i = b_i$ for all $i \in I$. One can think of a family $(a_i)_{i \in I}$, where $a_i$ lies in the set $A$ for all $i \in I$, as a function from $I$ to $A$: in this interpretation, the image of $i$ under the function is $a_i$.

On the other hand, curly braces { }, as in

$$\{d_1, \ldots, d_n\} \qquad \text{or} \qquad \{d \in D : \text{statement } P(d) \text{ is true}\},$$

will often be used to indicate sets. A set is completely determined by its members, and no concept of 'position' is involved when the members of the set are displayed within braces. The distinction between a family and a set parallels that between a function and its image. To illustrate the distinction, let $d_1 = d_2 = 1$ and $d_3 = 3$. Then the family $(d_i)_{i=1}^3$ can be thought of as the ordered triple $(1, 1, 3)$, whereas the set $\{d_1, d_2, d_3\}$ is just the 2-element set $\{1, 1, 3\} = \{1, 3\}$.

As we are going to regard the contents of [15] as typical preparation for the study of this book, we shall in the main follow the terminology of [15]. In particular, all the rings we study will have multiplicative identity elements. To be precise, by a *ring* we shall mean a set, $R$ say, furnished with two laws of composition, addition and multiplication, such that $R$ is an Abelian group with respect to addition, multiplication is associative and both right and left distributive over addition, and $R$ contains a multiplicative identity element $1_R$ (or simply 1) such that

$$1_R r = r = r 1_R \qquad \text{for all } r \in R.$$

If, in addition, the multiplication in $R$ is commutative, then we shall say that $R$ is a *commutative ring*. Virtually all the rings we shall study in this book will be commutative, although occasionally we shall focus attention on certain commutative subrings of rings which might not be commutative, such as endomorphism rings of modules. Thus we shall occasionally have to refer to non-commutative rings, and for this reason the word 'commutative' will always be inserted at appropriate places in hypotheses.

The reader should have a substantial fund of examples of commutative rings at his or her disposal, and we review some familiar examples now. We use this opportunity to introduce some more of the notation that will be employed in this book.

**1.2** EXAMPLES. (i) The ring of integers $\mathbf{Z}$ is an example of a commutative ring.

(ii) The ring of Gaussian integers will be denoted by $\mathbf{Z}[i]$. See [15, p. 18]. The ring $\mathbf{Z}[i]$ consists of all complex numbers of the form $a + ib$ where

$a, b \in \mathbf{Z}$, and the ring operations are ordinary addition and multiplication of complex numbers. This is, of course, an example of a commutative ring.

(iii) Let $n$ be an integer with $n > 1$. The ring of residue classes of integers modulo $n$ will (sometimes) be denoted by $\mathbf{Z}_n$. See [15, 1.7]. This ring has exactly $n$ elements, and so is an example of a finite commutative ring.

(iv) Another example of a commutative ring is given by the set $C[0,1]$ of all continuous real-valued functions defined on the closed interval $[0,1]$. See [15, p. 8]. In this ring, the operations of addition and multiplication are defined 'pointwise': thus, for $f, g \in C[0,1]$ we define $f + g$ and $fg$ by the rules

$$(f + g)(x) = f(x) + g(x) \quad \text{for all } x \in [0,1]$$

and

$$(fg)(x) = f(x)g(x) \quad \text{for all } x \in [0,1].$$

**1.3** REMARK. Let $R$ be a commutative ring. In our definition, there is no requirement that the multiplicative identity element $1_R$ of $R$ should be different from its zero element $0_R$ (or 0). (This is one way in which our approach differs from that of [15].) A ring $R$ in which $1_R = 0_R$ is called a *trivial* ring; such a ring consists of just one, necessarily zero, element.

Let $R$ be a commutative ring. Two new commutative rings which can be constructed from $R$ are the ring $R[X]$ of polynomials in the indeterminate $X$ with coefficients in $R$, and the ring $R[[X]]$ of formal power series in $X$ with coefficients in $R$. As both these methods of constructing new commutative rings from old are absolutely fundamental to the subject matter of this book, it is appropriate for us to review the ideas involved here. It is expected that the review will be revision (both $R[X]$ and $R[[X]]$ are discussed in [15]); this means that it is reasonable to take the neat approach of discussion of $R[[X]]$ before $R[X]$.

A typical element of $R[[X]]$ is a 'formal power series'

$$a_0 + a_1 X + \cdots + a_n X^n + \cdots,$$

where the coefficients $a_0, a_1, \ldots, a_n, \ldots \in R$. (For each non-negative integer $n$, we refer to $a_n$ as the *n-th coefficient* of the above formal power series.) Even though the symbol '+' is used, the reader should not think that, at this elementary stage, an addition is involved: the above expression is really just a convenient notation for the infinite sequence

$$(a_0, a_1, \ldots, a_n, \ldots),$$

and the alternative notation $\sum_{i=0}^{\infty} a_i X^i$ is preferable from some points of view.

Two formal power series $\sum_{i=0}^{\infty} a_i X^i$ and $\sum_{i=0}^{\infty} b_i X^i$ in $R[[X]]$ are considered equal precisely when $a_i = b_i$ for all integers $i \geq 0$. Addition and multiplication in $R[[X]]$ are defined as follows: for all $\sum_{i=0}^{\infty} a_i X^i$, $\sum_{i=0}^{\infty} b_i X^i \in R[[X]]$,

$$\sum_{i=0}^{\infty} a_i X^i + \sum_{i=0}^{\infty} b_i X^i = \sum_{i=0}^{\infty} (a_i + b_i) X^i,$$

and

$$\left( \sum_{i=0}^{\infty} a_i X^i \right) \left( \sum_{j=0}^{\infty} b_j X^j \right) = \sum_{k=0}^{\infty} c_k X^k,$$

where, for each integer $k \geq 0$,

$$c_k = a_0 b_k + a_1 b_{k-1} + \cdots + a_k b_0.$$

With these definitions, it turns out that $R[[X]]$ is a commutative ring, with zero element $\sum_{i=0}^{\infty} 0 X^i$ (abbreviated to 0, of course) and identity element

$$1 + 0X + \cdots + 0X^n + \cdots.$$

The subset of $R[[X]]$ consisting of all formal power series $\sum_{i=0}^{\infty} a_i X^i \in R[[X]]$ in which only finitely many of the coefficients $a_i$ are non-zero is also a commutative ring with respect to the above operations, having the same identity element as $R[[X]]$. It is called the *ring of polynomials in X with coefficients in R* and is denoted by $R[X]$. It is customary to omit a 'term' $a_n X^n$ from the formal expression

$$a_0 + a_1 X + \cdots + a_i X^i + \cdots$$

for a formal power series or polynomial when the coefficient $a_n$ is zero. Thus, with this convention, a typical polynomial in $R[X]$ has the form

$$a_0 + a_1 X + \cdots + a_d X^d$$

for some non-negative integer $d$, where $a_0, \ldots, a_d \in R$, and furthermore the '+' signs in the above expression really can now be interpreted as standing for addition. If we have $a_d \neq 0$ here, then we say that $d$ is the *degree* of the above polynomial. We define the degree of the zero polynomial to be $-\infty$.

With the convention just introduced, $R$ itself is regarded as a subset of $R[X]$ and of $R[[X]]$. It is time we had the concept of subring at our disposal.

**1.4 DEFINITION.** A subset $S$ of a ring $R$ is said to be a *subring* of $R$ precisely when $S$ is itself a ring with respect to the operations in $R$ and $1_S = 1_R$, that is, the multiplicative identity of $S$ is equal to that of $R$.

It should be clear to the reader that, if $R$ is a commutative ring, and $X$ is an indeterminate, then $R$ is a subring of $R[X]$ and also a subring of $R[[X]]$, and $R[X]$ is a subring of $R[[X]]$.

There is a simple criterion for a subset of a ring $R$ to be a subring of $R$.

**1.5** THE SUBRING CRITERION. (See [15, Theorem 1.4.4].) *Let $R$ be a ring and let $S$ be a subset of $R$. Then $S$ is a subring of $R$ if and only if the following conditions hold:*

(i) *$1_R \in S$;*
(ii) *whenever $a, b \in S$, then $a + b \in S$;*
(iii) *whenever $a \in S$, then $-a \in S$;*
(iv) *whenever $a, b \in S$, then $ab \in S$.* □

The notion of subring leads naturally to the concept of ring homomorphism.

**1.6** DEFINITION. Let $f : R \to S$ be a mapping from the ring $R$ to the ring $S$. Then $f$ is said to be a *homomorphism* (or *ring homomorphism* ) precisely when

(i) $f(a + b) = f(a) + f(b)$ for all $a, b \in R$,
(ii) $f(ab) = f(a)f(b)$ for all $a, b \in R$, and
(iii) $f(1_R) = 1_S$.

A bijective ring homomorphism is called an *isomorphism* (or *ring isomorphism*).

For example, if $R'$ is a subring of the ring $R$, then the inclusion mapping $i : R' \to R$ is an injective ring homomorphism. In fact, there are many situations where we use an injective ring homomorphism $f : T \to R$ from a ring $T$ to a ring $R$ to identify elements of $T$ as elements of $R$.

**1.7** ♯EXERCISE. Let $R, S$ be rings, and let $f : R \to S$ be an isomorphism of rings. Prove that the inverse mapping

$$f^{-1} : S \to R$$

is also a ring isomorphism.

In view of this result, we say, if there is a ring isomorphism from $R$ to $S$, that $R$ and $S$ are *isomorphic* rings, and we write $R \cong S$.

**1.8** LEMMA. (See [15, Theorem 1.4.5].) *Let $f : R \to S$ be a homomorphism of rings. Then $\operatorname{Im} f$, the image of $f$, is a subring of $S$.* □

**1.9** DEFINITION. Let $R$ be a commutative ring. By an *$R$-algebra* we shall mean a ring $S$ endowed with a ring homomorphism $f : R \to S$. Thus the homomorphism $f$ is to be regarded as part of the structure of the $R$-algebra $S$. When we have this situation, it is automatic that $S$ is an algebra over its subring $\operatorname{Im} f$ by virtue of the inclusion homomorphism.

We should point out at once that the concept of $R$-algebra introduced in 1.9 above occurs very frequently in ring theory, simply because every ring is automatically a $\mathbb{Z}$-algebra. We explain in 1.10 why this is the case.

**1.10** REMARK. Let $R$ be a ring. Then the mapping $f : \mathbb{Z} \to R$ defined by $f(n) = n(1_R)$ for all $n \in \mathbb{Z}$ is a ring homomorphism, and, in fact, is the only ring homomorphism from $\mathbb{Z}$ to $R$.

Here,

$$
n(1_R) = \begin{cases}
1_R + \cdots + 1_R & (n \text{ terms}) & \text{for } n > 0, \\
0_R & & \text{for } n = 0, \\
(-1_R) + \cdots + (-1_R) & (-n \text{ terms}) & \text{for } n < 0.
\end{cases}
$$

It should be clear from 1.5 that the intersection of the members of any non-empty family of subrings of a ring $R$ is again a subring of $R$. This observation leads to the following lemma. Before we state it, it is appropriate to point out the convention whereby, for $a \in R$, the symbol $a^0$ is interpreted as $1_R$.

**1.11** LEMMA. *Let $S$ be a subring of the ring $R$, and let $\Gamma$ be a subset of $R$. Then $S[\Gamma]$ is defined to be the intersection of all subrings of $R$ which contain both $S$ and $\Gamma$. (There certainly is one such subring, namely $R$ itself.) Thus $S[\Gamma]$ is a subring of $R$ which contains both $S$ and $\Gamma$, and it is the smallest such subring of $R$ in the sense that it is contained in every other subring of $R$ that contains both $S$ and $\Gamma$.*

*In the special case in which $\Gamma$ is a finite set $\{\alpha_1, \alpha_2, \ldots, \alpha_n\}$, we write $S[\Gamma]$ as $S[\alpha_1, \alpha_2, \ldots, \alpha_n]$.*

*In the special case in which $S$ is commutative, and $\alpha \in R$ is such that $\alpha s = s\alpha$ for all $s \in S$, we have*

$$
S[\alpha] = \left\{ \sum_{i=0}^{t} s_i \alpha^i : t \in \mathbb{N}_0, \ s_0, \ldots, s_t \in S \right\}.
$$

*Proof.* Only the claim in the last paragraph still requires proof. For this, let

$$
H = \left\{ \sum_{i=0}^{t} s_i \alpha^i : t \in \mathbb{N}_0, \ s_0, \ldots, s_t \in S \right\}.
$$

Since $S$ is commutative and $\alpha s = s\alpha$ for all $s \in S$, it is clear from the Subring Criterion 1.5 that $H$ is a subring of $R$; it also contains $S$ and $\alpha = (1_S)\alpha$. Hence

$$
S[\alpha] \subseteq H.
$$

On the other hand, it is clear that $H$ must be contained in every subring of $R$ which contains both $S$ and $\alpha$. Hence $S[\alpha] = H$. $\square$

Note that, when $R$ is a commutative ring and $X$ is an indeterminate, then it follows from 1.11 that our earlier use of $R[X]$ to denote the polynomial ring is consistent with this new use of $R[X]$ to denote 'ring adjunction'. A similar comment applies to our earlier notation $\mathbf{Z}[i]$ (of 1.2(ii)) for the ring of Gaussian integers: of course, the set $\mathbf{C}$ of all complex numbers is a ring with respect to ordinary addition and multiplication of complex numbers, and, since $i^2 = -1$, the ring of Gaussian integers is the smallest subring of $\mathbf{C}$ which contains both $\mathbf{Z}$ and $i$.

**1.12** ♯EXERCISE. Let $S$ be a subring of the commutative ring $R$, and let $\Gamma$, $\Delta$ be subsets of $R$. Show that $S[\Gamma \cup \Delta] = S[\Gamma][\Delta]$, and

$$S[\Gamma] = \bigcup_{\Omega \subseteq \Gamma, \ |\Omega| < \infty} S[\Omega].$$

(Here is a hint: show that the right-hand side in the above display is a subring of $R$ which contains both $S$ and $\Gamma$.)

The polynomial ring $R[X]$, where $R$ is a commutative ring, has the 'universal property' described in the following lemma.

**1.13** LEMMA. *Let $R$ be a commutative ring, and let $X$ be an indeterminate; let $T$ be a commutative $R$-algebra with structural ring homomorphism $f : R \to T$; and let $\alpha \in T$. Then there is a unique ring homomorphism $f_1 : R[X] \to T$ which extends $f$ (that is, is such that $f_1|_R = f$) and satisfies $f_1(X) = \alpha$.*

*Proof.* If $f_1 : R[X] \to T$ were a ring homomorphism which extends $f$ and satisfies $f_1(X) = \alpha$, then it would have to satisfy $f_1(rX^i) = f(r)\alpha^i$ for $r \in R$ and $i \in \mathbf{N}_0$, and it follows that the only possible candidate for $f_1$ is the mapping defined by

$$f_1\left(\sum_{i=0}^{n} r_i X^i\right) = \sum_{i=0}^{n} f(r_i)\alpha^i$$

for all $n \in \mathbf{N}_0$, $r_0, \ldots, r_n \in R$. It is completely straightforward to check that this mapping does indeed have all the desired properties. $\square$

Consider again the ring of polynomials $R[X]$ in the indeterminate $X$ with coefficients in the commutative ring $R$ (we sometimes say 'ring of polynomials over $R'$). What happens if we form the ring of polynomials over $R[X]$ in another indeterminate $Y$? The new ring can be denoted by $R[X][Y]$, and, in view of 1.12, also by $R[X, Y]$; but what can we say about its elements?

A typical element of $R[X][Y]$ has the form

$$f_0 + f_1 Y + \cdots + f_n Y^n$$

for some $n \in \mathbf{N}_0$ and $f_0, \ldots, f_n \in R[X]$, and so can be expressed as a finite sum of expressions of the form

$$r_{ij} X^i Y^j,$$

where $i, j \in \mathbf{N}_0$, $r_{ij} \in R$. Moreover, it is easy to see that an expression of the form

$$\sum_{i=0}^{n} \sum_{j=0}^{m} s_{ij} X^i Y^j$$

in $R[X][Y]$, where $n, m \in \mathbf{N}_0$ and $s_{ij} \in R$ for $i = 0, \ldots, n$, $j = 0, \ldots, m$, is zero if and only if $s_{ij} = 0$ for all $i = 0, \ldots, n$ and $j = 0, \ldots, m$. We describe this property of $X$ and $Y$ by saying that they are 'algebraically independent' over $R$.

The above ideas can easily be extended from 2 to any finite number of indeterminates.

**1.14** DEFINITION. Let $R$ be a commutative ring, and let $\alpha_1, \ldots, \alpha_n \in R$; let $R_0$ be a subring of $R$. Then $\alpha_1, \ldots, \alpha_n$ are said to be *algebraically independent over $R_0$* (strictly speaking, we should say *the family $(\alpha_i)_{i=1}^n$ is algebraically independent over $R_0$*) precisely when the following condition is satisfied: whenever $\Lambda$ is a finite subset of $\mathbf{N}_0{}^n$ and elements

$$r_{i_1, \ldots, i_n} \in R_0 \qquad ((i_1, \ldots, i_n) \in \Lambda)$$

are such that

$$\sum_{(i_1, \ldots, i_n) \in \Lambda} r_{i_1, \ldots, i_n} \alpha_1^{i_1} \ldots \alpha_n^{i_n} = 0,$$

then $r_i = 0$ for all $i \in \Lambda$.

**1.15** REMARK. Let $R$ be a commutative ring and let $n$ be a positive integer. Form polynomial rings successively by defining $R_0 = R$, $R_i = R_{i-1}[X_i]$ for $i = 1, \ldots, n$, where $X_1, \ldots, X_n$ are indeterminates. Then
  (i) $R_n = R[X_1, \ldots, X_n]$;
  (ii) $X_1, \ldots, X_n$ are algebraically independent over $R$;

(iii) a typical element of $R_n$ has the form

$$\sum_{(i_1,\ldots,i_n)\in\Lambda} r_{i_1,\ldots,i_n} X_1^{i_1} \ldots X_n^{i_n}$$

for some finite subset $\Lambda$ of $\mathbf{N}_0^n$ and some

$$r_{i_1,\ldots,i_n} \in R \qquad ((i_1,\ldots,i_n) \in \Lambda),$$

and, if it is non-zero, then its *(total) degree* is defined to be the greatest $d \in \mathbf{N}_0$ for which there exists $(i_1,\ldots,i_n) \in \Lambda$ such that $i_1 + \cdots + i_n = d$ and $r_{i_1,\ldots,i_n} \neq 0$; and

(iv) as in the case of one variable, the *(total) degree* of the zero element of $R_n$ is defined to be $-\infty$.

We shall refer to $R_n$ as the *ring of polynomials with coefficients in $R$ (or over $R$) in the $n$ indeterminates $X_1,\ldots,X_n$*.

The next exercise shows that the above polynomial ring $R[X_1,\ldots,X_n]$ has a universal property analogous to that described for $R[X]$ in 1.13.

**1.16 ♯EXERCISE.** Let $R'$ be a commutative ring, and let $\xi_1,\ldots,\xi_n \in R'$ be algebraically independent over the subring $R$ of $R'$. Let $T$ be a commutative $R$-algebra with structural ring homomorphism $f : R \to T$ and let $\alpha_1,\ldots,\alpha_n \in T$. Show that there is exactly one ring homomorphism

$$g : R[\xi_1,\ldots,\xi_n] \to T$$

which extends $f$ (that is, is such that $g|_R = f$) and is such that $g(\xi_i) = \alpha_i$ for all $i = 1,\ldots,n$.

Deduce that there is a (unique) ring isomorphism

$$h : R[\xi_1,\ldots,\xi_n] \longrightarrow R[X_1,\ldots,X_n],$$

where $R[X_1,\ldots,X_n]$ denotes the polynomial ring constructed in 1.15, such that $h(\xi_i) = X_i$ for all $i = 1,\ldots,n$ and $h|_R : R \to R$ is the identity map.

This exercise shows that, whenever $\xi_1,\ldots,\xi_n$ are elements of a commutative ring $R'$ and $\xi_1,\ldots,\xi_n$ are algebraically independent over the subring $R$ of $R'$, then $R[\xi_1,\ldots,\xi_n]$ is 'essentially' the ring of polynomials $R[X_1,\ldots,X_n]$ discussed in 1.15. Indeed, whenever we discuss such a ring of polynomials in the rest of the book, it will, of course, be (tacitly) understood that the family $(X_i)_{i=1}^n$ is algebraically independent over $R$. The reader is reminded (see 1.1) that the symbols $X, Y, X_1,\ldots,X_n$ always denote indeterminates in this book.

The above exercise leads to the idea of 'evaluation' of a polynomial.

**1.17** DEFINITION. Let $R$ be a subring of the commutative ring $S$, and consider the polynomial ring $R[X_1, \ldots, X_n]$ over $R$ in $n$ indeterminates $X_1, \ldots, X_n$. Let $\alpha_1, \ldots, \alpha_n \in S$. By 1.16, there is exactly one ring homomorphism $g : R[X_1, \ldots, X_n] \to S$ with the properties that

$$g(r) = r \quad \text{for all } r \in R$$

and

$$g(X_i) = \alpha_i \quad \text{for all } i = 1, \ldots, n.$$

This homomorphism $g$ is called *the evaluation homomorphism*, or just *evaluation*, at $\alpha_1, \ldots, \alpha_n$.

It is clear that, in the situation of 1.17, the effect of $g$ on an element $p \in R[X_1, \ldots, X_n]$ is worked out simply by replacing, for each $i = 1, \ldots, n$, each occurrence of $X_i$ by $\alpha_i$. For this reason, $g$ is sometimes referred to as 'the result of putting $X_i = \alpha_i$ for $i = 1, \ldots, n$'. This is perhaps unfortunate, because, although we shall certainly write the image of $p$ under the evaluation homomorphism $g$ as

$$p(\alpha_1, \ldots, \alpha_n)$$

on occasion, one should certainly not confuse the concept of polynomial with that of function. The following exercise illustrates the point.

**1.18** EXERCISE. Let $p = X^7 - X \in \mathbf{Z}_7[X]$. Show that $p(\alpha) = 0$ for all $\alpha \in \mathbf{Z}_7$.

**1.19** EXERCISE. Let $K$ be an infinite field, let $\Lambda$ be a finite subset of $K$, and let $f \in K[X_1, \ldots, X_n]$, the ring of polynomials over $K$ in the indeterminates $X_1, \ldots, X_n$. Suppose that $f \neq 0$. Show that there exist infinitely many choices of

$$(\alpha_1, \ldots, \alpha_n) \in (K \setminus \Lambda)^n$$

for which $f(\alpha_1, \ldots, \alpha_n) \neq 0$.

Again, let $R$ be a commutative ring, and let $X_1, \ldots, X_n$ be indeterminates. We can successively form power series rings by the following inductive procedure: set $R_0 = R$, and, for each $i \in \mathbf{N}$ with $0 < i \leq n$, let

$$R_i = R_{i-1}[[X_i]].$$

Such power series rings are very important in commutative algebra, and it is desirable that we have available a convenient description of them. To this end, we introduce the idea of *homogeneous polynomial* in $R[X_1, \ldots, X_n]$: a polynomial in this ring is said to be homogeneous, or to be a *form*, if it has the form

$$\sum_{i_1 + \cdots + i_n = d} r_{i_1, \ldots, i_n} X_1^{i_1} \ldots X_n^{i_n}$$

for some $d \in \mathbb{N}_0$ and some $r_{i_1, \ldots, i_n} \in R$. (Thus any non-zero term which actually appears in the polynomial has degree $d$.) Note that the zero polynomial is considered to be homogeneous.

We can use the concept of homogeneous polynomial to describe the elements of the ring $R_1 = R[[X_1]]$: an arbitrary element of $R_1$ can be expressed as a formal sum

$$\sum_{i=0}^{\infty} f_i,$$

where $f_i$ is a homogeneous polynomial in $R[X_1]$ which is either 0 or of degree $i$ (for each $i \in \mathbb{N}_0$).

With this in mind, we now introduce another ring which can be constructed from $R$ and the above indeterminates $X_1, \ldots, X_n$. This is the *ring of formal power series* in $X_1, \ldots, X_n$ with coefficients in $R$, and it is denoted by

$$R[[X_1, \ldots, X_n]].$$

The elements of this ring are formal sums of the form

$$\sum_{i=0}^{\infty} f_i,$$

where $f_i$ is, for each $i \in \mathbb{N}_0$, a homogeneous polynomial in $R[X_1, \ldots, X_n]$ which is either zero or of degree $i$. Two such 'formal power series' $\sum_{i=0}^{\infty} f_i$ and $\sum_{i=0}^{\infty} g_i$ are considered to be equal precisely when $f_i = g_i$ for all $i \in \mathbb{N}_0$. The operations of addition and multiplication are given by

$$\left( \sum_{i=0}^{\infty} f_i \right) + \left( \sum_{i=0}^{\infty} g_i \right) = \sum_{i=0}^{\infty} (f_i + g_i),$$

$$\left( \sum_{i=0}^{\infty} f_i \right) \left( \sum_{i=0}^{\infty} g_i \right) = \sum_{i=0}^{\infty} \left( \sum_{j=0}^{i} f_j g_{i-j} \right)$$

for all $\sum_{i=0}^{\infty} f_i, \sum_{i=0}^{\infty} g_i \in R[[X_1, \ldots, X_n]]$. It is straightforward to check that these definitions do indeed provide $R[[X_1, \ldots, X_n]]$ with the structure of a commutative ring.

The next exercise, although tedious to verify, concerns a fundamental fact about rings of formal power series.

**1.20** ♯EXERCISE. Let $R$ be a commutative ring and let $X_1, \ldots, X_n$ (where $n > 1$) be indeterminates. Define a map

$$\psi : R[[X_1, \ldots, X_{n-1}]][[X_n]] \longrightarrow R[[X_1, \ldots, X_{n-1}, X_n]]$$

as follows. For each

$$f = \sum_{i=0}^{\infty} f_{(i)} X_n^i \in R[[X_1, \ldots, X_{n-1}]][[X_n]],$$

where

$$f_{(i)} = \sum_{j=0}^{\infty} f_{(i)j} \in R[[X_1, \ldots, X_{n-1}]]$$

for each $i \in \mathbf{N}_0$ (so that each $f_{(i)j}$ is zero or a homogeneous polynomial of degree $j$ in $X_1, \ldots, X_{n-1}$), set

$$\psi(f) = \sum_{k=0}^{\infty} \left( \sum_{j=0}^{k} f_{(k-j)j} X_n^{k-j} \right).$$

Prove that $\psi$ is a ring isomorphism.

Readers will no doubt recall from their elementary studies of ring theory the special types of commutative ring called 'integral domain' and 'field'. It is appropriate for us to include a review of these concepts in this introductory chapter.

**1.21** DEFINITION. Let $R$ be a commutative ring. A *zerodivisor* in $R$ is an element $r \in R$ for which there exists $y \in R$ with $y \neq 0_R$ such that $ry = 0_R$. An element of $R$ which is not a zerodivisor in $R$ is called a *non-zerodivisor* of $R$. The set of zerodivisors in $R$ will often be denoted by $Z(R)$.

If $R$ is non-trivial, then $0_R$ is a zerodivisor in $R$, simply because

$$0_R 1_R = 0_R.$$

**1.22** DEFINITION. Let $R$ be a commutative ring. Then $R$ is said to be an *integral domain* precisely when
  (i) $R$ is not trivial, that is, $1_R \neq 0_R$, and
  (ii) $0_R$ is the only zerodivisor in $R$ (it *is* one, by 1.21).

Of course, the ring $\mathbf{Z}$ of integers and the ring $\mathbf{Z}[i]$ of Gaussian integers are examples of integral domains. The reader should also be familiar with the following general result, which allows us to produce new integral domains from old.

**1.23** PROPOSITION. (See [15, Theorem 1.5.2].) *Let $R$ be an integral domain and let $X$ be an indeterminate. Then the polynomial ring $R[X]$ is also an integral domain.* $\square$

In particular, this result shows that the ring of polynomials

$$Z[X_1, \ldots, X_n]$$

over $Z$ in indeterminates $X_1, \ldots, X_n$ is again an integral domain.

The reader should not be misled into believing that every non-trivial commutative ring is an integral domain: the ring $Z_6$ of residue classes of integers modulo 6 is not an integral domain; also the ring $C[0, 1]$ of continuous real-valued functions defined on the closed interval $[0, 1]$ is not an integral domain (see [15, p. 20]).

**1.24** DEFINITION. Let $R$ be a commutative ring. A *unit* of $R$ is an element $r \in R$ for which there exists $u \in R$ such that $ru = 1_R$. When $r \in R$ is a unit of $R$, then there is exactly one element $u \in R$ with the property that $ru = 1_R$; this element is called the *inverse* of $r$, and is denoted by $r^{-1}$.

The set of all units in $R$ is an Abelian group with respect to the multiplication of $R$.

**1.25** DEFINITION. Let $R$ be a commutative ring. Then we say that $R$ is a *field* precisely when

(i) $R$ is not trivial (that is, $1_R \neq 0_R$), and
(ii) every non-zero element of $R$ is a unit.

We remind the reader of some of the elementary interrelations between the concepts of field and integral domain.

**1.26** REMARK. (See [15, Theorem 1.6.3].) Every field is an integral domain. However, the converse statement is not true: $Z$ is an example of an integral domain which is not a field.

**1.27** LEMMA. (See [15, Theorem 1.6.4].) *Every finite integral domain is a field.* $\square$

**1.28** LEMMA. (See [15, Theorem 1.7.1 and Corollary 1.7.2].) *Let $n \in N$ with $n > 1$. Then the following statements, about the ring $Z_n$ of residue classes of integers modulo $n$, are equivalent:*

(i) *$Z_n$ is a field;*
(ii) *$Z_n$ is an integral domain;*
(iii) *$n$ is a prime number.* $\square$

Of course, $\mathbf{Q}$, $\mathbf{R}$ and $\mathbf{C}$ are examples of fields; note also that if $K$ is any field, then the ring $K[X_1, \ldots, X_n]$ of polynomials over $K$ in the $n$ indeterminates $X_1, \ldots, X_n$ is automatically an integral domain.

**1.29** EXERCISE. Let $R$ be a commutative ring, and let $X$ be an indeterminate. Show that $R[X]$ is never a field. Show also that if $R[X]$ is an integral domain, then so too is $R$.

**1.30** EXERCISE. Let $R$ be a commutative ring. Show that $R$ is an integral domain if and only if the ring of formal power series $R[[X_1, \ldots, X_n]]$ in $n$ indeterminates $X_1, \ldots, X_n$ is an integral domain.

Another important connection between the concepts of integral domain and field is that every integral domain can be embedded, as a subring, in a field: the construction that makes this possible is that of the so-called 'field of fractions' of an integral domain. We review this next.

**1.31** THE FIELD OF FRACTIONS OF AN INTEGRAL DOMAIN. (See [15, pp. 25-26].) *Let $R$ be an integral domain. Then there exist a field $F$ and an injective ring homomorphism $f : R \to F$ such that each element of $F$ can be written in the form $f(r)f(s)^{-1}$ for some $r, s \in R$ with $s \neq 0_R$.*

*Sketch.* Let $S = R \setminus \{0_R\}$. We set up an equivalence relation $\sim$ on $R \times S$ as follows: for $(a, b)$, $(c, d) \in R \times S$, we write

$$(a, b) \ \sim \ (c, d) \qquad \Longleftrightarrow \qquad ad = bc.$$

It is straightforward to check that $\sim$ is an equivalence relation on $R \times S$: for $(a, b) \in R \times S$, we denote the equivalence class which contains $(a, b)$ by $a/b$ or

$$\frac{a}{b} \, .$$

The set $F$ of all equivalence classes of $\sim$ can be given the structure of a field under operations of addition and multiplication for which

$$\frac{a}{b} + \frac{c}{d} = \frac{ad + bc}{bd}, \qquad \frac{a}{b}\frac{c}{d} = \frac{ac}{bd}$$

for all $a/b, c/d \in F$; the zero element of this field is $0/1$, and the identity element is $1/1$ (which is equal to $a/a$ for each $a \in R \setminus \{0\}$).

The mapping

$$f : R \to F$$

defined by $f(a) = a/1$ for all $a \in R$ is an injective ring homomorphism with all the desired properties. $\square$

*Note.* The field $F$ constructed from the integral domain $R$ in the above sketch of proof for 1.31 is referred to as the *field of fractions of the integral domain $R$*.

**1.32** DEFINITION. Let $R$ be a commutative ring. An element $r \in R$ is said to be *nilpotent* precisely when there exists $n \in \mathbb{N}$ such that $r^n = 0$.

**1.33** EXERCISE. Determine all the nilpotent elements and all the units in the ring $\mathbb{Z}_{12}$. For $n \in \mathbb{N}$ with $n > 1$, determine all the nilpotent elements and all the units in the ring $\mathbb{Z}_n$.

**1.34** ♯EXERCISE. Let $R$ be a commutative ring, and let $x, y \in R$. Show that, for $n \in \mathbb{N}$,

$$(x + y)^n = \sum_{i=0}^{n} \binom{n}{i} x^{n-i} y^i.$$

Deduce that the sum of two nilpotent elements of $R$ is again nilpotent.

**1.35** EXERCISE. Let $a$ be a nilpotent element of the commutative ring $R$. Show that $1 + a$ is a unit of $R$, and deduce that $u + a$ is a unit of $R$ for each unit $u$ of $R$.

**1.36** EXERCISE. Let $R$ be a commutative ring, and let $X$ be an indeterminate; let

$$f = r_0 + r_1 X + \cdots + r_n X^n \in R[X].$$

(i) Prove that $f$ is a unit of $R[X]$ if and only if $r_0$ is a unit of $R$ and $r_1, \ldots, r_n$ are all nilpotent. (Here is a hint: if $n > 0$ and $f$ is a unit of $R[X]$ with inverse $a_0 + a_1 X + \cdots + a_m X^m$, show by induction on $i$ that $r_n^{i+1} a_{m-i} = 0$ for each $i = 0, \ldots, m$.)

(ii) Prove that $f$ is nilpotent if and only if $r_0, \ldots, r_n$ are all nilpotent.

(iii) Prove that $f$ is a zerodivisor in $R[X]$ if and only if there exists $c \in R$ such that $c \neq 0$ but $cf = 0$. (Here is another hint: if $f$ is a zerodivisor in $R[X]$, choose a polynomial

$$0 \neq g = c_0 + c_1 X + \cdots + c_k X^k \in R[X]$$

of least degree $k$ such that $fg = 0$, and show by induction that $r_{n-i} g = 0$ for each $i = 0, \ldots, n$.)

**1.37** EXERCISE. Generalize the results of 1.36 to a polynomial ring

$$R[X_1, \ldots, X_n]$$

over the commutative ring $R$ in $n$ indeterminates $X_1, \ldots, X_n$.

We shall also need to assume that the reader is familiar with certain basic ideas in the theory of factorization in integral domains, including the concepts of 'Euclidean domain', 'irreducible element' of an integral domain, and 'unique factorization domain' (UFD for short). In keeping with the spirit of this book, we end this chapter with a brief review of some of the main points from this theory, and include appropriate references.

**1.38** DEFINITION. Let $R$ be an integral domain. An element $p \in R$ is said to be an *irreducible element of $R$* precisely when
  (i) $p \neq 0$ and $p$ is not a unit of $R$, and
  (ii) whenever $p$ is expressed as $p = ab$ with $a, b \in R$, then either $a$ or $b$ is a unit of $R$.

**1.39** DEFINITION. Let $R$ be an integral domain. We say that $R$ is a *unique factorization domain* (UFD for short) precisely when
  (i) each non-zero, non-unit element of $R$ can be expressed in the form $p_1 p_2 \ldots p_s$, where $p_1, \ldots, p_s$ are irreducible elements of $R$, and
  (ii) whenever $s, t \in \mathsf{N}$ and $p_1, \ldots, p_s, q_1, \ldots, q_t$ are irreducible elements of $R$ such that

$$p_1 p_2 \ldots p_s = q_1 q_2 \ldots q_t$$

then $s = t$ and there exist units $u_1, \ldots, u_s \in R$ such that, after a suitable renumbering of the $q_j$,

$$p_i = u_i q_i \qquad \text{for all } i = 1, \ldots, s.$$

**1.40** DEFINITION. The integral domain $R$ is said to be a *Euclidean domain* precisely when there is a function $\partial : R \setminus \{0\} \to \mathsf{N}_0$, called the *degree function of $R$*, such that
  (i) whenever $a, b \in R \setminus \{0\}$ and $a$ is a factor of $b$ in $R$ (that is, there exists $c \in R$ such that $ac = b$), then $\partial(a) \leq \partial(b)$, and
  (ii) whenever $a, b \in R$ with $b \neq 0$, then there exist $q, r \in R$ such that

$$a = qb + r \quad \text{with} \quad \text{either } r = 0 \ \text{ or } r \neq 0 \text{ and } \partial(r) < \partial(b).$$

(We could, of course, follow our practice with polynomials and define $\partial(0) = -\infty$; this would lead to a neater statement for (ii).)

We expect the reader to be familiar with the facts that the ring of integers $\mathsf{Z}$, the ring of Gaussian integers $\mathsf{Z}[i]$ and the ring of polynomials $K[X]$ in the indeterminate $X$ with coefficients in the field $K$ are all examples of Euclidean domains.

We state now, with references, two fundamental facts concerning Euclidean domains and UFDs.

**1.41** THEOREM. (See [15, Theorem 2.6.1].) *Every Euclidean domain is a unique factorization domain.* □

**1.42** THEOREM. (See [15, Theorem 2.8.12].) *If $R$ is a unique factorization domain, then so also is the polynomial ring $R[X]$, where $X$ is an indeterminate.* □

It is immediate from the last two theorems that, if $K$ is a field, then the ring $K[X_1, \ldots, X_n]$ of polynomials over $K$ in the $n$ indeterminates $X_1, \ldots, X_n$ is a UFD; also, the ring $\mathbf{Z}[Y_1, \ldots, Y_m]$ of polynomials over $\mathbf{Z}$ in the $m$ indeterminates $Y_1, \ldots, Y_m$ is another example of a UFD.

**1.43** EXERCISE. Let $R$ be a commutative ring, and consider the ring $R[[X_1, \ldots, X_n]]$ of formal power series over $R$ in indeterminates $X_1, \ldots, X_n$. Let

$$f = \sum_{i=0}^{\infty} f_i \in R[[X_1, \ldots, X_n]],$$

where $f_i$ is either zero or a form of degree $i$ in $R[X_1, \ldots, X_n]$ (for each $i \in \mathbf{N}_0$). Prove that $f$ is a unit of $R[[X_1, \ldots, X_n]]$ if and only if $f_0$ is a unit of $R$.

# Chapter 2

# Ideals

Some experienced readers will have found it amazing that a whole chapter of this book has been covered without mention of the absolutely fundamental concept of ideal in a commutative ring. Ideals constitute the most important substructure of such a ring: they are to commutative rings what normal subgroups are to groups. Furthermore, the concepts of prime ideal and maximal ideal are central to the applications of commutative ring theory to algebraic geometry.

We begin by putting some flesh on the above statement that ideals are to commutative rings what normal subgroups are to groups: just as, for a group $G$, a subset $N$ of $G$ is a normal subgroup of $G$ if and only if there exist a group $H$ and a group homomorphism $\theta : G \to H$ which has kernel equal to $N$, we shall see that, for a commutative ring $R$, a subset $I$ of $R$ is an ideal of $R$ if and only if there exist a commutative ring $S$ and a ring homomorphism $f : R \to S$ with kernel equal to $I$. Thus the first step is the definition, and examination of the properties, of the kernel of a homomorphism of commutative rings.

**2.1 DEFINITION and LEMMA.** *Let $R$ and $S$ be commutative rings, and let $f : R \to S$ be a ring homomorphism. Then we define the* kernel *of $f$, denoted $\operatorname{Ker} f$, by*

$$\operatorname{Ker} f := \{ r \in R : f(r) = 0_S \} .$$

*Note that*

    (i) *$0_R \in \operatorname{Ker} f$, so that $\operatorname{Ker} f \neq \emptyset$;*

    (ii) *whenever $a, b \in \operatorname{Ker} f$, then $a + b \in \operatorname{Ker} f$ also; and*

    (iii) *whenever $a \in \operatorname{Ker} f$ and $r \in R$, then $ra \in \operatorname{Ker} f$ also.* $\square$

The above lemma provides motivation for the definition of ideal in a commutative ring, but before we give the definition, we record a fundamental fact about kernels of homomorphisms of commutative rings.

**2.2** LEMMA. *Let $R$ and $S$ be commutative rings, and let $f : R \rightarrow S$ be a ring homomorphism. Then $\operatorname{Ker} f = \{0_R\}$ if and only if $f$ is injective.*

*Proof.* ($\Rightarrow$) Let $r, r' \in R$ be such that $f(r) = f(r')$. Then $r - r' \in \operatorname{Ker} f = \{0_R\}$.

($\Leftarrow$) Of course, $0_R \in \operatorname{Ker} f$, by 2.1(i). Let $r \in \operatorname{Ker} f$. Then $f(r) = 0_S = f(0_R)$, so that $r = 0_R$ since $f$ is injective. $\square$

**2.3** DEFINITION. Let $R$ be a commutative ring. A subset $I$ of $R$ is said to be an *ideal* of $R$ precisely when the following conditions are satisfied:
  (i) $I \neq \emptyset$;
  (ii) whenever $a, b \in I$, then $a + b \in I$ also; and
  (iii) whenever $a \in I$ and $r \in R$, then $ra \in I$ also.

It should be clear to the reader that an ideal of a commutative ring $R$ is closed under subtraction. Any reader experienced in non-commutative ring theory should note that we shall not discuss ideals of non-commutative rings in this book, and so we shall have no need of the concepts of left ideal and right ideal.

**2.4** EXERCISE. Let $X$ be an indeterminate and consider the ring $\mathbf{Q}[X]$ of polynomials in $X$ with coefficients in $\mathbf{Q}$. Give
  (i) an example of a subring of $\mathbf{Q}[X]$ which is not an ideal of $\mathbf{Q}[X]$, and
  (ii) an example of an ideal of $\mathbf{Q}[X]$ which is not a subring of $\mathbf{Q}[X]$.

**2.5** ♯EXERCISE. Let $R$ be a commutative ring, and let $I$ be an ideal of $R$. Show that the set $N$ of all nilpotent elements of $R$ is an ideal of $R$. (The ideal $N$ is often referred to as the *nilradical* of $R$.)

Show also that

$$\sqrt{I} := \{r \in R : \text{there exists } n \in \mathbf{N} \text{ with } r^n \in I\}$$

is an ideal of $R$. (We call $\sqrt{I}$ the *radical* of $I$; thus the nilradical $N$ of $R$ is just $\sqrt{\{0_R\}}$, the radical of the 'zero ideal' $\{0_R\}$ of $R$.)

**2.6** EXERCISE. Let $R_1, \ldots, R_n$ be commutative rings. Show that the Cartesian product set

$$\prod_{i=1}^{n} R_i = R_1 \times \cdots \times R_n$$

has the structure of a commutative ring under componentwise operations of addition and multiplication. (This just means that we define

$$(r_1, \ldots, r_n) + (s_1, \ldots, s_n) = (r_1 + s_1, \ldots, r_n + s_n)$$

and

$$(r_1, \ldots, r_n)(s_1, \ldots, s_n) = (r_1 s_1, \ldots, r_n s_n)$$

for all $r_i, s_i \in R_i$ $(i = 1, \ldots, n)$.) We call this new ring the *direct product* of $R_1, \ldots, R_n$.

Show that, if $I_i$ is an ideal of $R_i$ for each $i = 1, \ldots, n$, then $I_1 \times \cdots \times I_n$ is an ideal of $\prod_{i=1}^{n} R_i$. Show further that each ideal of $\prod_{i=1}^{n} R_i$ has this form.

In order to enable us to show that every ideal of a commutative ring $R$ is the kernel of a ring homomorphism from $R$ to some other commutative ring, we shall now produce a ring-theoretic analogue of the construction, for a normal subgroup $N$ of a group $G$, of the factor group $G/N$. As this construction is of fundamental importance for our subject, we shall give a thorough account of it.

**2.7** REMINDERS and NOTATION. Let $I$ be an ideal of the commutative ring $R$, and let $r \in R$. The *coset* of $I$ in $R$ determined by, or which contains, $r$ is the set

$$r + I = \{r + z : z \in I\}.$$

Note that, for $r, s \in R$, the cosets $r + I$ and $s + I$ are equal if and only if $r - s \in I$; in fact, the cosets of $I$ in $R$ are precisely the equivalence classes of the equivalence relation $\sim$ on $R$ defined by

$$a \sim b \quad \Longleftrightarrow \quad a - b \in I \quad \text{for } a, b \in R.$$

We denote the set of all cosets of $I$ in $R$ by $R/I$.

**2.8** REMINDERS. Let $I$ and $J$ be ideals of the commutative ring $R$ such that $I \subseteq J$. Now $R$ is, of course, an Abelian group with respect to addition, and $I, J$ are subgroups of $R$; thus $I$ is a subgroup of $J$ and we can form the factor group $J/I$. The elements of this group are the cosets of $I$ in $J$; thus

$$J/I = \{a + I : a \in J\}$$

and, as before, for $a, b \in J$, we have $a + I = b + I$ if and only if $a - b \in I$.

The addition in the group $J/I$ is such that

$$(a + I) + (b + I) = (a + b) + I$$

for all $a, b \in J$. Of course, one needs to verify that this formula is unambiguous, but this is straightforward: if $a, a', b, b' \in J$ are such that

$$a + I = a' + I, \qquad b + I = b' + I,$$

then $a - a', b - b' \in I$, so that $(a + b) - (a' + b') \in I$ and

$$(a + b) + I = (a' + b') + I.$$

It is routine to check that this operation of addition provides $J/I$ with the structure of an Abelian group. The resulting group is often referred to as the *factor group* or *residue class group* of $J$ modulo $I$.

**2.9** THE CONSTRUCTION OF RESIDUE CLASS RINGS. Let $I$ be an ideal of the commutative ring $R$. Of course, $R$ is an ideal of itself, and we can apply the construction of 2.8 to form the factor group $R/I$. We show now how to put a ring structure on this Abelian group.

Suppose that $r, r', s, s' \in R$ are such that

$$r + I = r' + I, \qquad s + I = s' + I.$$

Then we have $r - r', s - s' \in I$, so that

$$\begin{aligned} rs - r's' &= rs - rs' + rs' - r's' \\ &= r(s - s') + (r - r')s' \in I, \end{aligned}$$

and this implies that $rs + I = r's' + I$. We can therefore unambiguously define an operation of multiplication on $R/I$ by the rule

$$(r + I)(s + I) = rs + I$$

for all $r, s \in R$. It is now very straightforward to verify that this operation provides $R/I$ with the structure of a commutative ring. The identity of this ring is $1 + I$, while the zero element of $R/I$ is, of course, $0 + I = I$. A common abbreviation, for $r \in R$, for the coset $r + I$ is $\bar{r}$; this notation is particularly useful when there is no ambiguity about the ideal $I$ which is under consideration.

The ring $R/I$ is referred to as the *residue class ring*, or *factor ring*, of $R$ modulo $I$.

It is appropriate for us to mention a very familiar example of the above construction before we proceed with the general development.

**2.10** EXAMPLE. Let $n \in \mathbf{N}$ with $n > 1$. Then the set

$$n\mathbf{Z} := \{nr : r \in \mathbf{Z}\}$$

of all integer multiples of $n$ is an ideal of the ring $\mathbf{Z}$ of integers, as the reader can easily check. We can therefore use the construction of 2.9 and form the residue class ring $\mathbf{Z}/n\mathbf{Z}$. This ring is just the ring of residue classes of integers modulo $n$ mentioned in 1.2(iii).

Thus the reader is probably very familiar with one example of the construction of residue class ring!

We now continue with the general development.

**2.11** LEMMA. *Let $I$ be an ideal of the commutative ring $R$. Then the mapping $f : R \to R/I$ defined by $f(r) = r + I$ for all $r \in R$ is a surjective ring homomorphism with kernel $I$. The homomorphism $f$ is often referred to as the* natural *or* canonical *ring homomorphism from $R$ to $R/I$.*

*Proof.* It is immediate from the construction of $R/I$ in 2.9 that, for all $r, s \in R$, we have

$$f(r + s) = f(r) + f(s), \qquad f(rs) = f(r)f(s);$$

it is equally clear that $f(1_R) = 1_{R/I}$. Thus $f$ is a ring homomorphism, and it is clear that it is surjective. Also, for $r \in R$, we have $f(r) = 0_{R/I}$ precisely when $r + I = I$, that is, if and only if $r \in I$. $\square$

**2.12** COROLLARY. *Let $R$ be a commutative ring, and let $I$ be a subset of $R$. Then $I$ is an ideal of $R$ if and only if there exist a commutative ring $S$ and a ring homomorphism $f : R \to S$ such that $I = \operatorname{Ker} f$.*

*Proof.* This is now immediate from 2.1, 2.3 and 2.11. $\square$

There is another very important connection between ring homomorphisms, ideals and residue class rings: this is the so-called 'Isomorphism Theorem' for commutative rings.

**2.13** THE ISOMORPHISM THEOREM FOR COMMUTATIVE RINGS. *Let $R$ and $S$ be commutative rings, and let $f : R \to S$ be a ring homomorphism. Then $f$ induces a ring isomorphism $\bar{f} : R/\operatorname{Ker} f \to \operatorname{Im} f$ for which*

$$\bar{f}(r + \operatorname{Ker} f) = f(r) \qquad \text{for all } r \in R.$$

*Proof.* Set $K = \operatorname{Ker} f$. In order to establish the existence of a mapping $\bar{f}$ which satisfies the formula given in the last line of the statement, we must show that if $r, s \in R$ are such that $r + K = s + K$, then $f(r) = f(s)$. However, this is easy: the equation $r + K = s + K$ implies that $r - s \in K = \operatorname{Ker} f$, so that

$$f(r) - f(s) = f(r - s) = 0_S$$

and $f(r) = f(s)$. It follows that there is indeed a mapping

$$\bar{f} : R/K \longrightarrow \operatorname{Im} f$$

given by the formula in the last line of the statement, and it is clear that $\bar{f}$ is surjective. Next note that, for all $r, s \in R$, we have

$$\begin{aligned}
\bar{f}((r + K) + (s + K)) &= \bar{f}((r + s) + K) = f(r + s) \\
&= f(r) + f(s) = \bar{f}(r + K) + \bar{f}(s + K)
\end{aligned}$$

and

$$\begin{aligned}
\bar{f}((r + K)(s + K)) &= \bar{f}(rs + K) = f(rs) \\
&= f(r)f(s) = \bar{f}(r + K)\bar{f}(s + K).
\end{aligned}$$

Also, $\bar{f}(1_{R/K}) = \bar{f}(1_R + K) = f(1_R) = 1_S$, the identity element of the subring $\operatorname{Im} f$ of $S$. Finally, if $r, s \in R$ are such that $\bar{f}(r + K) = \bar{f}(s + K)$, then $f(r - s) = f(r) - f(s) = 0_S$, so that $r - s \in K$ and $r + K = s + K$. Hence $\bar{f}$ is injective and the proof that it is an isomorphism is complete. □

Of course, we shall study many examples of ideals in the course of this book. However, we should point out two rather uninteresting examples next.

**2.14** REMARK. Let $R$ be a commutative ring. Then $R$ itself is always an ideal of $R$. We say that an ideal $I$ of $R$ is *proper* precisely when $I \neq R$.

Observe that, for an ideal $I$ of $R$, the ring $R/I$ is trivial if and only if $1 + I = 0 + I$; this is the case if and only if $1 \in I$, and this in turn occurs precisely when $I = R$, that is $I$ is *improper*. (If $1 \in I$ then $r = r1 \in I$ for all $r \in R$.)

Another example of an ideal of $R$ is the set $\{0_R\}$: this is referred to as the *zero ideal* of $R$, and is usually denoted just by $0$.

We met another example of an ideal in 2.10: for a positive integer $n > 1$, the set

$$n\mathbf{Z} = \{nr : r \in \mathbf{Z}\}$$

is an ideal of the ring $\mathbf{Z}$ of integers. This is a particular example of the concept of 'principal ideal' in a commutative ring, which we now define.

**2.15** LEMMA and DEFINITION. *Let $R$ be a commutative ring; let $a \in R$.*
*Then the set*

$$aR := \{ar : r \in R\}$$

*is an ideal of $R$, called the* principal ideal of $R$ generated by $a$. *Alternative*
*notations for $aR$ are $(a)$ and $Ra$.* □

It should be noted that, with the notation of 2.15, the principal ideal
$1_R R$ of $R$ is just $R$ itself, while the principal ideal $0_R R$ of $R$ is the zero
ideal.

**2.16** ♯EXERCISE. Let $R$ be a commutative ring and let $r \in R$. Show that
$r$ is a unit of $R$ if and only if $(r) = (1_R)$.

It is now time that we explored the words 'generated by' that were used
in the definition of principal ideal given in 2.15.

**2.17** GENERATION OF IDEALS. Let $R$ be a commutative ring, and let
$(I_\lambda)_{\lambda \in \Lambda}$ be a non-empty family of ideals of $R$. (This just means that $\Lambda \neq \emptyset$
and $I_\lambda$ is an ideal of $R$ for each $\lambda \in \Lambda$.) It should be clear from the definition
of ideal in 2.3 that

$$\bigcap_{\lambda \in \Lambda} I_\lambda,$$

the intersection of our family of ideals, is again an ideal of $R$. We interpret
$\bigcap_{\lambda \in \Lambda'} I_\lambda$ as $R$ itself in the case where $\Lambda' = \emptyset$.

Let $H \subseteq R$. We define the *ideal of $R$ generated by $H$*, denoted by $(H)$
or $RH$ or $HR$, to be the intersection of the family of all ideals of $R$ which
contain $H$. Note that this family is certainly non-empty, since $R$ itself is
automatically an ideal of $R$ which contains $H$.

It thus follows that

(i) $(H)$ is an ideal of $R$ and $H \subseteq (H)$;

(ii) $(H)$ is the *smallest* ideal of $R$ which contains $H$ in the sense that if
$I$ is any ideal of $R$ for which $H \subseteq I$, then $(H) \subseteq I$ simply because $I$ must
be one of the ideals in the family that was used to define $(H)$.

What does $(H)$ look like? This is the next point we address.

**2.18** PROPOSITION. *Let $R$ be a commutative ring and let $\emptyset \neq H \subseteq R$.*
*Then*

$$(H) = \left\{ \sum_{i=1}^{n} r_i h_i : n \in \mathbb{N},\ r_1, \ldots, r_n \in R,\ h_1, \ldots, h_n \in H \right\}.$$

*Also, $(\emptyset)$, the ideal of $R$ generated by its empty subset, is just given by*
$(\emptyset) = 0$, *the zero ideal of $R$.*

*Proof.* Set

$$J = \left\{ \sum_{i=1}^{n} r_i h_i : n \in \mathbf{N},\ r_1, \ldots, r_n \in R,\ h_1, \ldots, h_n \in H \right\}.$$

It is clear that $H \subseteq J$: each $h \in H$ can be written as $h = 1_R h$ and the right-hand side here shows that $h$ can be expressed in the form necessary to ensure that it belongs to $J$. Since $H \neq \emptyset$, it follows that $J \neq \emptyset$. We show next that $J$ is an ideal of $R$.

If $r_1, \ldots, r_n,\ r_{n+1}, \ldots, r_{n+m} \in R$ and $h_1, \ldots, h_n,\ h_{n+1}, \ldots, h_{n+m} \in H$ (where $n$ and $m$ are positive integers), then

$$\sum_{i=1}^{n} r_i h_i + \sum_{i=n+1}^{n+m} r_i h_i = \sum_{i=1}^{n+m} r_i h_i$$

is again an element of $J$, and it is just as easy to see that $J$ is closed under multiplication by arbitrary elements of $R$. Thus $J$ is an ideal of $R$ which contains $H$, and is therefore one of the ideals in the family used to define $(H)$. It follows that $(H) \subseteq J$.

On the other hand, $(H)$ is, by 2.17, an ideal of $R$ which contains $H$. It therefore must contain all elements of $R$ of the form $\sum_{i=1}^{n} r_i h_i$, where $n \in \mathbf{N}$, $r_1, \ldots, r_n \in R$, and $h_1, \ldots, h_n \in H$. Hence $J \subseteq (H)$, so that $J = (H)$, as claimed.

To establish the final claim of the proposition, just note that $0$ is an ideal of $R$ which (of course!) contains $\emptyset$, so that $(\emptyset) \subseteq 0$. On the other hand, every ideal of $R$ must contain $0_R$, and so $0 \subseteq (\emptyset)$. $\square$

**2.19** REMARKS. Let the situation be as in 2.18. Simplifications occur in the theory described in 2.18 in the special case in which $H$ is finite and non-empty.

(i) Suppose that $H = \{h_1, \ldots, h_t\} \subseteq R$ (where $t > 0$). Since, for $r_1, \ldots, r_t,\ s_1, \ldots, s_t \in R$, we have

$$\sum_{i=1}^{t} r_i h_i + \sum_{i=1}^{t} s_i h_i = \sum_{i=1}^{t} (r_i + s_i) h_i,$$

it follows that in this case

$$(H) = \left\{ \sum_{i=1}^{t} r_i h_i : r_1, \ldots, r_t \in R \right\}.$$

In this situation, $(H)$ is usually written as $(h_1, \ldots, h_t)$ (instead of the more cumbersome $(\{h_1, \ldots, h_t\})$), and is referred to as the *ideal generated by* $h_1, \ldots, h_t$.

(ii) In particular, for $h \in R$, we have

$$(\{h\}) = \{rh : r \in R\},$$

the principal ideal of $R$ generated by $h$ which we discussed in 2.15. Thus our notation $(h)$ for this ideal does not conflict with the use of this notation in 2.15.

(iii) If the ideal $I$ of $R$ is equal to $(H)$ for some finite subset $H$ of $R$, then we say that $I$ is a *finitely generated* ideal of $R$.

**2.20** ♯EXERCISE. Consider the ring $R[X_1, \ldots, X_n]$ of polynomials over the commutative ring $R$ in indeterminates $X_1, \ldots, X_n$. Let $\alpha_1, \ldots, \alpha_n \in R$, and let

$$f : R[X_1, \ldots, X_n] \longrightarrow R$$

be the evaluation homomorphism (see 1.17) at $\alpha_1, \ldots, \alpha_n$. Prove that

$$\operatorname{Ker} f = (X_1 - \alpha_1, \ldots, X_n - \alpha_n).$$

The reader should not be misled into thinking that every ideal of a commutative ring must be finitely generated; nor must he or she think that every finitely generated ideal of such a ring must be principal. The next two exercises emphasize these points.

**2.21** ♯EXERCISE. Let $K$ be a field and let $(X_i)_{i \in \mathbb{N}}$ be a family of indeterminates such that, for all $n \in \mathbb{N}$, it is the case that $X_1, \ldots, X_n$ are algebraically independent over $K$. (In these circumstances, we say that the family $(X_i)_{i \in \mathbb{N}}$ is *algebraically independent over $K$*.)

For each $n \in \mathbb{N}$, set $R_n = K[X_1, \ldots, X_n]$; set $R_0 = K$. Thus, for each $n \in \mathbb{N}_0$, we can view $R_n$ as a subring of $R_{n+1}$ in the natural way. Show that $R_\infty := \bigcup_{n \in \mathbb{N}_0} R_n$ can be given the structure of a commutative ring in such a way that $R_n$ is a subring of $R_\infty$ for each $n \in \mathbb{N}_0$. Show that $R_\infty = K[\Gamma]$, where $\Gamma = \{X_i : i \in \mathbb{N}\}$. In fact, we shall sometimes denote $R_\infty$ by $K[X_1, \ldots, X_n, \ldots]$, and refer to it as the *the ring of polynomials with coefficients in $K$ in the countably infinite family of indeterminates $(X_i)_{i \in \mathbb{N}}$*.

Show that the ideal of $R_\infty$ generated by $\Gamma$ is not finitely generated.

**2.22** EXERCISE. Let $K$ be a field. Show that the ideal $(X_1, X_2)$ of the commutative ring $K[X_1, X_2]$ (of polynomials over $K$ in indeterminates $X_1, X_2$) is not principal.

The concept of the *sum* of a family of ideals of a commutative ring $R$ is intimately related to the idea of generation of an ideal by a subset of $R$.

**2.23** SUMS OF IDEALS. Let $(I_\lambda)_{\lambda \in \Lambda}$ be a family of ideals of the commutative ring $R$. We define the *sum* $\sum_{\lambda \in \Lambda} I_\lambda$ of this family to be the ideal of $R$ generated by $\bigcup_{\lambda \in \Lambda} I_\lambda$: thus

$$\sum_{\lambda \in \Lambda} I_\lambda = \left( \bigcup_{\lambda \in \Lambda} I_\lambda \right).$$

In particular, if $\Lambda = \emptyset$, then $\sum_{\lambda \in \Lambda} I_\lambda = 0$.

Since an arbitrary ideal of $R$ is closed under addition and under scalar multiplication by arbitrary elements of $R$, it follows from 2.18 that, in the case in which $\Lambda \neq \emptyset$, an arbitrary element of $\sum_{\lambda \in \Lambda} I_\lambda$ can be expressed in the form $\sum_{i=1}^{n} c_{\lambda_i}$, where $n \in \mathbb{N}$, $\lambda_1, \ldots, \lambda_n \in \Lambda$, and $c_{\lambda_i} \in I_{\lambda_i}$ for each $i = 1, \ldots, n$. Another way of denoting such an expression is as $\sum_{\lambda \in \Lambda} c_\lambda$, where $c_\lambda \in I_\lambda$ for all $\lambda \in \Lambda$ and $c_\lambda = 0$ for all except finitely many $\lambda \in \Lambda$.

We shall make a great deal of use of the results of the next exercise, but they are very easy to prove.

**2.24** ‡EXERCISE. Let $R$ be a commutative ring.

(i) Show that the binary operation on the set of all ideals of $R$ given by ideal sum is both commutative and associative.

(ii) Let $I_1, \ldots, I_n$ be ideals of $R$. Show that

$$\sum_{i=1}^{n} I_i = \left\{ \sum_{i=1}^{n} r_i : r_i \in I_i \text{ for } i = 1, \ldots, n \right\}.$$

We often denote $\sum_{i=1}^{n} I_i$ by $I_1 + \cdots + I_n$.

(iii) Let $h_1, \ldots, h_n \in R$. Show that

$$(h_1, \ldots, h_n) = Rh_1 + \cdots + Rh_n,$$

(that is, that $(h_1, \ldots, h_n) = (h_1) + \cdots + (h_n)$).

**2.25** ‡EXERCISE. Let $R$ be a commutative ring, and let $I, J$ be ideals of $R$. The radical $\sqrt{I}$ of $I$ was defined in 2.5. Show that

(i) $\sqrt{(I + J)} = \sqrt{(\sqrt{(I)} + \sqrt{(J)})}$;
(ii) $\sqrt{(\sqrt{(I)})} = \sqrt{I}$;
(iii) $\sqrt{(I)} = (1)$ if and only if $I = (1)$;
(iv) if $\sqrt{(I)} + \sqrt{(J)} = (1)$, then $I + J = (1)$.

Another concept which is related to the idea of generation of ideals is that of the product of two ideals of a commutative ring. We explain this next.

**2.26** DEFINITION and LEMMA. *Let $I$ and $J$ be ideals of the commutative ring $R$. The product of $I$ and $J$, denoted by $I.J$, or more usually by $IJ$, is defined to be the ideal of $R$ generated by the set $\{ab : a \in I,\ b \in J\}$.*
*We have*

$$IJ = \left\{ \sum_{i=1}^{n} a_i b_i : n \in \mathbf{N},\ a_1, \ldots, a_n \in I,\ b_1, \ldots, b_n \in J \right\}.$$

*Proof.* This is immediate from 2.18 simply because an element of $R$ of the form $rab$, where $r \in R$, $a \in I$ and $b \in J$, can be written as $(ra)b$ and $ra \in I$. $\square$

The reader should note that, in the situation of 2.26, the product ideal $IJ$ is *not* in general equal to the subset $\{ab : a \in I,\ b \in J\}$ of $R$. The following exercise illustrates this point.

**2.27** EXERCISE. Let $K$ be a field, and let $R = K[X_1, X_2, X_3, X_4]$, the ring of polynomials over $K$ in indeterminates $X_1, X_2, X_3, X_4$. Set

$$I = RX_1 + RX_2, \qquad J = RX_3 + RX_4.$$

Show that $IJ \neq \{fg : f \in I,\ g \in J\}$.

We develop next the elementary properties of products of ideals.

**2.28** REMARKS. Let $R$ be a commutative ring, and let $I, J, K, I_1, \ldots, I_n$ be ideals of $R$.
(i) Clearly $IJ = JI \subseteq I \cap J$.
(ii) It is easy to check that $(IJ)K = I(JK)$ and that both are equal to the ideal $RH$ of $R$ generated by the set

$$H = \{abc : a \in I,\ b \in J,\ c \in K\}.$$

Thus a typical element of $(IJ)K = I(JK) =: IJK$ has the form

$$\sum_{i=1}^{t} a_i b_i c_i,$$

where $t \in \mathbf{N}$, $a_1, \ldots, a_t \in I$, $b_1, \ldots, b_t \in J$ and $c_1, \ldots, c_t \in K$.
(iii) It follows from (i) and (ii) above that we can unambiguously define the product $\prod_{i=1}^{n} I_i$ of the ideals $I_1, \ldots, I_n$ of $R$: we have

$$\prod_{i=1}^{n} I_i = I_1 \ldots I_n = RL,$$

where

$$L = \{a_1 \ldots a_n : a_1 \in I_1, \ldots, a_n \in I_n\}.$$

We therefore see that a typical element of $I_1 \ldots I_n$ is a sum of finitely many elements of $L$.

(iv) The reader should find it easy to check that $I(J + K) = IJ + IK$.

(v) Note in particular that the powers $I^m$, for $m \in \mathbb{N}$, of $I$ are defined; we adopt the convention that $I^0 = R$. Note that, by (iii), a general element of $I^m$ (for positive $m$) has the form

$$a_{11}a_{12} \ldots a_{1m} + a_{21}a_{22} \ldots a_{2m} + \cdots + a_{n1}a_{n2} \ldots a_{nm},$$

where $n \in \mathbb{N}$ and $a_{ij} \in I$ for all $i = 1, \ldots, n$ and $j = 1, \ldots, m$.

**2.29** EXERCISE. Let $R$ be a commutative ring and let $m \in \mathbb{N}$. Describe the ideal $(X_1, \ldots, X_n)^m$ of the ring $R[X_1, \ldots, X_n]$ of polynomials over $R$ in indeterminates $X_1, \ldots, X_n$.

**2.30** ♯EXERCISE. Let $I, J$ be ideals of the commutative ring $R$. Show that

$$\sqrt{(IJ)} = \sqrt{(I \cap J)} = \sqrt{(I)} \cap \sqrt{(J)}.$$

We have now built up quite a sort of 'arithmetic' of ideals of a commutative ring $R$: we have discussed the intersection and sum of an arbitrary family of ideals of $R$, and also the product of a non-empty finite family of ideals of $R$. There is yet another binary operation on the set of all ideals of $R$, namely 'ideal quotient'.

For an ideal $I$ of $R$ and $a \in R$, the notation $aI$ will denote the set $\{ac : c \in I\}$. This is again an ideal of $R$, because it is just the product ideal $(a)I$.

**2.31** DEFINITIONS. Let $I, J$ be ideals of the commutative ring $R$. We define the *ideal quotient* $(I : J)$ by

$$(I : J) = \{a \in R : aJ \subseteq I\};$$

clearly this is another ideal of $R$ and $I \subseteq (I : J)$.

In the special case in which $I = 0$, the ideal quotient

$$(0 : J) = \{a \in R : aJ = 0\} = \{a \in R : ab = 0 \text{ for all } b \in J\}$$

is called the *annihilator of $J$* and is also denoted by $\mathrm{Ann}\, J$ or $\mathrm{Ann}_R J$.

**2.32** ♯EXERCISE. Let $H$ be a subset of the commutative ring $R$, and let $I$ be an ideal of $R$. Show that

$$(I : RH) = \{a \in R : ah \in I \text{ for all } h \in H\}.$$

The ideal $(I : RH)$ is sometimes denoted by $(I : H)$; also, $(0 : H)$ is referred to as the annihilator of $H$. For $d \in R$, we write $(0 : d)$ rather than the more cumbersome $(0 : \{d\})$, and we similarly abbreviate $(I : \{d\})$ to $(I : d)$.

**2.33** ♯EXERCISE. Let $I, J, K$ be ideals of the commutative ring $R$, and let $(I_\lambda)_{\lambda \in \Lambda}$ be a family of ideals of $R$. Show that
  (i) $((I : J) : K) = (I : JK) = ((I : K) : J)$;
  (ii) $\left(\bigcap_{\lambda \in \Lambda} I_\lambda : K\right) = \bigcap_{\lambda \in \Lambda} (I_\lambda : K)$;
  (iii) $\left(J : \sum_{\lambda \in \Lambda} I_\lambda\right) = \bigcap_{\lambda \in \Lambda} (J : I_\lambda)$.

It is time we investigated some of our arithmetic of ideals in some familiar examples of commutative rings. Many elementary examples of commutative rings which come to mind when one starts out on the study of the subject are, in fact, Euclidean domains. There is one fact which greatly simplifies the ideal theory of Euclidean domains.

**2.34** THEOREM. *Each ideal $I$ in a Euclidean domain $R$ is principal.*

*Proof.* The zero ideal of $R$ is principal, and so we can, and do, assume that $I \neq 0$. Thus there exists $a \in I$ with $a \neq 0$.

We use $\partial : R \setminus \{0\} \to \mathbf{N}_0$ to denote the degree function of the Euclidean domain $R$. The set

$$\{\partial(a) : a \in I \setminus \{0\}\}$$

is a non-empty set of non-negative integers, and so has a smallest element: let this be $\partial(h)$, where $h \in I \setminus \{0\}$.

Since $I$ is an ideal of $R$, it is clear that $hR \subseteq I$. To establish the reverse inclusion, let $b \in I$. By the definition of Euclidean domain (see 1.40), there exist $q, r \in R$ such that

$$b = qh + r \quad \text{with} \quad \text{either } r = 0 \quad \text{or } r \neq 0 \text{ and } \partial(r) < \partial(h).$$

But $r = b - qh \in I$, and so we must have $r = 0$ or else there results a contradiction to the definition of $h$. Hence $b = qh \in hR$, and we have proved that $I \subseteq hR$. □

**2.35** EXERCISE. Let $R$ be a Euclidean domain, and let $a, b \in R \setminus \{0\}$. Let $h = \text{GCD}(a, b)$.
  (i) Show that $aR + bR = hR$.

(ii) Describe each of $(a) + (b)$, $(a) \cap (b)$, $(a)(b)$ and $(aR : b)$: your description should consist of a single generator for the (necessarily principal) ideal, with your generator given in terms of the factorizations of $a$ and $b$ into irreducible elements of $R$.

**2.36** DEFINITION. An integral domain $R$ is said to be a *principal ideal domain* (PID for short) precisely when every ideal of $R$ is principal.

It thus follows from 2.34 that every Euclidean domain is a PID. Thus $\mathbf{Z}, \mathbf{Z}[i]$ and $K[X]$, where $K$ is any field and $X$ is an indeterminate over $K$, are all examples of principal ideal domains.

It is important for us to be able to describe the ideals of a residue class ring.

**2.37** THE IDEALS OF A RESIDUE CLASS RING. *Let $I$ be an ideal of the commutative ring $R$.*

(i) *If $J$ is an ideal of $R$ such that $J \supseteq I$, then the Abelian group $J/I$ is an ideal of $R/I$, and, furthermore, for $r \in R$, we have $r + I \in J/I$ if and only if $r \in J$.*

(ii) *Each ideal $\mathcal{J}$ of $R/I$ can be expressed as $K/I$ for exactly one ideal $K$ of $R$ having the property that $K \supseteq I$; in fact, the unique ideal $K$ of $R$ which satisfies these conditions is given by*

$$K = \{a \in R : a + I \in \mathcal{J}\}.$$

*Proof.* (i) It is clear from 2.8 that the residue class group

$$J/I = \{a + I : a \in J\} \subseteq \{r + I : r \in R\} = R/I,$$

so that $J/I$ is a subgroup of the additive group $R/I$. Furthermore, since, for all $r \in R$ and $a \in J$, we have

$$(r + I)(a + I) = ra + I \in J/I,$$

it follows that $J/I$ is an ideal of $R/I$.

For the other claim, just observe that, if $r \in R$ is such that $r + I = j + I$ for some $j \in J$, then $r = (r - j) + j$ and $r - j \in I \subseteq J$.

(ii) Let $\mathcal{J}$ be an ideal of $R/I$. Set

$$K = \{a \in R : a + I \in \mathcal{J}\}.$$

Clearly $I \subseteq K$, since $a + I = I \in \mathcal{J}$ for all $a \in I$. Let $a, b \in K$ and $r \in R$. Then we have $a + I, b + I \in \mathcal{J}$, so that $(a + b) + I, ra + I \in \mathcal{J}$; hence $a + b, ra \in K$ and $K$ is an ideal of $R$. We have already remarked that $K \supseteq I$. It is clear from the definition of $K$ that $K/I = \mathcal{J}$.

Now suppose that $L$ is another ideal of $R$ with the properties that $L \supseteq I$ and $L/I = \mathcal{J}$. Let $a \in L$. Then

$$a + I \in L/I = \mathcal{J}$$

and so $a \in K$ by definition of $K$. On the other hand, if $b \in K$, then $b + I \in \mathcal{J} = L/I$, and so, by (i), we have $b \in L$. Hence $K \subseteq L$, and we have proved that $L = K$. $\square$

**2.38** ♯EXERCISE. Let $I$ be an ideal of the commutative ring $R$, and let $J, K$ be ideals of $R$ which contain $I$. Let $a_1, \ldots, a_h \in R$. For each of the following choices of the ideal $\mathcal{J}$ of $R/I$, determine the unique ideal $L$ of $R$ which has the properties that $L \supseteq I$ and $L/I = \mathcal{J}$.

(i) $\mathcal{J} = J/I + K/I$.
(ii) $\mathcal{J} = (J/I)(K/I)$.
(iii) $\mathcal{J} = (J/I : K/I)$.
(iv) $\mathcal{J} = (J/I)^n$, where $n \in \mathbb{N}$.
(v) $\mathcal{J} = (J/I) \cap (K/I)$.
(vi) $\mathcal{J} = 0$.
(vii) $\mathcal{J} = \sum_{i=1}^{h}(R/I)(a_i + I)$.
(viii) $\mathcal{J} = R/I$.

**2.39** REMARK. The results of 2.37 are so important that it is worth our while to dwell on them a little longer. So let, once again, $R$ be a commutative ring and let $I$ be an ideal of $R$. Let us use $\mathcal{I}_R$ to denote the set of all ideals of $R$, so that, in this notation, $\mathcal{I}_{R/I}$ denotes the set of all ideals of $R/I$. One way of describing the results of 2.37 is to say that there is a mapping

$$\theta \; : \; \{J \in \mathcal{I}_R : J \supseteq I\} \; \longrightarrow \; \mathcal{I}_{R/I}$$
$$J \; \longmapsto \; J/I$$

which is *bijective*. Note also that both $\theta$ and its inverse preserve inclusion relations: this means that, for $J_1, J_2 \in \mathcal{I}_R$ with $J_i \supseteq I$ for $i = 1, 2$, we have

$$J_1 \subseteq J_2 \quad \text{if and only if} \quad J_1/I \subseteq J_2/I.$$

**2.40** ♯EXERCISE. Let $I, J$ be ideals of the commutative ring $R$ such that $J \supseteq I$. Show that there is a ring isomorphism

$$\xi : (R/I)/(J/I) \longrightarrow R/J$$

for which $\xi\big((r + I) + J/I\big) = r + J$ for all $r \in R$. (Here is a hint: try using the Isomorphism Theorem for commutative rings.)

The results described in 2.37 and 2.39 are actually concerned with the effect on ideals produced by the natural ring homomorphism $f : R \to R/I$: they show that if $J$ is an ideal of $R$ which contains $I$, then $f(J)$ is actually an ideal of $R/I$, and that if $\mathcal{J}$ is an ideal of $R/I$, then

$$f^{-1}(\mathcal{J}) = \{r \in R : f(r) \in \mathcal{J}\}$$

is an ideal of $R$ which contains $I$.

We have also considered earlier the inverse image under a ring homomorphism of a particular ideal of a commutative ring: if $R$ and $S$ are both commutative rings and $g : R \to S$ is a ring homomorphism, then we can view $\operatorname{Ker} g$ as the inverse image under $g$ of the zero ideal of $S$. Of course, $\operatorname{Ker} g$ is an ideal of $R$.

The facts outlined in the preceding two paragraphs lead naturally to consideration of general phenomena, often described using the expressions 'extension' and 'contraction' of ideals under a ring homomorphism.

**2.41** LEMMA and DEFINITIONS. *Let $R$ and $S$ be commutative rings and let $f : R \to S$ be a ring homomorphism.*

(i) *Whenever $J$ is an ideal of $S$, then $f^{-1}(J) := \{r \in R : f(r) \in J\}$ is an ideal of $R$, called the* contraction *of $J$ to $R$. When there is no possibility of confusion over which ring homomorphism is under discussion, $f^{-1}(J)$ is often denoted by $J^c$.*

(ii) *For each ideal $I$ of $R$, the ideal $f(I)S$ of $S$ generated by $f(I)$ is called the* extension *of $I$ to $S$. Again, when no confusion is possible, $f(I)S$ is often denoted by $I^e$.*

*Proof.* Only part (i) needs proof. For this, consider the composite ring homomorphism

$$g : R \xrightarrow{f} S \longrightarrow S/J,$$

where the second map is the natural surjective ring homomorphism. The composition of two ring homomorphisms is, of course, again a ring homomorphism, and $\operatorname{Ker} g = \{r \in R : f(r) \in J\} = f^{-1}(J)$; it thus follows from 2.1 that $f^{-1}(J)$ is an ideal of $R$. $\square$

**2.42** ♯EXERCISE. Let the situation be as in 2.41 and suppose that the ideal $I$ of $R$ is generated by the set $H$. Show that the extension $I^e$ of $I$ to $S$ under $f$ is generated by $f(H) = \{f(h) : h \in H\}$.

**2.43** ♯EXERCISE. Let the situation be as in 2.41 and let $I_1, I_2$ be ideals of $R$ and $J_1, J_2$ be ideals of $S$. Prove that
  (i) $(I_1 + I_2)^e = I_1^e + I_2^e$;
  (ii) $(I_1 I_2)^e = I_1^e I_2^e$;

(iii) $(J_1 \cap J_2)^c = J_1^c \cap J_2^c$;

(iv) $(\sqrt{J_1})^c = \sqrt{(J_1^c)}$.

**2.44** LEMMA. *Let $R$ and $S$ be commutative rings, and let $f : R \to S$ be a ring homomorphism. Let $I$ be an ideal of $R$, and let $J$ be an ideal of $S$. Use the notation $I^e$ and $J^c$ of 2.41. Then*

(i) $I \subseteq I^{ec}$,

(ii) $J^{ce} \subseteq J$,

(iii) $I^e = I^{ece}$, *and*

(iv) $J^{cec} = J^c$.

*Proof.* (i) Let $r \in I$. Since $f(r) \in f(I) \subseteq f(I)S = I^e$, it is immediate that $r \in I^{ec}$.

(ii) By definition, $J^{ce}$ is the ideal of $S$ generated by $f(f^{-1}(J))$. But

$$f(f^{-1}(J)) \subseteq J,$$

an ideal of $S$, and so we must have $J^{ce} \subseteq J$.

(iii) By (i), we have $I \subseteq I^{ec}$. Take extensions to see that $I^e \subseteq I^{ece}$. The reverse inclusion is obtained by application of (ii) to the ideal $I^e$ of $S$.

(iv) This is proved in a way similar to that just used for (iii). By (ii), we have $J^{ce} \subseteq J$, so that $J^{cec} \subseteq J^c$ on contraction back to $R$. On the other hand, (i) applied to the ideal $J^c$ of $R$ yields that $J^c \subseteq J^{cec}$. $\square$

**2.45** COROLLARY. *Let the situation be as in 2.44, and use the notation $\mathcal{I}_R$ of 2.39 to denote the set of all ideals of $R$. Furthermore, set*

$$\mathcal{C}_R = \{J^c : J \in \mathcal{I}_S\}, \qquad \mathcal{E}_S = \{I^e : I \in \mathcal{I}_R\}.$$

*(We sometimes refer to $\mathcal{C}_R$ as the set of ideals of $R$ which are contracted from $S$ under $f$, or, more loosely, as the set of contracted ideals of $R$; similarly $\mathcal{E}_S$ is the set of ideals of $S$ which are extended from $R$ under $f$, or the set of extended ideals of $S$.)*

*It follows from 2.44(iv) that $I^{ec} = I$ for all $I \in \mathcal{C}_R$, and from 2.44(iii) that $J^{ce} = J$ for all $J \in \mathcal{E}_S$. Hence extension and contraction give us bijective mappings*

$$\begin{array}{ccc} \mathcal{C}_R & \longrightarrow & \mathcal{E}_S \\ I & \longmapsto & I^e \end{array} \qquad and \qquad \begin{array}{ccc} \mathcal{E}_S & \longrightarrow & \mathcal{C}_R \\ J & \longmapsto & J^c \end{array}$$

*which are inverses of each other.* $\square$

**2.46** ♯EXERCISE. We use the notation of 2.45. Suppose that the ring homomorphism $f : R \to S$ is surjective. Show that

$$\mathcal{C}_R = \{I \in \mathcal{I}_R : I \supseteq \mathrm{Ker}\, f\} \qquad and \qquad \mathcal{E}_S = \mathcal{I}_S.$$

Deduce that there is a bijective mapping

$$
\begin{aligned}
\{I \in \mathcal{I}_R : I \supseteq \operatorname{Ker} f\} &\longrightarrow \mathcal{I}_S \\
I &\longmapsto f(I)
\end{aligned}
$$

whose inverse is given by contraction.

Thus the above exercise shows that there is, for a *surjective* homomorphism $f : R \to S$ of commutative rings, a bijective mapping from the set of ideals of $R$ which contain $\operatorname{Ker} f$ to the set of all ideals of $S$ (and both this bijection and its inverse preserve inclusion relations). This is not surprising, since, by the Isomorphism Theorem for commutative rings 2.13, $R/\operatorname{Ker} f \cong S$, and we obtained similar results about the ideals of a residue class ring like $R/\operatorname{Ker} f$ in 2.39.

Our ideas about extension and contraction of ideals can throw light on some of the ideals in a polynomial ring.

**2.47** ♯EXERCISE. Let $R$ be a commutative ring and let $X$ be an indeterminate over $R$. Let $f : R \to R[X]$ denote the natural ring homomorphism, and use the extension and contraction terminology and notation of 2.41 with reference to $f$.

Let $I$ be an ideal of $R$, and for $r \in R$, denote the natural image of $r$ in $R/I$ by $\bar{r}$. By 1.13, there is a ring homomorphism

$$
\eta : R[X] \longrightarrow (R/I)[X]
$$

for which

$$
\eta\left(\sum_{i=0}^{n} r_i X^i\right) = \sum_{i=0}^{n} \overline{r_i} X^i
$$

for all $n \in \mathsf{N}_0$ and $r_0, r_1, \ldots, r_n \in R$. Show that
  (i) $I^e = \operatorname{Ker} \eta$, that is,

$$
I^e = \left\{ \sum_{i=0}^{n} r_i X^i \in R[X] : n \in \mathsf{N}_0,\ r_i \in I \text{ for all } i = 0, \ldots, n \right\};
$$

  (ii) $I^{ec} = I$, so that $\mathcal{C}_R = \mathcal{I}_R$;
  (iii) $R[X]/I^e = R[X]/IR[X] \cong (R/I)[X]$; and
  (iv) if $I_1, \ldots, I_n$ are ideals of $R$, then

$$
(I_1 \cap \ldots \cap I_n) R[X] = I_1 R[X] \cap \ldots \cap I_n R[X].
$$

**2.48** EXERCISE. Extend the results of Exercise 2.47 to the polynomial ring $R[X_1, \ldots, X_n]$ over the commutative ring $R$ in indeterminates $X_1, \ldots, X_n$.

**2.49** EXERCISE. Let $R$ be a commutative ring and let $f : R \to R[[X]]$ be the natural ring homomorphism, where $X$ is an indeterminate. To what extent can you imitate the results of 2.47 for this ring homomorphism?

**2.50** EXERCISE. Find an ideal of $\mathbf{Z}[X]$ (where $X$ is an indeterminate) which is not extended from $\mathbf{Z}$ under the natural ring homomorphism.

Is every ideal of $\mathbf{Z}[X]$ principal? Is $\mathbf{Z}[X]$ a Euclidean domain? Justify your responses.

# Chapter 3

# Prime ideals and maximal ideals

It was mentioned at the beginning of the last chapter that the concepts of prime ideal and maximal ideal are central to the applications of commutative ring theory to algebraic geometry. This chapter is concerned with the development of the theory of these two topics. It is reasonable to take the view that prime ideals form the most important class of ideals in commutative ring theory: one of our tasks in this chapter is to show that there is always an adequate supply of them.

But we must begin at the beginning, with the basic definitions. A good starting point is a discussion of the ideals of a field $K$. Since $K$ is, in particular, a non-trivial ring, we have $K \neq 0$, and so $K$ certainly has two (distinct) ideals, namely itself and its zero ideal. However, these are the only ideals of $K$, since if $I$ is an ideal of $K$ and $I \neq 0$, then there exists $r \in I$ with $r \neq 0$, and since $r$ is a unit of $K$, it follows that

$$K = Kr \subseteq I \subseteq K,$$

so that $I = K$. Thus we see that a field has exactly two ideals. In fact, this property serves to characterize fields among commutative rings, as we now show.

**3.1 Lemma.** *Let $R$ be a commutative ring. Then $R$ is a field if and only if $R$ has exactly two ideals.*

*Proof.* ($\Rightarrow$) This was proved in the paragraph immediately preceding the statement of the lemma.

37

($\Leftarrow$) Since a trivial commutative ring has just one ideal (namely itself), it follows that $R$ is not trivial. Let $r \in R$ with $r \neq 0$: we must show that $r$ is a unit of $R$. Now the principal ideal $Rr$ of $R$ is not zero, since $r \in Rr$. As $R$ and $0$ are two ideals of $R$ and are now known to be different, we must have $Rr = R$, and so there exists $u \in R$ with $ur = 1$. Thus $r$ is a unit of $R$; it follows that $R$ is a field. $\square$

Thus if $K$ is a field, then the zero ideal of $K$ is *maximal* with respect to inclusion among the set of proper ideals of $K$. (Recall from 2.14 that an ideal $I$ of a commutative ring $R$ is said to be proper precisely when $I \neq R$.)

**3.2** DEFINITION. An ideal $M$ of a commutative ring $R$ is said to be *maximal* precisely when $M$ is a maximal member, with respect to inclusion, of the set of proper ideals of $R$.

In other words, the ideal $M$ of $R$ is maximal if and only if
(i) $M \subset R$, and
(ii) there does not exist an ideal $I$ of $R$ with $M \subset I \subset R$.

**3.3** LEMMA. *Let $I$ be an ideal of the commutative ring $R$. Then $I$ is maximal if and only if $R/I$ is a field.*

*Proof.* This is immediate from 2.37 and 2.39, in which we established that there are bijective mappings

$$\begin{array}{ccc} \mathcal{C}_R & \longrightarrow & \mathcal{I}_{R/I} \\ I' & \longmapsto & I'/I = I'^e \end{array} \quad \text{and} \quad \begin{array}{ccc} \mathcal{I}_{R/I} & \longrightarrow & \mathcal{C}_R \\ J & \longmapsto & J^c \end{array}$$

which preserve inclusion relations. Here, we are using the extension and contraction notation of 2.41 and 2.45 in relation to the natural surjective ring homomorphism $R \rightarrow R/I$: thus, in particular, $\mathcal{C}_R$ denotes

$$\{I' \in \mathcal{I}_R : I' \supseteq I\},$$

the set of all ideals of $R$ which contain $I$.

It therefore follows from these results that $I$ is a maximal ideal of $R$ if and only if $0$ is a maximal ideal of $R/I$, and, by 3.1, this is the case if and only if $R/I$ is a field. $\square$

**3.4** ♯EXERCISE. Let $I, M$ be ideals of the commutative ring $R$ such that $M \supseteq I$. Show that $M$ is a maximal ideal of $R$ if and only if $M/I$ is a maximal ideal of $R/I$.

**3.5** EXAMPLE. The maximal ideals of the ring $\mathbf{Z}$ of integers are precisely the ideals of the form $\mathbf{Z}p$ where $p \in \mathbf{Z}$ is a prime number.

*Proof.* In 1.28 we saw that, for $n \in \mathbf{N}$ with $n > 1$, the residue class ring $\mathbf{Z}/\mathbf{Z}n$ is a field if and only if $n$ is a prime number, that is (in view of 3.3) the ideal $\mathbf{Z}n$ of $\mathbf{Z}$ is maximal if and only if $n$ is a prime number. Note that, by 2.34, each ideal of $\mathbf{Z}$ is principal, that $(-m)\mathbf{Z} = m\mathbf{Z}$ for each $m \in \mathbf{Z}$, and that the ideals $1\mathbf{Z}$ and $0\mathbf{Z}$ are not maximal (since $1\mathbf{Z} = \mathbf{Z}$ and $0\mathbf{Z} \subset 2\mathbf{Z} \subset \mathbf{Z}$); the result follows. $\square$

**3.6** ♯EXERCISE. Determine all the maximal ideals of the ring $K[X]$, where $K$ is a field and $X$ is an indeterminate.

The results of 3.5 and 3.6, together with the above-noted fact that the zero ideal of a field $K$ is a maximal ideal of $K$, do give us some examples of maximal ideals. However, at the moment we have no guarantee that an arbitrary non-trivial commutative ring has any maximal ideals. (It is clear that a trivial commutative ring has no maximal ideal since it does not even have any proper ideal.) To address this problem in a general commutative ring, we are going to use Zorn's Lemma. A few brief reminders about the terminology and ideas needed for Zorn's Lemma are therefore in order at this point.

**3.7** REMINDERS. Let $V$ be a non-empty set. A relation $\preceq$ on $V$ is said to be a *partial order* on $V$ precisely when it is reflexive (that is $u \preceq u$ for all $u \in V$), transitive (that is $u \preceq v$ and $v \preceq w$ for $u, v, w \in V$ imply $u \preceq w$) and antisymmetric (that is $u \preceq v$ and $v \preceq u$ for $u, v \in V$ imply $u = v$). If $\preceq$ is a partial order on $V$, then we write that $(V, \preceq)$ is a *partially ordered set*.

The partially ordered set $(V, \preceq)$ is said to be *totally ordered* precisely when, for each $u, v \in V$, it is the case that at least one of $u \preceq v$, $v \preceq u$ holds. Of course, each non-empty subset $W$ of our partially ordered set $(V, \preceq)$ is again partially ordered by $\preceq$, and we can discuss whether or not $W$ is totally ordered.

Let $W$ be a non-empty subset of the partially ordered set $(V, \preceq)$. An element $u \in V$ is said to be an *upper bound* of $W$ precisely when $w \preceq u$ for all $w \in W$.

If $(V, \preceq)$ is a partially ordered set, then, for $u, v \in V$, we write $u \prec v$ precisely when $u \preceq v$ and $u \neq v$. An element $m \in V$ is said to be a *maximal element of* $V$ precisely when there does not exist $w \in V$ with $m \prec w$. Thus $m \in V$ is a maximal element of $V$ if and only if $m \preceq v$ with $v \in V$ implies that $m = v$.

We now have available all the terminology needed for the statement of Zorn's Lemma.

**3.8** ZORN'S LEMMA. *Let $(V, \preceq)$ be a (non-empty) partially ordered set which has the property that every non-empty totally ordered subset of $V$ has an upper bound in $V$. Then $V$ has at least one maximal element.* □

For the purposes of this book, we should regard Zorn's Lemma as an axiom. Any reader who would like to learn more about it, such as its equivalence to the Axiom of Choice, is referred to Halmos' book [3].

Our first use of Zorn's Lemma in this book is to establish the existence of at least one maximal ideal in an arbitrary non-trivial commutative ring. However, there will be other uses of Zorn's Lemma later in the book.

**3.9** PROPOSITION. *Let $R$ be a non-trivial commutative ring. Then $R$ has at least one maximal ideal.*

*Proof.* Since $R$ is not trivial, the zero ideal $0$ is proper, and so the set $\Omega$ of all proper ideals of $R$ is not empty. Of course, the relation of inclusion, $\subseteq$, is a partial order on $\Omega$, and a maximal ideal of $R$ is just a maximal member of the partially ordered set $(\Omega, \subseteq)$. We therefore apply Zorn's Lemma to this partially ordered set.

Let $\Delta$ be a non-empty totally ordered subset of $\Omega$. Set

$$J = \bigcup_{I \in \Delta} I;$$

it is clear that $J$ is a non-empty subset of $R$ with the property that $ra \in J$ for all $a \in J$ and $r \in R$. Let $a, b \in J$; thus there exist $I_1, I_2 \in \Delta$ with $a \in I_1, b \in I_2$. Since $\Delta$ is totally ordered with respect to inclusion, either $I_1 \subseteq I_2$ or $I_2 \subseteq I_1$, and so $a + b$ belongs to the larger of the two. Hence $J$ is an ideal of $R$; furthermore, $J$ is proper since, for each $I \in \Delta$, we have $1 \notin I$.

Thus we have shown that $J \in \Omega$; it is clear that $J$ is an upper bound for $\Delta$ in $\Omega$. Since the hypotheses of Zorn's Lemma are satisfied, it follows that the partially ordered set $(\Omega, \subseteq)$ has a maximal element, and $R$ has a maximal ideal. □

A variant of the above result is very important.

**3.10** COROLLARY. *Let $I$ be a proper ideal of the commutative ring $R$. Then there exists a maximal ideal $M$ of $R$ with $I \subseteq M$.*

*Proof.* By 2.14, the residue class ring $R/I$ is non-trivial, and so, by 3.9, has a maximal ideal, which, by 2.37, will have to have the form $M/I$ for exactly one ideal $M$ of $R$ for which $M \supseteq I$. It now follows from 2.39 that $M$ is a maximal ideal of $R$.

Alternatively, one can modify the proof of 3.9 above and apply Zorn's Lemma to the set

$$\Omega' := \{K \in \mathcal{I}_R : R \supset K \supseteq I\}$$

(where, as usual, $\mathcal{I}_R$ denotes the set of all ideals of $R$); it is recommended that the reader tries this. □

**3.11** COROLLARY. *Let $R$ be a commutative ring, and let $a \in R$. Then $a$ is a unit of $R$ if and only if, for each maximal ideal $M$ of $R$, it is the case that $a \notin M$, that is, if and only if $a$ lies outside each maximal ideal of $R$.*

*Proof.* By 2.16, $a$ is a unit of $R$ if and only if $aR = R$.

($\Rightarrow$) If we had $a \in M$ for some maximal ideal $M$ of $R$, then we should have $aR \subseteq M \subset R$, so that $a$ could not be a unit of $R$.

($\Leftarrow$) If $a$ were not a unit of $R$, then $aR$ would be a proper ideal of $R$, and it would follow from 3.10 that $aR \subseteq M$ for some maximal ideal $M$ of $R$; but this would contradict the fact that $a$ lies outside each maximal ideal of $R$. □

A field is an example of a commutative ring which has exactly one maximal ideal, for its zero ideal is its unique proper ideal. There is a special name for a commutative ring which has exactly one maximal ideal.

**3.12** DEFINITION. A commutative ring $R$ which has exactly one maximal ideal, $M$ say, is said to be *quasi-local*. In these circumstances, the field $R/M$ is called the *residue field* of $R$.

**3.13** LEMMA. *Let $R$ be a commutative ring. Then $R$ is quasi-local if and only if the set of non-units of $R$ is an ideal.*

*Proof.* ($\Rightarrow$) Assume that $R$ is quasi-local with maximal ideal $M$. By 3.11, $M$ is precisely the set of non-units of $R$.

($\Leftarrow$) Assume that the set of non-units of $R$ is an ideal $I$ of $R$. Since $0 \in I$, we see that $0$ is a non-unit of $R$, and so $0 \neq 1$. Thus $R$ is not trivial, and so, by 3.9, has at least one maximal ideal: let $M$ be one such. By 3.11, $M$ consists of non-units of $R$, and so $M \subseteq I \subset R$. (It should be noted that $1 \notin I$ because $1$ is a unit of $R$.) Since $M$ is a maximal ideal of $R$, we have $M = I$. We have thus shown that $R$ has at least one maximal ideal, and any maximal ideal of $R$ must be equal to $I$. Hence $R$ is quasi-local. □

**3.14** REMARK. Suppose that the commutative ring $R$ is quasi-local. Then it follows from 3.11 that the unique maximal ideal of $R$ is precisely the set of non-units of $R$.

**3.15** ♯EXERCISE. Let $K$ be a field and let $a_1, \ldots, a_n \in K$. Show that the ideal

$$(X_1 - a_1, \ldots, X_n - a_n)$$

of the ring $K[X_1, \ldots, X_n]$ (of polynomials with coefficients in $K$ in indeterminates $X_1, \ldots, X_n$) is maximal.

The concept of maximal ideal in a commutative ring immediately leads to the very important idea of the Jacobson radical of such a ring.

**3.16** DEFINITION. Let $R$ be a commutative ring. We define the *Jacobson radical* of $R$, sometimes denoted by $\mathrm{Jac}(R)$, to be the intersection of all the maximal ideals of $R$.

Thus $\mathrm{Jac}(R)$ is an ideal of $R$: even in the case when $R$ is trivial, our convention concerning the intersection of the empty family of ideals of a commutative ring means that $\mathrm{Jac}(R) = R$.

Note that when $R$ is quasi-local, $\mathrm{Jac}(R)$ is the unique maximal ideal of $R$.

We can provide a characterization of the Jacobson radical of a commutative ring.

**3.17** LEMMA. *Let $R$ be a commutative ring, and let $r \in R$. Then $r \in \mathrm{Jac}(R)$ if and only if, for every $a \in R$, the element $1 - ra$ is a unit of $R$.*

*Proof.* ($\Rightarrow$) Suppose that $r \in \mathrm{Jac}(R)$. Suppose that, for some $a \in R$, it is the case that $1 - ra$ is not a unit of $R$. Then, by 3.11, there exists a maximal ideal $M$ of $R$ such that $1 - ra \in M$. But $r \in M$ by definition of $\mathrm{Jac}(R)$, and so

$$1 = (1 - ra) + ra \in M,$$

a contradiction.

($\Leftarrow$) Suppose that, for each $a \in R$, it is the case that $1 - ra$ is a unit of $R$. Let $M$ be a maximal ideal of $R$: we shall show that $r \in M$. If this were not the case, then we should have

$$M \subset M + Rr \subseteq R.$$

Hence, by the maximality of $M$, we deduce that $M + Rr = R$, so that there exist $b \in M$ and $a \in R$ with $b + ar = 1$. Hence $1 - ra \in M$, and so cannot be a unit of $R$. This contradiction shows that $r \in M$, as claimed. As this is true for each maximal ideal of $R$, we have $r \in \mathrm{Jac}(R)$. □

**3.18** EXERCISE. Consider the commutative ring $C[0,1]$ of all continuous real-valued functions defined on the closed interval $[0,1]$: see 1.2(iv). Let $z \in [0,1]$. Show that

$$M_z := \{f \in C[0,1] : f(z) = 0\}$$

is a maximal ideal of $C[0,1]$. Show further that every maximal ideal of $C[0,1]$ has this form. (Here is a hint for the second part. Let $M$ be a maximal ideal of $C[0,1]$. Argue by contradiction to show that the set

$$\{a \in [0,1] : f(a) = 0 \text{ for all } f \in M\}$$

is non-empty: remember that $[0,1]$ is a compact subset of **R**.)

**3.19** EXERCISE. Let $R$ be a quasi-local commutative ring with maximal ideal $M$. Show that the ring $R[[X_1, \ldots, X_n]]$ of formal power series in indeterminates $X_1, \ldots, X_n$ with coefficients in $R$ is again a quasi-local ring, and that its maximal ideal is generated by $M \cup \{X_1, \ldots, X_n\}$.

We now introduce the concept of prime ideal in a commutative ring.

**3.20** DEFINITION. Let $P$ be an ideal in a commutative ring $R$. We say that $P$ is a *prime ideal* of $R$ precisely when
  (i) $P \subset R$, that is, $P$ is a proper ideal of $R$, and
  (ii) whenever $a, b \in R$ with $ab \in P$, then either $a \in P$ or $b \in P$.

**3.21** REMARKS. Let $R$ be a commutative ring.
  (i) Note that $R$ itself is *not* considered to be a prime ideal of $R$.
  (ii) When $R$ is an integral domain, its zero ideal $0$ is a prime ideal of $R$.

**3.22** EXERCISE. (i) Determine all the prime ideals of the ring **Z** of integers.
  (ii) Determine all the prime ideals of the ring $K[X]$, where $K$ is a field and $X$ is an indeterminate.

The observation in 3.21(ii) provides a clue to a characterization of prime ideals in terms of residue class rings.

**3.23** LEMMA. *Let $I$ be an ideal of the commutative ring $R$. Then $I$ is prime if and only if the residue class ring $R/I$ is an integral domain.*

*Proof.* ($\Rightarrow$) Assume that $I$ is prime. Since $I$ is proper, $R/I$ is not trivial. Suppose that $a \in R$ is such that, in $R/I$, the element $a + I$ is a zerodivisor. Thus there exists $b \in R$ such that $b + I \neq 0_{R/I}$ but

$$(a + I)(b + I) = 0_{R/I} = 0 + I.$$

Then $ab \in I$ but $b \notin I$, so that, since $I$ is prime, we must have $a \in I$. Thus $a + I = 0_{R/I}$, and it follows that $R/I$ is an integral domain.

($\Leftarrow$) Assume that $R/I$ is an integral domain. Then $I \neq R$. Let $a, b \in R$ be such that $ab \in I$. Then, in $R/I$, we have

$$(a + I)(b + I) = ab + I = 0 + I = 0_{R/I};$$

since $R/I$ is an integral domain, either $a + I = 0_{R/I}$ or $b + I = 0_{R/I}$, so that either $a \in I$ or $b \in I$. Hence $I$ is prime. $\square$

**3.24** EXERCISE. Show that the residue class ring $S$ of the ring of polynomials $\mathbf{R}[X_1, X_2, X_3]$ over the real field $\mathbf{R}$ in indeterminates $X_1, X_2, X_3$ given by

$$S = \mathbf{R}[X_1, X_2, X_3]/(X_1^2 + X_2^2 + X_3^2)$$

is an integral domain.

**3.25** REMARKS. Let $R$ be a commutative ring.

(i) Since every field is an integral domain, it is immediate from 3.3 and 3.23 that every maximal ideal of $R$ is prime.

(ii) However, the converse of (i) is not true, since, for example, the zero ideal 0 of $\mathbf{Z}$ is prime, but $0 \subset 2\mathbf{Z} \subset \mathbf{Z}$.

**3.26** DEFINITION. Let $R$ be a commutative ring. We define the *prime spectrum*, or just the *spectrum*, of $R$ to be the set of all prime ideals of $R$. The spectrum of $R$ is denoted by $\mathrm{Spec}(R)$.

**3.27** REMARKS. Let $R$ be a commutative ring.

(i) It is immediate from 3.9 and the observation in 3.25(i) that each maximal ideal of $R$ is prime that $R$ is non-trivial if and only if $\mathrm{Spec}(R) \neq \emptyset$.

(ii) Let $f : R \to S$ be a homomorphism of commutative rings and let $Q \in \mathrm{Spec}(S)$. Then the composite ring homomorphism

$$R \xrightarrow{f} S \longrightarrow S/Q$$

(in which the second homomorphism is the natural surjective one) has kernel $f^{-1}(Q) = \{r \in R : f(r) \in Q\}$; hence, by the Isomorphism Theorem 2.13, $R/f^{-1}(Q)$ is isomorphic to a subring of the integral domain $S/Q$, and so must itself be an integral domain. Hence $f^{-1}(Q) \in \mathrm{Spec}(R)$, by 3.23. Thus the ring homomorphism $f : R \to S$ induces a mapping

$$\begin{array}{ccc} \mathrm{Spec}(S) & \longrightarrow & \mathrm{Spec}(R) \\ Q & \longmapsto & f^{-1}(Q) \end{array}.$$

(iii) However, if, in the situation of (ii) above, $N$ is a maximal ideal of $S$, then, although it is automatic that $f^{-1}(N) \in \operatorname{Spec}(R)$, it is not necessarily the case that $f^{-1}(N)$ is a maximal ideal of $R$. To see this, consider the inclusion ring homomorphism $f : \mathbf{Z} \to \mathbf{Q}$, and take $N = 0$, the zero ideal of $\mathbf{Q}$. Then $f^{-1}(N) = 0$, which is *not* a maximal ideal of $\mathbf{Z}$.

It is very important that the reader should have at his fingertips a description of the prime ideals of a residue class ring. We discuss this next.

**3.28** LEMMA. *Let $I$ be an ideal of the commutative ring $R$; let $J$ be an ideal of $R$ with $J \supseteq I$. Then the ideal $J/I$ of the residue class ring $R/I$ is prime if and only if $J$ is a prime ideal of $R$.*

*In other words, $J/I \in \operatorname{Spec}(R/I)$ if and only if $J \in \operatorname{Spec}(R)$.*

*Proof.* By 2.40,
$$(R/I) / (J/I) \cong R/J;$$

thus one of these rings is an integral domain if and only if the other is, and so the result follows from 3.23. $\square$

**3.29** EXERCISE. Determine the prime ideals of the ring $\mathbf{Z}/60\mathbf{Z}$ of residue classes of integers modulo 60.

**3.30** ♯EXERCISE. Let $R$ and $S$ be commutative rings, and let $f : R \to S$ be a surjective ring homomorphism. Use the extension and contraction notation of 2.41 and 2.45 in conjunction with $f$, so that, by 2.46, $\mathcal{C}_R = \{I \in \mathcal{I}_R : I \supseteq \operatorname{Ker} f\}$ and $\mathcal{E}_S = \mathcal{I}_S$.

Let $I \in \mathcal{C}_R$. Show that $I$ is a prime (respectively maximal) ideal of $R$ if and only if $I^e$ is a prime (respectively maximal) ideal of $S$.

Now we have available the concept of prime ideal, we can deal efficiently with one important elementary result. Recall from 2.36 that we say that an integral domain $R$ is a principal ideal domain (PID for short) precisely when every ideal of $R$ is principal, and that we saw in 2.34 that a Euclidean domain is a PID. A basic result about Euclidean domains is (see 1.41) that every Euclidean domain is a unique factorization domain (UFD for short). An important and related fact is that every PID is a UFD, and we now turn our attention to a proof of this.

**3.31** EXERCISE. Let $R$ be an integral domain. Recall (see [15, Definition 2.4.1]) that, for $a_1, \ldots, a_n \in R$, where $n \in \mathbf{N}$, a *greatest common divisor* (GCD for short) or *highest common factor* of $a_1, \ldots, a_n$ is an element $d \in R$ such that

(i) $d \mid a_i$ for all $i = 1, \ldots, n$, and

(ii) whenever $c \in R$ is such that $c \mid a_i$ for all $i = 1, \ldots, n$, then $c \mid d$.

Show that every non-empty finite set of elements in a PID has a GCD.

**3.32** Lemma. *Let $R$ be an integral domain, and let $a, b \in R \setminus \{0\}$. Then $aR = bR$ if and only if $a$ and $b$ are associates (that is (see [15, Definition 2.2.2]), $a = ub$ for some unit $u$ of $R$.*

*Proof.* ($\Rightarrow$) Suppose that $aR = bR$. Then $a = ub$ and $b = va$ for some $u, v \in R$. It follows that $a = uva$, so that, since $R$ is an integral domain and $a \neq 0$, we have $1 = uv$.

($\Leftarrow$) Suppose that $a = ub$ for some unit $u$ of $R$. Then $a \in bR$, so that $aR \subseteq bR$. Similarly, since $b = u^{-1}a$, we have $bR \subseteq aR$. $\square$

The reader will perhaps recall that the concept of prime element is very relevant to the theory of UFDs. We recall the definition, and, now that we have discussed prime ideals, it is desirable that we establish quickly the relationship between prime elements and prime ideals.

**3.33** Definition. (See [15, Definition 2.5.1].) Let $R$ be an integral domain and let $p \in R$. We say that $p$ is a *prime element* of $R$ precisely when $p$ is a non-zero, non-unit of $R$ with the property that, whenever $a, b \in R$ are such that $p \mid ab$, then either $p \mid a$ or $p \mid b$.

Some basic facts about prime elements in integral domains are established in [15, Theorem 2.5.2]: there it is shown that every prime element in an integral domain is irreducible, and that, in a UFD (and in particular in a Euclidean domain), every irreducible element is prime. It is clear from the definition that, if $R$ is an integral domain and $p \in R$, then $pR$ is a non-zero prime ideal if and only if $p$ is a prime element of $R$. We can say more when $R$ is a PID.

**3.34** Lemma. *Let $R$ be a PID, and let $p \in R \setminus \{0\}$. Then the following statements are equivalent:*

(i) *$pR$ is a maximal ideal of $R$;*
(ii) *$pR$ is a non-zero prime ideal of $R$;*
(iii) *$p$ is a prime element of $R$;*
(iv) *$p$ is an irreducible element of $R$.*

*Proof.* (i) $\Rightarrow$ (ii) This is clear because $p \neq 0$ and every maximal ideal of $R$ is prime.

(ii) $\Rightarrow$ (iii) As commented above, this is clear from the definitions.

(iii) $\Rightarrow$ (iv) This is easy, and proved in [15, Theorem 2.5.2], as was mentioned in the paragraph immediately preceding this lemma.

(iv) $\Rightarrow$ (i) Since $p$ is not a unit of $R$, it follows from 2.16 that $pR \subset R$. Let $I$ be an ideal of $R$ for which $pR \subseteq I \subset R$. Since $R$ is a PID, there exists $a \in R$ such that $I = aR$, and $a$ is a non-unit of $R$ since $I$ is proper. Now $p \in I$, and so $p = ab$ for some $b \in R$; since $p$ is irreducible and $a$ is a non-unit, it follows that $b$ is a unit of $R$, so that $pR = aR = I$ by 3.32. Thus $pR$ is maximal. $\square$

A consequence of 3.34 is that, in a PID which is not a field, an ideal is maximal if and only if it is a non-zero prime ideal. However, the reader should not lose sight of the fact that a field is automatically a PID, and the same statement is not true in a field!

In order to approach the result that, in a principal ideal domain $R$, each non-zero, non-unit of $R$ can be expressed as a product of a finite number of irreducible elements of $R$, it is convenient for us to introduce some general considerations about partially ordered sets.

**3.35** DEFINITIONS. Let $(V, \preceq)$ be a non-empty partially ordered set.

(i) We say that $(V, \preceq)$ satisfies the *ascending chain condition* if (and only if), whenever $(v_i)_{i \in \mathbb{N}}$ is a family of elements of $V$ such that

$$v_1 \preceq v_2 \preceq \ldots \preceq v_i \preceq v_{i+1} \preceq \ldots,$$

then there exists $k \in \mathbb{N}$ such that $v_k = v_{k+i}$ for all $i \in \mathbb{N}$.

(ii) We say that $(V, \preceq)$ satisfies the *maximal condition* if (and only if) every non-empty subset of $V$ contains a maximal element (with respect to $\preceq$).

It is a fundamental fact of commutative algebra that the ascending chain condition and the maximal condition are equivalent.

**3.36** LEMMA. *Let $(V, \preceq)$ be a non-empty partially ordered set. Then $(V, \preceq)$ satisfies the ascending chain condition if and only if it satisfies the maximal condition.*

*Proof.* Recall from 3.7 that, for $v, w \in V$, we write '$v \prec w$' to denote that $v \preceq w$ and $v \neq w$.

($\Rightarrow$) Let $T$ be a non-empty subset of $V$, and suppose that $T$ does not possess a maximal element. There exists $t_1 \in T$; since $T$ does not have a maximal element, there exists $t_2 \in T$ with $t_1 \prec t_2$. Continue in this way: if we have found $t_n \in T$, then there exists $t_{n+1} \in T$ such that $t_n \prec t_{n+1}$. In this manner we construct an infinite strictly ascending chain

$$t_1 \prec t_2 \prec \ldots \prec t_n \prec t_{n+1} \prec \ldots$$

of elements of $T \subseteq V$.

($\Leftarrow$) Now assume that $(V, \preceq)$ satisfies the maximal condition. Let

$$v_1 \preceq \ldots \preceq v_n \preceq v_{n+1} \preceq \ldots$$

be an ascending chain of elements of $V$. By the maximal condition applied to the subset $\{v_i : i \in \mathbf{N}\} =: T$ consisting of all terms in the chain, there exists $k \in \mathbf{N}$ such that $v_k$ is a maximal element of $T$. Then $v_k = v_{k+i}$ for all $i \in \mathbf{N}$. $\square$

**3.37** DEFINITIONS. Let $R$ be a commutative ring. We denote by $\mathcal{I}_R$ the set of all ideals of $R$. (We shall consistently use this notation throughout the book.) We say that $R$ is *Noetherian* precisely when the partially ordered set $(\mathcal{I}_R, \subseteq)$ satisfies the conditions of 3.35 (which are equivalent, by 3.36).

In other words, $R$ is Noetherian if and only if every ascending chain

$$I_1 \subseteq \ldots \subseteq I_n \subseteq I_{n+1} \subseteq \ldots$$

of ideals of $R$ is 'eventually stationary', and this is the case if and only if every non-empty set of ideals of $R$ has a maximal member with respect to inclusion.

We shall have a great deal to say about commutative Noetherian rings later in the book. Indeed, the development of the properties of commutative Noetherian rings is one of the major aims of commutative algebra. For our present purposes though, we just want to point out that a PID is Noetherian.

**3.38** PROPOSITION. *Let $R$ be a principal ideal domain. Then $R$ is Noetherian.*

*Proof.* Let

$$I_1 \subseteq \ldots \subseteq I_n \subseteq I_{n+1} \subseteq \ldots$$

be an ascending chain of ideals of $R$. It is easy to see that

$$J := \bigcup_{i \in \mathbf{N}} I_i$$

is an ideal of $R$: it is clearly non-empty and closed under multiplication by arbitrary elements of $R$, and, if $a \in I_n, b \in I_m$ where, for the sake of argument, $n \leq m$, then $a + b \in I_m$. Thus, since $R$ is a PID, there exists $a \in R$ such that $J = aR$. By definition of $J$, there exists $k \in \mathbf{N}$ such that $a \in I_k$. But then we have

$$J = aR \subseteq I_k \subseteq I_{k+i} \subseteq J$$

for all $i \in \mathbf{N}$. Thus our ascending chain of ideals must be stationary. $\square$

We are now in a position to prove that every PID is a UFD. Readers familiar with, say, the proof that every Euclidean domain is a UFD in

[15, Theorem 2.6.1] will perhaps realise that the argument used there to establish uniqueness can be used in a PID, now that we know from 3.34 that every irreducible element in a PID is a prime element. To establish the 'existence', we shall use the fact, just established in 3.38, that a principal ideal domain $R$ is Noetherian, so that the partially ordered set $(\mathcal{I}_R, \subseteq)$ satisfies the maximal condition.

**3.39** THEOREM. *Every principal ideal domain is a unique factorization domain.*

*Proof.* Let $R$ be a PID. We first show that every non-zero, non-unit element of $R$ can be factorized into the product of finitely many irreducible elements of $R$. Suppose that this is not the case. Then the set $\Omega$ of all ideals of $R$ of the form $aR$, where $a$ is a non-zero, non-unit element of $R$ which does *not* have a factorization of the above type, is non-empty; hence, by 3.38, the set $\Omega$ has a maximal element with respect to inclusion, $bR$ say, where $b$ is, in particular, a non-zero, non-unit element of $R$.

Now $b$ itself cannot be irreducible, for if it were, $b = b$ would be a factorization of the desired kind (with just one factor). Thus $b = cd$ for some $c, d \in R$, neither of which is a unit. It follows easily that

$$bR \subset cR \subset R \quad \text{and} \quad bR \subset dR \subset R.$$

(Bear in mind 3.32.) Hence, by the maximality of $bR$ in $\Omega$, we have $cR \notin \Omega$ and $dR \notin \Omega$. Neither $c$ nor $d$ is zero; neither is a unit. Therefore each of $c, d$ can be expressed as a product of finitely many irreducible elements of $R$, and so the same is true of $b = cd$. This is a contradiction. Hence every non-zero, non-unit element of $R$ can be factorized as a product of finitely many irreducible elements of $R$.

The uniqueness of such factorizations can now be established by an argument entirely similar to that used in [15, Theorem 2.6.1], and the details are left as an exercise for the reader. $\square$

**3.40** EXERCISE. Complete the proof of Theorem 3.39.

**3.41** EXERCISE. Show that the subring $\mathbf{Z}[\sqrt{-5}]$ of the field $\mathbf{C}$ is not a PID. Find an ideal in $\mathbf{Z}[\sqrt{-5}]$ which is not principal.

**3.42** ♯EXERCISE. Show that an irreducible element in a unique factorization domain $R$ generates a prime ideal of $R$.

We have already used Zorn's Lemma once in this chapter, in 3.9, where we showed that each non-trivial commutative ring possesses at least one maximal ideal; this shows, in particular, that a non-trivial commutative ring has at least one prime ideal. We have two more uses of Zorn's Lemma planned for this chapter, both of which are concerned with existence of prime ideals. The first can be regarded as a sharpening of 3.9. It is concerned with a multiplicatively closed subset in a commutative ring.

**3.43** DEFINITION. We say that a subset $S$ of a commutative ring $R$ is *multiplicatively closed* precisely when
  (i) $1 \in S$, and
  (ii) whenever $s_1, s_2 \in S$, then $s_1 s_2 \in S$ too.

The concept of multiplicatively closed subset of $R$ introduced in 3.43 is of fundamental importance in the subject. Two crucial examples of the idea are $R \setminus P$, where $P \in \operatorname{Spec}(R)$, and $\{f^n : n \in \mathbf{N}_0\}$, where $f$ is a (fixed) element of $R$. (Recall that $f^0$ is interpreted as 1.)

**3.44** THEOREM. *Let $I$ be an ideal of the commutative ring $R$, and let $S$ be a multiplicatively closed subset of $R$ such that $I \cap S = \emptyset$. Then the set*

$$\Psi := \{J \in \mathcal{I}_R : J \supseteq I \text{ and } J \cap S = \emptyset\}$$

*of ideals of $R$ (partially ordered by inclusion) has at least one maximal element, and any such maximal element of $\Psi$ is a prime ideal of $R$.*

*Proof.* Clearly $I \in \Psi$, and so $\Psi \neq \emptyset$. The intention is to apply Zorn's Lemma to the partially ordered set $\Psi$. So let $\Delta$ be a non-empty totally ordered subset of $\Psi$. Then

$$Q := \bigcup_{J \in \Delta} J$$

is an ideal of $R$ such that $Q \supseteq I$ and $Q \cap S = \emptyset$. (To see that $Q$ is closed under addition, note that, for $J, J' \in \Delta$, we have either $J \subseteq J'$ or $J' \subseteq J$.) Thus $Q$ is an upper bound for $\Delta$ in $\Psi$, and so it follows from Zorn's Lemma that $\Psi$ has at least one maximal element.

Let $P$ be an arbitrary maximal element of $\Psi$. Since $P \cap S = \emptyset$ and $1 \in S$, we see that $1 \notin P$ and $P \subset R$. Now let $a, a' \in R \setminus P$: we must show that $aa' \notin P$.

Since $a \notin P$, we have

$$I \subseteq P \subset P + Ra.$$

By the maximality of $P$ in $\Psi$, we must have $(P + Ra) \cap S \neq \emptyset$, and so there exist $s \in S, r \in R$ and $u \in P$ such that

$$s = u + ra;$$

similarly, there exist $s' \in S, r' \in R$ and $u' \in P$ such that

$$s' = u' + r'a'.$$

But then

$$ss' = (u + ra)(u' + r'a') = (uu' + rau' + r'a'u) + rr'aa'.$$

Since $ss' \in S$ (because $S$ is multiplicatively closed) and

$$uu' + rau' + r'a'u \in P,$$

we must have $aa' \notin P$ because $P \cap S = \emptyset$. Thus $P \in \mathrm{Spec}(R)$. □

The reader should note that 3.44 can be used to provide another proof of the result, already proved in 3.10, that a proper ideal $I$ of a commutative ring $R$ is contained in a maximal ideal of $R$: just take, in 3.44, $S = \{1\}$, which is certainly a multiplicatively closed subset of $R$ such that $I \cap S = \emptyset$, and note that a maximal member of

$$\{J \in \mathcal{I}_R : J \supseteq I \text{ and } J \cap S = \emptyset\}$$

must actually be a maximal ideal of $R$.

**3.45** REMARK. Let $P$ be a prime ideal of the commutative ring $R$. Note that $P$ is maximal if and only if $P$ is a maximal member of $\mathrm{Spec}(R)$ (with respect to inclusion).

We do have another use of 3.44 in mind in addition to a second proof of one of our earlier results. We use it now in connection with the idea of the radical of an ideal, introduced in Exercise 2.5. As this idea is of great importance in commutative algebra, we shall essentially provide a solution for Exercise 2.5 now.

**3.46** LEMMA and DEFINITION. *Let $R$ be a commutative ring and let $I$ be an ideal of $R$. Then*

$$\sqrt{I} := \{r \in R : \text{there exists } n \in \mathbf{N} \text{ with } r^n \in I\}$$

*is an ideal of $R$ which contains $I$, and is called the* radical *of $I$.*

*Alternative notation for $\sqrt{I}$, to be used when it is necessary to specify the ring under consideration, is $\mathrm{rad}_R I$.*

*Proof.* It is clear that $I \subseteq \sqrt{I}$, and that for $r \in R$ and $a \in \sqrt{I}$, we have $ra \in \sqrt{I}$. Let $a, b \in \sqrt{I}$, so that there exist $n, m \in \mathbf{N}$ such that $a^n, b^m \in I$. By 1.34,

$$(a + b)^{n+m-1} = \sum_{i=0}^{n+m-1} \binom{n + m - 1}{i} a^{n+m-1-i} b^i.$$

Now for each $i = 0, \ldots, n + m - 1$,

$$\text{either} \quad n + m - 1 - i \geq n \quad \text{or} \quad i \geq m,$$

so that either $a^{n+m-1-i} \in I$ or $b^i \in I$. Hence $(a + b)^{n+m-1} \in I$ and $a + b \in \sqrt{I}$. Thus $\sqrt{I}$ is an ideal of $R$. □

**3.47** ♯EXERCISE. Let $P$ be a prime ideal of the commutative ring $R$. Show that $\sqrt{(P^n)} = P$ for all $n \in \mathbf{N}$.

We are now ready to give another application of 3.44.

**3.48** NOTATION and LEMMA. *Let $I$ be an ideal of the commutative ring $R$. Define the* variety of $I$, *denoted* $\text{Var}(I)$, *to be the set*

$$\{P \in \text{Spec}(R) : P \supseteq I\}.$$

*Then*

$$\sqrt{I} = \bigcap_{P \in \text{Var}(I)} P = \bigcap_{\substack{P \in \text{Spec}(R) \\ P \supseteq I}} P.$$

*Proof.* Let $a \in \sqrt{I}$ and let $P \in \text{Var}(I)$. Then there exists $n \in \mathbf{N}$ such that $a^n \in I \subseteq P$, so that, since $P$ is prime, $a \in P$. Hence

$$\sqrt{I} \subseteq \bigcap_{P \in \text{Var}(I)} P.$$

To establish the reverse inclusion, let $b \in \bigcap_{P \in \text{Var}(I)} P$. We suppose that $b \notin \sqrt{I}$, and look for a contradiction. Our supposition means that $I \cap S = \emptyset$, where $S = \{b^h : h \in \mathbf{N}_0\}$, a multiplicatively closed subset of $R$. Hence, by 3.44, there exists a prime ideal $P'$ of $R$ such that $I \subseteq P'$ and $P' \cap S = \emptyset$. It follows that $P' \in \text{Var}(I)$, so that $b \in P' \cap S$. With this contradiction, the proof is complete. □

**3.49** COROLLARY. *The nilradical $\sqrt{0}$ of the commutative ring $R$ satisfies*

$$\sqrt{0} = \bigcap_{P \in \text{Spec}(R)} P.$$

*Proof.* This is immediate from 3.48 because every prime ideal of $R$ contains the zero ideal. □

**3.50** EXERCISE. Let $R$ be a commutative ring, and let $N$ be the nilradical of $R$. Show that the ring $R/N$ has zero nilradical. (A commutative ring is said to be *reduced* if and only if it has zero nilradical.)

**3.51** EXERCISE. Let $R$ be a non-trivial commutative ring. Show that $R$ has exactly one prime ideal if and only if each element of $R$ is either a unit or nilpotent.

We are now ready for the other application of Zorn's Lemma to an existence result about prime ideals which was hinted at earlier in the chapter. This one is concerned with the set $\text{Var}(I)$ of prime ideals of the commutative ring $R$ which contain the ideal $I$ of $R$. We know, from 3.10, that, if $I$ is proper, then $\text{Var}(I) \neq \emptyset$; what we are going to establish next is that $\text{Var}(I)$ actually contains *minimal* members with respect to inclusion. One nice aspect of the use of Zorn's Lemma here is that $\text{Var}(I)$ is regarded as a partially ordered set by *reverse* inclusion (that is, we write, for $P_1, P_2 \in \text{Var}(I)$,

$$P_1 \preceq P_2 \quad \text{if and only if} \quad P_1 \supseteq P_2),$$

so that a maximal member of this partially ordered set is just a minimal member of $\text{Var}(I)$ with respect to inclusion.

**3.52** THEOREM and DEFINITIONS. *Let $I$ be a proper ideal of the commutative ring $R$. Then*

$$\text{Var}(I) := \{P \in \text{Spec}(R) : P \supseteq I\}$$

*has at least one minimal member with respect to inclusion. Such a minimal member is called a* minimal prime ideal of $I$ *or a* minimal prime ideal containing $I$. *In the case when $R$ is not trivial, the minimal prime ideals of the zero ideal $0$ of $R$ are sometimes referred to as the* minimal prime ideals of $R$.

*Proof.* By 3.10, we have $\text{Var}(I) \neq \emptyset$. Partially order $\text{Var}(I)$ by reverse inclusion in the manner described just before the statement of the theorem. We are thus trying to establish the existence of a maximal element of our partially ordered set, and we use Zorn's Lemma for this purpose.

Let $\Omega$ be a non-empty subset of $\text{Var}(I)$ which is totally ordered with respect to the above partial order. Then

$$Q := \bigcap_{P \in \Omega} P$$

is a proper ideal of $R$, since $\Omega \neq \emptyset$. We show that $Q \in \text{Spec}(R)$. Let $a \in R \setminus Q, b \in R$ be such that $ab \in Q$. We must show that $b \in Q$. Let $P \in \Omega$. There exists $P_1 \in \Omega$ such that $a \notin P_1$.

Since $\Omega$ is totally ordered, either $P_1 \subseteq P$ or $P \subseteq P_1$. In the first case, the facts that $ab \in P_1$ and $a \notin P_1$ imply that $b \in P_1 \subseteq P$; in the second case, we must have $a \notin P$ and $ab \in P$, so that $b \in P$. Thus $b \in P$ in any event, and, since $P$ was any arbitrary member of $\Omega$, it follows that $b \in Q$. Therefore $Q \in \text{Spec}(R)$. Since $Q \supseteq I$, we have $Q \in \text{Var}(I)$, and $Q$ is an upper bound for $\Omega$ in our partially ordered set. We now use Zorn's Lemma to complete the proof. $\square$

In fact, a variation of the above result is perhaps needed more than the result itself: we often need to know that, if $P$ is a prime ideal of the commutative ring $R$ and $P$ contains the ideal $I$ of $R$, then there exists a minimal prime ideal $P'$ of $I$ with $P \supseteq P'$. This can be achieved with a modification of the above argument, and this modification is left as an exercise for the reader.

**3.53** ♯EXERCISE. Let $P, I$ be ideals of the commutative ring $R$ with $P$ prime and $P \supseteq I$. Show that the non-empty set

$$\Theta := \{P' \in \text{Spec}(R) : P \supseteq P' \supseteq I\}$$

has a minimal member with respect to inclusion (by partially ordering $\Theta$ by reverse inclusion and using Zorn's Lemma). Note that a minimal member of $\Theta$ is a minimal prime ideal of $I$, and so deduce that there exists a minimal prime ideal $P''$ of $I$ with $P'' \subseteq P$.

**3.54** COROLLARY. *Let $I$ be a proper ideal of the commutative ring $R$, and let $\text{Min}(I)$ denote the set of minimal prime ideals of $I$. Then*

$$\sqrt{I} = \bigcap_{P \in \text{Min}(I)} P.$$

*Proof.* By 3.48, $\sqrt{I} = \bigcap_{P \in \text{Var}(I)} P$, and, since $\text{Min}(I) \subseteq \text{Var}(I)$, it is clear that

$$\bigcap_{P \in \text{Var}(I)} P \subseteq \bigcap_{P \in \text{Min}(I)} P.$$

However, the reverse inclusion is immediate from 3.53, which shows that every prime in $\text{Var}(I)$ contains a minimal prime ideal of $I$. $\square$

The final few results in this chapter are concerned with properties of prime ideals. The most important is probably the Prime Avoidance Theorem because it is, among other things, absolutely fundamental to the theory of regular sequences in commutative algebra.

**3.55** LEMMA. *Let $P$ be a prime ideal of the commutative ring $R$, and let $I_1, \ldots, I_n$ be ideals of $R$. Then the following statements are equivalent:*

(i) $P \supseteq I_j$ *for some $j$ with* $1 \leq j \leq n$;
(ii) $P \supseteq \bigcap_{i=1}^{n} I_i$;
(iii) $P \supseteq \prod_{i=1}^{n} I_i$.

*Proof.* It is clear that (i) $\Rightarrow$ (ii) and (ii) $\Rightarrow$ (iii).

(iii) $\Rightarrow$ (i) Suppose that, for all $j$ with $1 \leq j \leq n$, it is the case that $P \not\supseteq I_j$. Then, for each such $j$, there exists $a_j \in I_j \setminus P$; but then

$$a_1 \ldots a_n \in \prod_{i=1}^{n} I_i \setminus P$$

(because $P$ is prime), and this contradicts the statement of (iii). $\square$

**3.56** COROLLARY. *Let $I_1, \ldots, I_n$ be ideals of the commutative ring $R$, and suppose that $P$ is a prime ideal of $R$ such that $P = \bigcap_{i=1}^{n} I_i$. Then $P = I_j$ for some $j$ with $1 \leq j \leq n$.* $\square$

The next proposition gives an illustration of the sort of use to which 3.55 can be put. It is concerned with comaximal ideals.

**3.57** DEFINITION. *Let $I, J, I_1, \ldots, I_n$, where $n \in \mathbb{N}$ with $n \geq 2$, be ideals of the commutative ring $R$. We say that $I$ and $J$ are comaximal (or coprime) precisely when $I + J = R$; also, we say that the family $(I_i)_{i=1}^{n}$ is pairwise comaximal if and only if $I_i + I_j = R$ whenever $1 \leq i, j \leq n$ and $i \neq j$.*

**3.58** LEMMA. *Let $I, J$ be comaximal ideals of the commutative ring $R$. Then $I \cap J = IJ$.*

*Proof.* Of course, $IJ \subseteq I \cap J$. By hypothesis, $I + J = R$; hence

$$I \cap J = (I \cap J)R = (I \cap J)(I + J) = (I \cap J)I + (I \cap J)J$$

by 2.28(iv). But $(I \cap J)I \subseteq JI$ and $(I \cap J)J \subseteq IJ$. It follows that $I \cap J \subseteq IJ$, and the proof is complete. $\square$

**3.59** PROPOSITION. *Let $(I_i)_{i=1}^{n}$ (where $n \geq 2$) be a pairwise comaximal family of ideals of the commutative ring $R$. Then*
(i) $I_1 \cap \ldots \cap I_{n-1}$ *and $I_n$ are comaximal, and*
(ii) $I_1 \cap \ldots \cap I_n = I_1 \ldots I_n$.

*Proof.* (i) Set $J := \bigcap_{i=1}^{n-1} I_i$. Suppose that $M$ is a maximal ideal of $R$ such that $J + I_n \subseteq M$. Then $I_n \subseteq M$ and

$$J = I_1 \cap \ldots \cap I_{n-1} \subseteq M;$$

hence, by 3.55, there is a $j \in \mathbb{N}$ with $1 \le j \le n - 1$ such that $I_j \subseteq M$, so that

$$I_j + I_n \subseteq M.$$

But this is a contradiction because $I_j$ and $I_n$ are comaximal. Hence there is no maximal ideal of $R$ that contains $J + I_n$, and so, by 3.10, $J + I_n = R$.

(ii) We prove this by induction on $n$, the case in which $n = 2$ having been dealt with in 3.58. So we suppose that $n = k \ge 3$ and that the result has been proved for smaller values of $n$. We see immediately from this induction hypothesis that

$$J := \bigcap_{i=1}^{k-1} I_i = \prod_{i=1}^{k-1} I_i.$$

By part (i) above, $J$ and $I_k$ are comaximal, so that, by 3.58, we have $J \cap I_k = JI_k$. It therefore follows from the above displayed equation that

$$\bigcap_{i=1}^{k} I_i = J \cap I_k = JI_k = \prod_{i=1}^{k} I_i.$$

This completes the inductive step, and the proof. $\square$

**3.60** EXERCISE. Let $I_1, \ldots, I_n$, where $n \ge 2$, be ideals of the commutative ring $R$. Recall the construction of the direct product $\prod_{i=1}^{n} R/I_i$ of the rings $R/I_1, \ldots, R/I_n$ from 2.6.

(i) Show that there is a ring homomorphism

$$f : R \longrightarrow R/I_1 \times \cdots \times R/I_n$$

given by $f(r) = (r + I_1, \ldots, r + I_n)$ for all $r \in R$.

(ii) Show that $f$ is injective if and only if $\bigcap_{i=1}^{n} I_i = 0$.

(iii) Show that $f$ is surjective if and only if the family $(I_i)_{i=1}^{n}$ is pairwise comaximal.

**3.61** THE PRIME AVOIDANCE THEOREM. *Let $P_1, \ldots, P_n$, where $n \ge 2$, be ideals of the commutative ring $R$ such that at most 2 of $P_1, \ldots, P_n$ are not prime. Let $S$ be an additive subgroup of $R$ which is closed under multiplication. (For example, $S$ could be an ideal of $R$, or a subring of $R$.) Suppose that*

$$S \subseteq \bigcup_{i=1}^{n} P_i.$$

*Then $S \subseteq P_j$ for some $j$ with $1 \le j \le n$.*

*Proof.* We use induction on $n$.

Consider first the case in which $n = 2$. Here we have $S \subseteq P_1 \cup P_2$, and we assume merely that $P_1$ and $P_2$ are ideals. Suppose that $S \nsubseteq P_1$ and $S \nsubseteq P_2$ and look for a contradiction. Thus there exists, for $j = 1, 2$, an element $a_j \in S \setminus P_j$; the hypotheses therefore imply that

$$a_1 \in P_2 \quad \text{and} \quad a_2 \in P_1.$$

Now $a_1 + a_2 \in S \subseteq P_1 \cup P_2$, and so $a_1 + a_2$ belongs to either $P_1$ or $P_2$. In the former case, we have

$$a_1 = (a_1 + a_2) - a_2 \in P_1,$$

which is a contradiction; the second possibility leads to a contradiction in a similar way. Thus we must have $S \subseteq P_j$ for $j = 1$ or $j = 2$.

We now turn to the inductive step. Assume, inductively, that $n = k+1$, where $k \geq 2$, and that the result has been proved in the case where $n = k$. Thus we have $S \subseteq \bigcup_{i=1}^{k+1} P_i$ and, since at most 2 of the $P_i$ are not prime, we can, and do, assume that they have been indexed in such a way that $P_{k+1}$ is prime.

Suppose that, for each $j = 1, \ldots, k+1$, it is the case that

$$S \nsubseteq \bigcup_{\substack{i=1 \\ i \neq j}}^{k+1} P_i.$$

Thus, for each $j = 1, \ldots, k+1$, there exists

$$a_j \in S \setminus \bigcup_{\substack{i=1 \\ i \neq j}}^{k+1} P_i.$$

The hypotheses imply that $a_j \in P_j$ for all $j = 1, \ldots, k+1$. Also, since $P_{k+1} \in \text{Spec}(R)$, we have $a_1 \ldots a_k \notin P_{k+1}$. Thus

$$a_1 \ldots a_k \in \bigcap_{i=1}^{k} P_i \setminus P_{k+1} \quad \text{and} \quad a_{k+1} \in P_{k+1} \setminus \bigcup_{i=1}^{k} P_i.$$

Now consider the element $b := a_1 \ldots a_k + a_{k+1}$: we cannot have $b \in P_{k+1}$, for that would imply

$$a_1 \ldots a_k = b - a_{k+1} \in P_{k+1},$$

a contradiction; also, we cannot have $b \in P_j$ for some $j$ with $1 \leq j \leq k$, for that would imply

$$a_{k+1} = b - a_1 \ldots a_k \in P_j,$$

again a contradiction. But $b \in S$ since $a_j \in S$ for $j = 1, \ldots, k+1$, and so we have a contradiction to the hypothesis that $S \subseteq \bigcup_{i=1}^{k+1} P_i$.

It follows that there is at least one $j$ with $1 \leq j \leq k+1$ for which

$$S \subseteq \bigcup_{\substack{i=1 \\ i \neq j}}^{k+1} P_i,$$

so that we can now use the inductive hypothesis to deduce that $S \subseteq P_i$ for some $i$ with $1 \leq i \leq k+1$.

This completes the inductive step, and so the theorem has been proved by induction. $\square$

**3.62** REMARKS. The notation is as in 3.61.

(i) The Prime Avoidance Theorem is most frequently used in situations where $S$ is actually an ideal of $R$ and $P_1, \ldots, P_n$ are all prime ideals of $R$. However, there are some occasions when it is helpful to have more of the full force of our statement of 3.61 available.

(ii) Why is 3.61 called the 'Prime Avoidance Theorem'? The name is explained by the following reformulation of its statement. If $P_1, \ldots, P_n$ are ideals of $R$, where $n \geq 2$, and at most 2 of $P_1, \ldots, P_n$ are not prime, and if, for each $i = 1, \ldots, n$, we have $S \not\subseteq P_i$, then there exists

$$c \in S \setminus \bigcup_{i=1}^{n} P_i,$$

so that $c$ 'avoids' all the ideals $P_1, \ldots, P_n$, 'most' of which are prime.

**3.63** EXERCISE. Let $R$ be a commutative ring which contains an infinite field as a subring. Let $I$ and $J_1, \ldots, J_n$, where $n \geq 2$, be ideals of $R$ such that

$$I \subseteq \bigcup_{i=1}^{n} J_i.$$

Prove that $I \subseteq J_j$ for some $j$ with $1 \leq j \leq n$.

There is a refinement of the Prime Avoidance Theorem that is sometimes extremely useful.

—

**3.64** THEOREM. *Let $P_1, \ldots, P_n$, where $n \geq 1$, be prime ideals of the commutative ring $R$, let $I$ be an ideal of $R$, and let $a \in R$ be such that*

$$aR + I \not\subseteq \bigcup_{i=1}^{n} P_i.$$

*Then there exists $c \in I$ such that*

$$a + c \notin \bigcup_{i=1}^{n} P_i.$$

*Proof.* First note that, if $P_i \subseteq P_j$ for some $i, j$ with $1 \leq i, j \leq n$ and $i \neq j$, then we can discard $P_i$ from our list of prime ideals without changing the problem. We can, and do, therefore assume that, for all $i, j = 1, \ldots, n$ with $i \neq j$, we have $P_i \not\subseteq P_j$ and $P_j \not\subseteq P_i$.

Now suppose that the $P_1, \ldots, P_n$ have been numbered (renumbered if necessary) so that $a$ lies in all of $P_1, \ldots, P_k$ but in none of $P_{k+1}, \ldots, P_n$. If $k = 0$, then $a = a + 0 \notin \bigcup_{i=1}^{n} P_i$ and we have an element of the desired form. We therefore assume henceforth in this proof that $k \geq 1$.

Now $I \not\subseteq \bigcup_{i=1}^{k} P_i$, for otherwise, by the Prime Avoidance Theorem 3.61, we would have $I \subseteq P_j$ for some $j$ with $1 \leq j \leq k$, which would imply that

$$aR + I \subseteq P_j \subseteq \bigcup_{i=1}^{n} P_i,$$

contrary to hypothesis. Thus there exists $d \in I \setminus (P_1 \cup \ldots \cup P_k)$.

Next, note that

$$P_{k+1} \cap \ldots \cap P_n \not\subseteq P_1 \cup \ldots \cup P_k:$$

this is clearly so if $k = n$, for then the left-hand side should be interpreted as $R$; and if the above claim were false in the case in which $k < n$, then it would follow from the Prime Avoidance Theorem 3.61 that

$$P_{k+1} \cap \ldots \cap P_n \subseteq P_j$$

for some $j$ with $1 \leq j \leq k$, and it would then follow from 3.55 that $P_h \subseteq P_j$ for some $h$ with $k+1 \leq h \leq n$, contrary to the arrangements that we made. Thus there exists

$$b \in P_{k+1} \cap \ldots \cap P_n \setminus (P_1 \cup \ldots \cup P_k).$$

Now define $c := db \in I$, and note that

$$c \in P_{k+1} \cap \ldots \cap P_n \setminus (P_1 \cup \ldots \cup P_k)$$

since $P_1, \ldots, P_k \in \operatorname{Spec}(R)$. Since

$$a \in P_1 \cap \ldots \cap P_k \setminus (P_{k+1} \cup \ldots \cup P_n),$$

it follows that $a + c \notin \bigcup_{i=1}^n P_i$. $\square$

**3.65** ♯EXERCISE. Let $R$ be a commutative ring and let $X$ be an indeterminate; use the extension and contraction notation of 2.41 in conjunction with the natural ring homomorphism $f : R \to R[X]$; and let $I$ be an ideal of $R$.

(i) Show that $I \in \operatorname{Spec}(R)$ if and only if $I^e \in \operatorname{Spec}(R[X])$.

(ii) Prove that

$$\sqrt{(I^e)} = (\sqrt{I})^e.$$

(iii) Let $M$ be a maximal ideal of $R$. Decide whether it is (a) always, (b) sometimes, or (c) never true that $M^e$ is a maximal ideal of $R[X]$, and justify your response.

**3.66** ♯EXERCISE. Let $K$ be a field, and let $R = K[X_1, \ldots, X_n]$ be the ring of polynomials over $K$ in indeterminates $X_1, \ldots, X_n$; let $\alpha_1, \ldots, \alpha_n \in K$. Show that, in $R$,

$$0 \subset (X_1 - \alpha_1) \subset (X_1 - \alpha_1, X_2 - \alpha_2) \subset \ldots$$
$$\subset (X_1 - \alpha_1, \ldots, X_i - \alpha_i) \subset \ldots$$
$$\subset (X_1 - \alpha_1, \ldots, X_n - \alpha_n)$$

is a (strictly) ascending chain of prime ideals.

**3.67** EXERCISE. Let $t \in \mathbb{N}$ and let $p_1, \ldots, p_t$ be $t$ distinct prime numbers. Show that

$$R = \{\alpha \in \mathbb{Q} : \alpha = m/n \text{ for some } m \in \mathbb{Z} \text{ and } n \in \mathbb{N} \text{ such that}$$
$$n \text{ is divisible by none of } p_1, \ldots, p_t\}$$

is a subring of $\mathbb{Q}$ which has exactly $t$ maximal ideals.

**3.68** EXERCISE. Let $R$ be a commutative ring, and let $f = \sum_{i=0}^{\infty} f_i \in R[[X]]$, the ring of formal power series over $R$ in the indeterminate $X$, where, for each $i \in \mathbb{N}_0$, $f_i$ is a form in $R[X]$ which is either 0 or of degree $i$. Use the contraction notation of 2.41 with reference to the natural inclusion ring homomorphism from $R$ to $R[[X]]$.

(i) Show that $f \in \operatorname{Jac}(R[[X]])$ if and only if $f_0 \in \operatorname{Jac}(R)$.

(ii) Let $\mathcal{M}$ be a maximal ideal of $R[[X]]$. Show that $\mathcal{M}$ is generated by $\mathcal{M}^c \cup \{X\}$, and that $\mathcal{M}^c$ is a maximal ideal of $R$.

(iii) Show that each prime ideal of $R$ is the contraction of a prime ideal of $R[[X]]$.

# Chapter 4

# Primary decomposition

One of the really satisfactory aspects of a Euclidean domain is that it is a unique factorization domain (UFD). We have also seen in 3.39 that every principal ideal domain is a UFD. It is natural to ask to what extent these results can be generalized. In fact, there is available a very elegant theory which can be viewed as providing a generalization of the fact that a PID is a UFD. This is the theory of primary decomposition of proper ideals in a commutative Noetherian ring, and we are going to provide an introduction to this theory in this chapter.

For motivation, let us temporarily consider a principal ideal domain $R$, which is not a field. The theory of primary decomposition is more concerned with ideals than elements, and so let us consider a non-zero, proper ideal $I$ of $R$. Of course, $I$ will be principal, and so there exists a non-zero, non-unit $a \in R$ such that $I = aR$. Since $R$ is, by 3.39, a UFD, there exist $s \in \mathsf{N}$, irreducible elements $p_1, \ldots, p_s \in R$ such that $p_i$ and $p_j$ are not associates whenever $i \neq j$ ($1 \leq i, j \leq s$), a unit $u$ of $R$, and $t_1, \ldots, t_s \in \mathsf{N}$ such that

$$a = u p_1^{t_1} \ldots p_s^{t_s}.$$

However, we are interested in the ideal $I = aR$: we can use the idea of the product of finitely many ideals of $R$ and the comments in 2.28 to deduce that

$$I = aR = \prod_{i=1}^{s} R p_i^{t_i}.$$

We can now use some of the results of Chapter 3 concerning comaximal ideals to deduce, from the above equation, another expression for $I$ as an intersection of ideals of a certain type. Let $i, j \in \mathsf{N}$ with $1 \leq i, j \leq s$ and $i \neq j$. By 3.34, $Rp_i$ and $Rp_j$ are maximal ideals of $R$, and since $p_i$ and $p_j$

61

are not associates, we can deduce from 3.32 that these two maximal ideals of $R$ are different. Hence

$$Rp_i \subset Rp_i + Rp_j \subseteq R,$$

since if $Rp_i = Rp_i + Rp_j$ were the case then we should have $Rp_j \subseteq Rp_i \subset R$, which would imply that $Rp_i = Rp_j$. It follows that $Rp_i + Rp_j = R$, so that $Rp_i$ and $Rp_j$ are comaximal. Hence $Rp_i^{t_i}$ and $Rp_j^{t_j}$ are also comaximal, because

$$\sqrt{(Rp_i^{t_i})} = Rp_i \qquad \text{and} \qquad \sqrt{(Rp_j^{t_j})} = Rp_j$$

by 3.47, so that $Rp_i^{t_i} + Rp_j^{t_j} = R$ by 2.25(iv). It therefore follows from 3.59(ii) that

$$I = Ra = Rp_1^{t_1} \cap \ldots \cap Rp_s^{t_s}.$$

Now, for each $i = 1, \ldots, s$, the ideal $Rp_i^{t_i} = (Rp_i)^{t_i}$ is a positive power of a maximal ideal of $R$, and we shall see from the definition and results below that a positive power of a maximal ideal of a commutative ring is an example of what is known as a 'primary ideal'. Thus we have expressed our ideal $I$ of $R$ as an intersection of finitely many primary ideals of $R$; such an expression is known as a 'primary decomposition' of $I$.

One of the main aims of this chapter is to show that every proper ideal in a commutative Noetherian ring has a primary decomposition, that is, can be expressed as an intersection of finitely many primary ideals. In view of the observations in the above paragraph, this result can be viewed as a generalization of the fact that a PID is a UFD.

But we must begin with the basic definitions, such as that of primary ideal.

**4.1** DEFINITION. Let $Q$ be an ideal of a commutative ring $R$. We say that $Q$ is a *primary ideal* of $R$ precisely when

  (i) $Q \subset R$, that is $Q$ is a proper ideal of $R$, and

  (ii) whenever $a, b \in R$ with $ab \in Q$ but $a \notin Q$, then there exists $n \in \mathbf{N}$ such that $b^n \in Q$.

Condition (ii) in 4.1 can be rephrased as follows: $a, b \in R$ and $ab \in Q$ imply $a \in Q$ or $b \in \sqrt{Q}$, where $\sqrt{Q}$ denotes the radical of $Q$ (see 3.46).

**4.2** REMARK. It should be clear to the reader that every prime ideal in a commutative ring $R$ is a primary ideal of $R$.

Recall from 3.23 that, for an ideal $I$ of $R$, we have that $I$ is prime if and only if $R/I$ is an integral domain. We used this in 3.27(ii) to deduce that, if $f : R \to S$ is a homomorphism of commutative rings and $P' \in \mathrm{Spec}(S)$, then $P'^c := f^{-1}(P') \in \mathrm{Spec}(R)$. There is a similar circle of ideas concerning primary ideals.

**4.3** LEMMA. (i) *Let $I$ be an ideal of the commutative ring $R$. Then $I$ is primary if and only if the ring $R/I$ is not trivial and has the property that every zerodivisor in $R/I$ is nilpotent.*

(ii) *Let $f : R \to S$ be a homomorphism of commutative rings, and let $Q$ be a primary ideal of $S$. Then $Q^c := f^{-1}(Q)$ is a primary ideal of $R$.*

*Proof.* (i) ($\Rightarrow$) Suppose that $I$ is primary. Since $I \neq R$ we deduce that $R/I$ is not trivial, by 2.14. Let $b \in R$ be such that the element $b + I$ in $R/I$ is a zerodivisor, so that there exists $a \in R$ such that $a + I \neq 0_{R/I}$ but $(a + I)(b + I) = 0_{R/I}$. These conditions mean that $a \notin I$ but $ab \in I$, so that, since $I$ is primary, there exists $n \in \mathbb{N}$ such that $b^n \in I$. Hence $(b + I)^n = b^n + I = 0_{R/I}$.

($\Leftarrow$) This is just as straightforward, and will be left as an exercise for the reader.

(ii) The composite ring homomorphism

$$R \xrightarrow{\ f\ } S \longrightarrow S/Q$$

(in which the second homomorphism is the natural surjective one) has kernel $Q^c$, and so it follows from the Isomorphism Theorem 2.13 that $R/Q^c$ is isomorphic to a subring of $S/Q$. Now if a commutative ring $R'$ is non-trivial and has the property that every zerodivisor in it is nilpotent, then each subring of $R'$ has the same two properties. Hence it follows from part (i) that $Q^c$ is a primary ideal of $R$. $\square$

**4.4** ♯EXERCISE. Complete the proof of 4.3(i).

Primary ideals have very nice radicals, as we now show.

**4.5** LEMMA and DEFINITION. *Let $Q$ be a primary ideal of the commutative ring $R$. Then $P := \sqrt{Q}$ is a prime ideal of $R$, and we say that $Q$ is $P$-primary.*

*Furthermore, $P$ is the smallest prime ideal of $R$ which contains $Q$, in that every prime ideal of $R$ which contains $Q$ must also contain $P$. Thus (see 3.52) $P$ is the unique minimal prime ideal of $Q$.*

*Proof.* Since $1 \notin Q$, we must have $1 \notin \sqrt{Q} = P$, so that $P$ is proper. Suppose that $a, b \in R$ with $ab \in \sqrt{Q}$ but $a \notin \sqrt{Q}$. Thus there exists $n \in \mathbb{N}$ such that $(ab)^n = a^n b^n \in Q$; however, no positive power of $a$ belongs to $Q$, and so no positive power of $a^n$ lies in $Q$. Since $Q$ is primary, it follows from the definition that $b^n \in Q$, so that $b \in \sqrt{Q}$. Hence $P = \sqrt{Q}$ is prime.

To prove the claim in the last paragraph, note that, if $P' \in \operatorname{Spec}(R)$ and $P' \supseteq Q$, then we can take radicals and use 3.47 to see that

$$P' = \sqrt{P'} \supseteq \sqrt{Q} = P.$$

Hence $P$ is the one and only minimal prime ideal of $Q$. $\square$

**4.6** REMARK. Let $f : R \to S$ be a homomorphism of commutative rings, and let $Q'$ be a $P'$-primary ideal of $S$. We saw in 4.3(ii) that $Q'^c := f^{-1}(Q')$ is a primary ideal of $R$. It follows from 2.43(iv) that $\sqrt{(Q'^c)} = P'^c$, so that $Q'^c$ is actually a $P'^c$-primary ideal of $R$.

**4.7** ♯EXERCISE. Let $f : R \to S$ be a surjective homomorphism of commutative rings. Use the extension and contraction notation of 2.41 and 2.45 in conjunction with $f$. Note that, by 2.46, $\mathcal{C}_R = \{I \in \mathcal{I}_R : I \supseteq \mathrm{Ker}\, f\}$ and $\mathcal{E}_S = \mathcal{I}_S$.

Let $I \in \mathcal{C}_R$. Show that

(i) $I$ is a primary ideal of $R$ if and only if $I^e$ is a primary ideal of $S$; and

(ii) when this is the case, $\sqrt{I} = (\sqrt{(I^e)})^c$ and $\sqrt{(I^e)} = (\sqrt{I})^e$.

**4.8** ♯EXERCISE. Let $I$ be a proper ideal of the commutative ring $R$, and let $P, Q$ be ideals of $R$ which contain $I$. Prove that $Q$ is a $P$-primary ideal of $R$ if and only if $Q/I$ is a $P/I$-primary ideal of $R/I$.

We have already mentioned in 4.2 that a prime ideal of a commutative ring is automatically primary. However, it is time that we had some further examples of primary ideals. It was hinted in the introduction to this chapter that each positive power of a maximal ideal in a commutative ring is primary: this fact will be a consequence of our next result.

**4.9** PROPOSITION. *Let $Q$ be an ideal of the commutative ring $R$ such that $\sqrt{Q} = M$, a maximal ideal of $R$. Then $Q$ is a primary (in fact $M$-primary) ideal of $R$.*

*Consequently, all positive powers $M^n$ ($n \in \mathbf{N}$) of the maximal ideal $M$ are $M$-primary.*

*Proof.* Since $Q \subseteq \sqrt{Q} = M \subset R$, it is clear that $Q$ is proper. Let $a, b \in R$ be such that $ab \in Q$ but $b \notin \sqrt{Q}$. Since $\sqrt{Q} = M$ is maximal and $b \notin M$, we must have $M + Rb = R$, so that

$$\sqrt{Q} + \sqrt{(Rb)} = R.$$

Hence, by 2.25(iv), $Q + Rb = R$. Thus there exist $d \in Q$, $c \in R$ such that $d + cb = 1$, and

$$a = a1 = a(d + cb) = ad + c(ab) \in Q$$

because $d, ab \in Q$. Hence $Q$ is $M$-primary.

The last claim is now an immediate consequence, because $\sqrt{(M^n)} = M$ for all $n \in \mathbf{N}$, by 3.47. $\square$

Proposition 4.9 enables us to increase our fund of examples of primary ideals.

**4.10** EXAMPLE. Let $R$ be a PID which is not a field. Then the set of all primary ideals of $R$ is

$$\{0\} \cup \{Rp^n : p \text{ an irreducible element of } R, \ n \in \mathbf{N}\}.$$

*Proof.* Since $0 \in \operatorname{Spec}(R)$ because $R$ is a domain, and, for an irreducible element $p$ of $R$ and $n \in \mathbf{N}$, the ideal $Rp^n$ is a power of a maximal ideal of $R$ by 3.34 and so is a primary ideal of $R$ by 4.9, we see that each member of the displayed set is indeed a primary ideal of $R$.

On the other hand, a non-zero primary ideal of $R$ must have the form $Ra$ for some non-zero $a \in R$, and $a$ cannot be a unit since a primary ideal is proper. By 3.39, we can express $a$ as a product of irreducible elements of $R$. If $a$ were divisible by two irreducible elements $p, q$ of $R$ which are not associates, then $Rp$ and $Rq$ would be distinct maximal ideals of $R$ by 3.32 and 3.34, and they would both be minimal prime ideals of $Ra$, in contradiction to 4.5. It follows that $Ra$ is generated by a positive power of some irreducible element of $R$. $\square$

The reader should not be misled into thinking that every $M$-primary ideal, where $M$ is a maximal ideal of a commutative ring $R$, has to be a power of $M$. The next example illustrates this point.

**4.11** EXAMPLE. Let $K$ be a field and let $R$ denote the ring $K[X, Y]$ of polynomials over $K$ in the indeterminates $X, Y$. Let $M = RX + RY$, a maximal ideal of $R$ by 3.15. Then $(X, Y^2)$ is an $M$-primary ideal of $R$ which is not a power of a prime ideal of $R$.

*Proof.* We have

$$M^2 = (X^2, XY, Y^2) \subseteq (X, Y^2) \subseteq (X, Y) = M,$$

so that, on taking radicals, we deduce that

$$M = \sqrt{(M^2)} \subseteq \sqrt{(X, Y^2)} \subseteq \sqrt{M} = M$$

with the aid of 3.47. Hence $\sqrt{(X, Y^2)} = M$, a maximal ideal of $R$, and so it follows from 4.9 that $(X, Y^2)$ is $M$-primary.

Furthermore, $(X, Y^2)$ is not a positive power of a prime ideal $P$ of $R$, because, if it were, we should have to have $P = M$ by 3.47, and, since the powers of $M$ form a descending chain

$$M \supseteq M^2 \supseteq \ldots \supseteq M^i \supseteq M^{i+1} \supseteq \ldots,$$

we should have to have $(X, Y^2) = M$ or $M^2$; neither of these is correct because $X \notin M^2$ (since every non-zero term which actually appears in a polynomial in $M^2$ has total degree at least 2), while $Y \notin (X, Y^2)$ (since otherwise

$$Y = Xf + Y^2 g$$

for some $f, g \in R$, and evaluation of $X, Y$ at $0, Y$ (see 1.17) leads to a contradiction). $\square$

Even though we have seen in 4.9 that every positive power of a maximal ideal of a commutative ring $R$ is a primary ideal of $R$, it is not necessarily the case that every positive power of a prime ideal of $R$ has to be primary. We give next an example which illustrates this point.

**4.12** EXAMPLE. Let $K$ be a field, and consider the residue class ring $R$ of the ring $K[X_1, X_2, X_3]$ of polynomials over $K$ in indeterminates $X_1, X_2, X_3$ given by

$$R = K[X_1, X_2, X_3]/(X_1 X_3 - X_2^2).$$

For each $i = 1, 2, 3$, let $x_i$ denote the natural image of $X_i$ in $R$. Then $P := (x_1, x_2)$ is a prime ideal of $R$, but $P^2$ is not primary.

Since $\sqrt{(P^2)} = P \in \mathrm{Spec}(R)$, this example also shows that an ideal of a commutative ring which has prime radical need not necessarily be primary.

*Proof.* By 3.15, the ideal of $K[X_1, X_2]$ generated by $X_1$ and $X_2$ is maximal. By 3.65(i), its extension to $K[X_1, X_2][X_3] = K[X_1, X_2, X_3]$ is a prime ideal, and, by 2.42, this extension is also generated by $X_1$ and $X_2$. Now, in $K[X_1, X_2, X_3]$, we have

$$(X_1, X_2) \supseteq (X_1 X_3 - X_2^2),$$

so that, by 3.28,

$$P = (x_1, x_2) = (X_1, X_2) / (X_1 X_3 - X_2^2) \in \mathrm{Spec}(R).$$

We show now that $P^2$ is not primary. Note that, by 3.47, $\sqrt{(P^2)} = P$. Now $x_1 x_3 = x_2^2 \in P^2$. However, we have $x_1 \notin P^2$ and $x_3 \notin P = \sqrt{(P^2)}$ (as is explained below), and so it follows that $P^2$ is not primary.

The claim that $x_1 \notin P^2$ is proved as follows. If this were not the case, then we should have

$$X_1 = X_1^2 f + X_1 X_2 g + X_2^2 h + (X_1 X_3 - X_2^2)d$$

for some $f, g, h, d \in K[X_1, X_2, X_3]$, and this is not possible since every term which actually appears in the right-hand side of the above equation has degree at least 2.

Similarly, if we had $x_3 \in P$, then we should have

$$X_3 = X_1 a + X_2 b + (X_1 X_3 - X_2^2)c$$

for some $a, b, c \in K[X_1, X_2, X_3]$, and we can obtain a contradiction by evaluating $X_1, X_2, X_3$ at $0, 0, X_3$. $\square$

We are going to study presentations of ideals in a commutative ring $R$ as intersections of finitely many primary ideals of $R$. We need some preliminary lemmas.

**4.13** LEMMA. *Let $P$ be a prime ideal of the commutative ring $R$, and let $Q_1, \ldots, Q_n$ (where $n \geq 1$) be $P$-primary ideals of $R$. Then $\bigcap_{i=1}^{n} Q_i$ is also $P$-primary.*

*Proof.* By repeated use of 2.30, we have

$$\sqrt{(Q_1 \cap \ldots \cap Q_n)} = \sqrt{Q_1} \cap \ldots \cap \sqrt{Q_n} = P \subset R.$$

This shows, among other things, that $\bigcap_{i=1}^{n} Q_i$ is proper. Suppose that $a, b \in R$ are such that $ab \in \bigcap_{i=1}^{n} Q_i$ but $b \notin \bigcap_{i=1}^{n} Q_i$. Then there exists an integer $j$ with $1 \leq j \leq n$ such that $b \notin Q_j$. Since $ab \in Q_j$ and $Q_j$ is $P$-primary, it follows that

$$a \in P = \sqrt{(Q_1 \cap \ldots \cap Q_n)}.$$

Hence $\bigcap_{i=1}^{n} Q_i$ is $P$-primary. $\square$

**4.14** LEMMA. *Let $Q$ be a $P$-primary ideal of the commutative ring $R$, and let $a \in R$.*
  (i) *If $a \in Q$, then $(Q : a) = R$.*
  (ii) *If $a \notin Q$, then $(Q : a)$ is $P$-primary, so that, in particular,*

$$\sqrt{(Q : a)} = P.$$

  (iii) *If $a \notin P$, then $(Q : a) = Q$.*

*Proof.* (i) This is immediate from the definition: see 2.31 and 2.32.

(ii) Let $b \in (Q : a)$. Then we have $ab \in Q$ and $a \notin Q$, so that, since $Q$ is $P$-primary, $b \in P = \sqrt{Q}$. Hence

$$Q \subseteq (Q : a) \subseteq P,$$

so that, on taking radicals, we see that

$$P = \sqrt{Q} \subseteq \sqrt{(Q : a)} \subseteq \sqrt{P} = P.$$

Hence $\sqrt{(Q : a)} = P$.

Now suppose that $c, d \in R$ are such that $cd \in (Q : a)$ but $d \notin P$. Then $cda \in Q$ but $d \notin P$ and $Q$ is $P$-primary. Hence $ca \in Q$ and $c \in (Q : a)$. It follows that $(Q : a)$ is $P$-primary.

(iii) This is immediate from the definition of $P$-primary ideal: we have $Q \subseteq (Q : a)$, of course, while if $b \in (Q : a)$ then $ab \in Q$, $a \notin P$ and $Q$ is $P$-primary, so that $b \in Q$. □

We are now ready to introduce formally the concept of primary decomposition.

**4.15** DEFINITION. Let $I$ be a proper ideal of the commutative ring $R$. A *primary decomposition* of $I$ is an expression for $I$ as an intersection of finitely many primary ideals of $R$. Such a primary decomposition

$$I = Q_1 \cap \ldots \cap Q_n \quad \text{with } \sqrt{Q_i} = P_i \text{ for } i = 1, \ldots, n$$

of $I$ (and it is to be understood that $Q_i$ is $P_i$-primary for all $i = 1, \ldots, n$ whenever we use this type of terminology) is said to be a *minimal primary decomposition* of $I$ precisely when

(i) $P_1, \ldots, P_n$ are $n$ different prime ideals of $R$, and

(ii) for all $j = 1, \ldots, n$, we have

$$Q_j \not\supseteq \bigcap_{\substack{i=1 \\ i \neq j}}^{n} Q_i.$$

We say that $I$ is a *decomposable* ideal of $R$ precisely when it has a primary decomposition.

Observe that condition 4.15(ii) can be rephrased as follows: for all $j = 1, \ldots, n$, we have

$$I \neq \bigcap_{\substack{i=1 \\ i \neq j}}^{n} Q_i,$$

so that $Q_j$ is not redundant and really is needed in the primary decomposition $I = \bigcap_{i=1}^{n} Q_i$.

**4.16** REMARKS. Let $I$ be a proper ideal of the commutative ring $R$, and let

$$I = Q_1 \cap \ldots \cap Q_n \quad \text{with } \sqrt{Q_i} = P_i \text{ for } i = 1, \ldots, n$$

be a primary decomposition of $I$.

(i) If two of the $P_i$, say $P_j$ and $P_k$ where $1 \leq j, k \leq n$ and $j \neq k$, are equal, then we can use 4.13 to combine together the terms $Q_j$ and $Q_k$ in our primary decomposition to obtain another primary decomposition of $I$ with $n-1$ terms. In fact, we can use 4.13 repeatedly in this way in order to produce a primary decomposition of $I$ in which the radicals of the primary terms are all different.

(ii) We can refine our given primary decomposition to produce one in which no term is redundant as follows. Firstly, discard $Q_1$ if and only if $I = \bigcap_{i=2}^{n} Q_i$, that is, if and only if $Q_1 \supseteq \bigcap_{i=2}^{n} Q_i$; then consider in turn $Q_2, \ldots, Q_n$; at the $j$-th stage, discard $Q_j$ if and only if it contains the intersection of those $Q_i$ with $i \neq j$ that have not yet been discarded. Observe that if $Q_j$ is not discarded at the $j$-th stage, then, at the end of the $n$-th stage, $Q_j$ will not contain the intersection of those $Q_i$ with $i \neq j$ that survive to the end. In this way we can refine our original primary decomposition of $I$ to obtain one in which every term present is irredundant.

(iii) Thus, starting with a given primary decomposition of $I$, we can first use the process described in (i) above and then use the refinement technique of (ii) in order to arrive at a minimal primary decomposition of $I$.

(iv) Thus every decomposable ideal of $R$ actually has a minimal primary decomposition.

(v) Note that, if $I$ has a primary decomposition with $t$ terms which is not minimal, then it follows from (i), (ii) and (iii) above that $I$ has a minimal primary decomposition with fewer than $t$ terms.

(vi) The phrases 'normal primary decomposition' and 'reduced primary decomposition' are alternatives, employed in some books, for 'minimal primary decomposition'.

Minimal primary decompositions have certain uniqueness properties.

**4.17** THEOREM. *Let $I$ be a decomposable ideal of the commutative ring $R$, and let*

$$I = Q_1 \cap \ldots \cap Q_n \quad \text{with } \sqrt{Q_i} = P_i \text{ for } i = 1, \ldots, n$$

*be a minimal primary decomposition of $I$. Let $P \in \mathrm{Spec}(R)$. Then the following statements are equivalent:*

(i) $P = P_i$ for some $i$ with $1 \leq i \leq n$;

(ii) there exists $a \in R$ such that $(I : a)$ is $P$-primary;

(iii) there exists $a \in R$ such that $\sqrt{(I : a)} = P$.

*Proof.* (i) $\Rightarrow$ (ii) Suppose that $P = P_i$ for some $i$ with $1 \leq i \leq n$. Since the primary decomposition $I = \bigcap_{i=1}^{n} Q_i$ is minimal, there exists

$$a_i \in \bigcap_{\substack{j=1 \\ j \neq i}}^{n} Q_j \setminus Q_i.$$

By 2.33(ii),

$$(I : a_i) = \left( \bigcap_{j=1}^{n} Q_j : a_i \right) = \bigcap_{j=1}^{n} (Q_j : a_i).$$

But, by 4.14(i) and (ii), $(Q_j : a_i) = R$ for $j \neq i$ $(1 \leq j \leq n)$, while $(Q_i : a_i)$ is $P_i$-primary. Since $P = P_i$, it follows that $(I : a_i)$ is $P$-primary.

(ii) $\Rightarrow$ (iii) This is immediate from 4.5 since the radical of a $P$-primary ideal is equal to $P$.

(iii) $\Rightarrow$ (i) Suppose that $a \in R$ is such that $\sqrt{(I : a)} = P$. By 2.33(ii),

$$(I : a) = \left( \bigcap_{i=1}^{n} Q_i : a \right) = \bigcap_{i=1}^{n} (Q_i : a).$$

By 4.14(i) and (ii), we have $(Q_i : a) = R$ if $a \in Q_i$, while $(Q_i : a)$ is $P_i$-primary if $a \notin Q_i$. Hence, on use of 2.30, we see that

$$P = \sqrt{(I : a)} = \bigcap_{\substack{i=1 \\ a \notin Q_i}}^{n} \sqrt{(Q_i : a)} = \bigcap_{\substack{i=1 \\ a \notin Q_i}}^{n} P_i.$$

Since $P$ is a proper ideal of $R$, it follows that there is at least one integer $i$ with $1 \leq i \leq n$ for which $a \notin Q_i$, and, by 3.56, $P = P_i$ for one such $i$. $\square$

**4.18** COROLLARY: THE FIRST UNIQUENESS THEOREM FOR PRIMARY DECOMPOSITION. *Let $I$ be a decomposable ideal of the commutative ring $R$, and let*

$$I = Q_1 \cap \ldots \cap Q_n \qquad \text{with } \sqrt{Q_i} = P_i \text{ for } i = 1, \ldots, n$$

*and*

$$I = Q_1' \cap \ldots \cap Q_{n'}' \qquad \text{with } \sqrt{Q_i'} = P_i' \text{ for } i = 1, \ldots, n'$$

*be two minimal primary decompositions of $I$. Then $n = n'$, and we have*

$$\{P_1, \ldots, P_n\} = \{P'_1, \ldots, P'_n\}.$$

*In other words, the number of terms appearing in a minimal primary decomposition of $I$ is independent of the choice of minimal primary decomposition, as also is the set of prime ideals which occur as the radicals of the primary terms.*

*Proof.* This is now immediate from 4.17, because that result shows that, for $P \in \text{Spec}(R)$, we have that $P$ is equal to one of $P_1, \ldots, P_n$ if and only if there exists $a \in R$ for which $\sqrt{(I : a)} = P$. Since this second statement is completely independent of any choice of minimal primary decomposition of $I$, the former statement must be similarly independent. □

The above theorem is one of the cornerstones of commutative algebra. It leads to the concept of 'associated prime ideal' of a decomposable ideal.

**4.19** DEFINITION. Let $I$ be a decomposable ideal of the commutative ring $R$, and let

$$I = Q_1 \cap \ldots \cap Q_n \quad \text{with } \sqrt{Q_i} = P_i \text{ for } i = 1, \ldots, n$$

be a minimal primary decomposition of $I$. Then the $n$-element set

$$\{P_1, \ldots, P_n\},$$

which is independent of the choice of minimal primary decomposition of $I$ by 4.18, is called *the set of associated prime ideals of $I$* and denoted by ass $I$ or ass$_R I$. The members of ass $I$ are referred to as the *associated prime ideals* or the *associated primes* of $I$, and are said to *belong* to $I$.

**4.20** REMARK. Let $I$ be a decomposable ideal of the commutative ring $R$, and let $P \in \text{Spec}(R)$. It follows from 4.17 that $P \in$ ass $I$ if and only if there exists $a \in R$ such that $(I : a)$ is $P$-primary, and that this is the case if and only if there exists $b \in R$ such that $\sqrt{(I : b)} = P$.

**4.21** ‡EXERCISE. Let $f : R \rightarrow S$ be a homomorphism of commutative rings, and use the contraction notation of 2.41 in conjunction with $f$. Let $\mathcal{I}$ be a decomposable ideal of $S$.

(i) Let

$$\mathcal{I} = \mathcal{Q}_1 \cap \ldots \cap \mathcal{Q}_n \quad \text{with } \sqrt{\mathcal{Q}_i} = \mathcal{P}_i \text{ for } i = 1, \ldots, n$$

be a primary decomposition of $\mathcal{I}$. Show that

$$\mathcal{I}^c = \mathcal{Q}_1^c \cap \ldots \cap \mathcal{Q}_n^c \quad \text{with } \sqrt{\mathcal{Q}_i^c} = \mathcal{P}_i^c \text{ for } i = 1, \ldots, n$$

is a primary decomposition of $\mathcal{I}^c$. (Note that $\sqrt{(Q_i^c)} = (\sqrt{Q_i})^c$ for $i = 1, \ldots, n$, by 2.43(iv).) Deduce that $\mathcal{I}^c$ is a decomposable ideal of $R$ and that

$$\mathrm{ass}_R(\mathcal{I}^c) \subseteq \{\mathcal{P}^c : \mathcal{P} \in \mathrm{ass}_S \mathcal{I}\}.$$

(ii) Now suppose that $f$ is surjective. Show that, if the first primary decomposition in (i) is minimal, then so too is the second, and deduce that, in these circumstances,

$$\mathrm{ass}_R(\mathcal{I}^c) = \{\mathcal{P}^c : \mathcal{P} \in \mathrm{ass}_S \mathcal{I}\}.$$

**4.22** ♯EXERCISE. Let $f : R \to S$ be a surjective homomorphism of commutative rings; use the extension notation of 2.41 in conjunction with $f$. Let $I, Q_1, \ldots, Q_n, P_1, \ldots, P_n$ be ideals of $R$ all of which contain $\mathrm{Ker}\, f$. Show that

$$I = Q_1 \cap \ldots \cap Q_n \quad \text{with } \sqrt{Q_i} = P_i \text{ for } i = 1, \ldots, n$$

is a primary decomposition of $I$ if and only if

$$I^e = Q_1^e \cap \ldots \cap Q_n^e \quad \text{with } \sqrt{(Q_i^e)} = P_i^e \text{ for } i = 1, \ldots, n$$

is a primary decomposition of $I^e$, and that, when this is the case, the first of these is minimal if and only if the second is.

Deduce that $I$ is a decomposable ideal of $R$ if and only if $I^e$ is a decomposable ideal of $S$, and, when this is the case,

$$\mathrm{ass}_S(I^e) = \{P^e : P \in \mathrm{ass}_R I\}.$$

**4.23** REMARK. Let $I$ be a proper ideal of the commutative ring $R$. The reader should notice the consequences of 4.22 for the natural ring homomorphism from $R$ to $R/I$. For instance, that exercise shows that, if $J$ is an ideal of $R$ such that $J \supseteq I$, then $J$ is a decomposable ideal of $R$ if and only if $J/I$ is a decomposable ideal of $R/I$, and, when this is the case,

$$\mathrm{ass}_{R/I}(J/I) = \{P/I : P \in \mathrm{ass}_R J\}.$$

It is obvious that, in the situation of 4.20, every associated prime of $I$ contains $I$, and so belongs to the set $\mathrm{Var}(I)$ of 3.48. We discussed minimal members of $\mathrm{Var}(I)$ in 3.52, and it is now appropriate for us to consider them once more in the context of primary decomposition.

**4.24** PROPOSITION. *Let $I$ be a decomposable ideal of the commutative ring $R$, and let $P \in \mathrm{Spec}(R)$. Then $P$ is a minimal prime ideal of $I$ (that is (see 3.52), $P$ is a minimal member with respect to inclusion of the set $\mathrm{Var}(I)$ of*

*all prime ideals of R which contain I) if and only if P is a minimal member
(again with respect to inclusion) of* ass $I$.

In particular, all the minimal prime ideals of $I$ belong to ass $I$, so that
$I$ has only finitely many minimal prime ideals, and if $P_1 \in \operatorname{Spec}(R)$ with
$P_1 \supseteq I$, then there exists $P_2 \in$ ass $I$ with $P_1 \supseteq P_2$.

*Proof.* Let

$$I = Q_1 \cap \ldots \cap Q_n \quad \text{with } \sqrt{Q_i} = P_i \text{ for } i = 1, \ldots, n$$

be a minimal primary decomposition of $I$. Note that $P \supseteq I$ if and only if
$P = \sqrt{P} \supseteq \sqrt{I}$, and, by 2.30,

$$\sqrt{I} = \bigcap_{i=1}^{n} \sqrt{Q_i} = \bigcap_{i=1}^{n} P_i.$$

It thus follows from 3.55 that $P \supseteq I$ if and only if $P \supseteq P_j$ for some $j$ with
$1 \le j \le n$, that is, if and only if $P \supseteq P'$ for some $P' \in$ ass $I$.

($\Rightarrow$) Assume that $P$ is a minimal prime ideal of $I$. Then by the above
argument, $P \supseteq P'$ for some $P' \in$ ass $I$. But ass $I \subseteq \operatorname{Var}(I)$, and so $P = P'$
must be a minimal member of ass $I$ with respect to inclusion.

($\Leftarrow$) Assume that $P$ is a minimal member of ass $I$. Thus $P \supseteq I$, and
so, by 3.53, there exists a minimal prime ideal $P'$ of $I$ such that $P \supseteq P'$.
Hence, by the first paragraph of this proof, there exists $P'' \in$ ass $I$ such
that $P' \supseteq P''$. But then,

$$P \supseteq P' \supseteq P'',$$

and since $P$ is a minimal member of ass $I$, we must have $P = P' = P''$.
Hence $P = P'$ is a minimal prime ideal of $I$.

The remaining outstanding claims follow from 3.53 and the fact that
ass $I$ is a finite set. $\Box$

**4.25** TERMINOLOGY. Let $I$ be a decomposable ideal of the commutative
ring $R$. We have just seen in 4.24 that the minimal members of ass $I$ are
precisely the minimal prime ideals of $I$: these prime ideals are called the
*minimal* or *isolated* primes of $I$. The remaining associated primes of $I$,
that is, the associated primes of $I$ which are not minimal, are called the
*embedded* primes of $I$.

Observe that a decomposable ideal in a commutative ring $R$ need not
have any embedded prime: a primary ideal $Q$ of $R$ is certainly decomposable, because '$Q = Q$' is a minimal primary decomposition of $Q$, so that
$\sqrt{Q}$ is the only associated prime of $Q$.

**4.26** EXERCISE. Suppose that the decomposable ideal $I$ of the commutative ring $R$ satisfies $\sqrt{I} = I$. Show that $I$ has no embedded prime.

The First Uniqueness Theorem for Primary Decomposition 4.18, together with the motivation for primary decomposition from the theory of unique factorization in a PID which was given at the beginning of this chapter, raise another question about uniqueness aspects of minimal primary decompositions: is a minimal primary decomposition of a decomposable ideal $I$ in a commutative ring $R$ uniquely determined by $I$? To see that this is not always the case, consider the following example.

**4.27** EXAMPLE. Let $K$ be a field and let $R = K[X, Y]$ be the ring of polynomials over $K$ in indeterminates $X, Y$. In $R$, let

$$M = (X, Y), \quad P = (Y), \quad Q = (X, Y^2), \quad I = (XY, Y^2).$$

Note that $M$ is a maximal ideal of $R$ by 3.15, $P$ is a prime ideal of $R$ by 3.66, and $Q$ is an $M$-primary ideal of $R$ different from $M^2$ by 4.11. We have that
$$I = Q \cap P \quad \text{and} \quad I = M^2 \cap P$$

are two minimal primary decompositions of $I$ with distinct $M$-primary terms.

*Proof.* It is clear that $I \subseteq P$ and $I \subseteq M^2 \subseteq Q$; hence

$$I \subseteq M^2 \cap P \subseteq Q \cap P.$$

Let $f \in Q \cap P$. Since $f \in P$, every monomial term which actually appears in $f$ involves $Y$; add together all these monomial terms which have degree at least 2 to form a polynomial $g \in I$ such that $f - g = cY$ for some $c \in K$. We claim that $c = 0$: if this were not the case, then we should have

$$Y = c^{-1}cY \in (Q \cap P) + I = Q \cap P \subseteq Q,$$

so that $Y = hX + eY^2$ for some $h, e \in R$, which is impossible. Thus $f = g \in I$ and we have proved that $I = M^2 \cap P = Q \cap P$. Furthermore, these equations give two primary decompositions of $I$ because $P \in \mathrm{Spec}(R)$ and $M^2$ is $M$-primary by 4.9. Finally, both these primary decompositions are minimal because

$$X^2 \in M^2 \setminus P, \quad X^2 \in Q \setminus P, \quad Y \in P \setminus Q, \quad Y \in P \setminus M^2.$$

We have thus produced two minimal primary decompositions of $I$ with different $M$-primary terms. $\square$

**4.28** EXERCISE. Let $K$ be a field and let $R = K[X, Y]$ be the ring of polynomials over $K$ in indeterminates $X, Y$. In $R$, let $I = (X^3, XY)$.

(i) Show that, for every $n \in \mathbf{N}$, the ideal $(X^3, XY, Y^n)$ of $R$ is primary.

(ii) Show that $I = (X) \cap (X^3, Y)$ is a minimal primary decomposition of $I$.

(iii) Construct infinitely many different minimal primary decompositions of $I$.

In spite of the above example and exercise, there is a positive result in this direction: it turns out that, for a decomposable ideal $I$ in a commutative ring $R$ and for any *minimal* prime ideal $P$ belonging to $I$, the $P$-primary term in a minimal primary decomposition of $I$ is uniquely determined by $I$ and is independent of the choice of minimal primary decomposition. (This does not conflict with the example in 4.27 because for the $I$ in that example we have ass $I = \{P, M\}$ and, since $P \subset M$, there is just one minimal prime of $I$, namely $P$, while $M$ is an embedded prime of $I$.)

This second uniqueness result is the subject of the Second Uniqueness Theorem for Primary Decomposition, to which we now turn.

**4.29** THE SECOND UNIQUENESS THEOREM FOR PRIMARY DECOMPOSITION. *Let $I$ be a decomposable ideal of the commutative ring $R$, and let* ass $I = \{P_1, \ldots, P_n\}$. *Let*

$$I = Q_1 \cap \ldots \cap Q_n \quad \text{with } \sqrt{Q_i} = P_i \text{ for } i = 1, \ldots, n$$

*and*

$$I = Q_1' \cap \ldots \cap Q_n' \quad \text{with } \sqrt{Q_i'} = P_i \text{ for } i = 1, \ldots, n$$

*be two minimal primary decompositions of $I$. (Of course, we are here making free use of the First Uniqueness Theorem for Primary Decomposition (4.18) and its consequences.) Then, for each $i$ with $1 \le i \le n$ for which $P_i$ is a minimal prime ideal belonging to $I$, we have*

$$Q_i = Q_i'.$$

*In other words, in a minimal primary decomposition of $I$, the primary term corresponding to an isolated prime ideal of $I$ is uniquely determined by $I$ and is independent of the choice of minimal primary decomposition.*

*Proof.* If $n = 1$, there is nothing to prove; we therefore suppose that $n > 1$.

Let $P_i$ be a minimal prime ideal belonging to $I$. Now there exists

$$a \in \bigcap_{\substack{j=1 \\ j \ne i}}^{n} P_j \setminus P_i,$$

for otherwise it would follow from 3.55 that $P_j \subset P_i$ for some $j \in \mathbb{N}$ with $1 \leq j \leq n$ and $j \neq i$, contrary to the fact that $P_i$ is a minimal prime ideal belonging to $I$.

For each $j = 1, \ldots, n$ with $j \neq i$, there exists $h_j \in \mathbb{N}$ such that $a^{h_j} \in Q_j$. Let $t \in \mathbb{N}$ be such that

$$t \geq \max \{h_1, \ldots, h_{i-1}, h_{i+1}, \ldots, h_n\}.$$

Then $a^t \notin P_i$, and so it follows that

$$(I : a^t) = \left( \bigcap_{j=1}^{n} Q_j : a^t \right) = \bigcap_{j=1}^{n} (Q_j : a^t) = Q_i$$

since $Q_i$ is $P_i$-primary. Thus we have shown that $Q_i = (I : a^t)$ whenever the integer $t$ is sufficiently large. In the same way, $Q_i' = (I : a^t)$ whenever $t$ is sufficiently large. Hence $Q_i = Q_i'$, as claimed. $\square$

We shall, in fact, give another proof of the above Second Uniqueness Theorem in the next chapter (see 5.42), because an illuminating way to approach its proof is by use of the theory of ideals in rings of fractions: we have not yet discussed these in this book but they will be one of the principal topics of the next chapter.

Another important topic which we have as yet hardly touched upon is the question of existence of primary decompositions for proper ideals in a commutative ring. An alert reader will probably realise from the comments at the very beginning of this chapter that every proper ideal in a PID does possess a primary decomposition. Any reader who is hoping that every proper ideal in every commutative ring has a primary decomposition will be disappointed by the following exercise.

**4.30** Exercise. Show that the zero ideal in the ring $C[0,1]$ of all continuous real-valued functions defined on the closed interval $[0,1]$ is not decomposable, that is, it does not have a primary decomposition.

(Here is a hint. Suppose that the zero ideal is decomposable, and look for a contradiction. Let $P \in \mathrm{ass}_{C[0,1]} 0$, so that, by 4.17, there exists $f \in C[0,1]$ such that $\sqrt{(0 : f)} = P$. Show that $(0 : f) = P$ and that there exists at most one real number $a \in [0,1]$ for which $f(a) \neq 0$.)

However, there is one beautiful existence result concerning primary decompositions which shows that every proper ideal in a commutative Noetherian ring possesses a primary decomposition. Commutative Noetherian rings were introduced in 3.37: a commutative ring $R$ is Noetherian precisely when every ascending chain of ideals of $R$ is eventually stationary.

Although these rings will be examined in greater detail in later chapters, we can already establish the existence of primary decompositions in this type of ring.

**4.31** DEFINITION. Let $I$ be an ideal of the commutative ring $R$. We say that $I$ is *irreducible* precisely when $I$ is proper and $I$ cannot be expressed as the intersection of two strictly larger ideals of $R$.

Thus $I$ is irreducible if and only if $I \subset R$ and, whenever $I = I_1 \cap I_2$ with $I_1, I_2$ ideals of $R$, then $I = I_1$ or $I = I_2$.

**4.32** EXERCISE. Let $f : R \to S$ be a surjective homomorphism of commutative rings, and use the extension notation of 2.41 in conjunction with $f$. Let $I$ be an ideal of $R$ which contains Ker $f$. Show that $I$ is an irreducible ideal of $R$ if and only if $I^e$ is an irreducible ideal of $S$.

The important ingredients in our proof of the existence of primary decompositions of proper ideals in a commutative Noetherian ring $R$ are that every proper ideal of $R$ can be expressed as an intersection of finitely many irreducible ideals of $R$, and that an irreducible ideal of $R$ is necessarily primary.

**4.33** PROPOSITION. *Let $R$ be a commutative Noetherian ring. Then every proper ideal of $R$ can be expressed as an intersection of finitely many irreducible ideals of $R$.*

*Proof.* Let $\Sigma$ denote the set of all proper ideals of $R$ that cannot be expressed as an intersection of finitely many irreducible ideals of $R$. Our aim is to show that $\Sigma = \emptyset$. Suppose that this is not the case. Then, since $R$ is Noetherian, it follows from 3.37 that $\Sigma$ has a maximal member, $I$ say, with respect to inclusion.

Then $I$ itself is not irreducible, for otherwise we could write $I = I \cap I$ and $I$ would not be in $\Sigma$. Since $I$ is proper, it therefore follows that $I = I_1 \cap I_2$ for some ideals $I_1, I_2$ of $R$ for which

$$I \subset I_1 \quad \text{and} \quad I \subset I_2.$$

Note that this implies that both $I_1$ and $I_2$ are proper ideals. By choice of $I$, we must have $I_i \notin \Sigma$ for $i = 1, 2$. Since both $I_1$ and $I_2$ are proper, it follows that both can be expressed as intersections of finitely many irreducible ideals of $R$; hence $I = I_1 \cap I_2$ has the same property, and this is a contradiction.

Hence $\Sigma = \emptyset$, and the proof is complete. $\square$

**4.34** PROPOSITION. *Let $R$ be a commutative Noetherian ring and let $I$ be an irreducible ideal of $R$. Then $I$ is primary.*

*Proof.* By definition of irreducible ideal (4.31), $I \subset R$. Suppose that $a, b \in R$ are such that $ab \in I$ but $b \notin I$. Now

$$(I : a) \subseteq (I : a^2) \subseteq \ldots \subseteq (I : a^i) \subseteq \ldots$$

is an ascending chain of ideals of $R$, so that, since $R$ is Noetherian, there exists $n \in \mathbf{N}$ such that $(I : a^n) = (I : a^{n+i})$ for all $i \in \mathbf{N}$.

We show that $I = (I + Ra^n) \cap (I + Rb)$. It is clear that

$$I \subseteq (I + Ra^n) \cap (I + Rb).$$

Let $r \in (I + Ra^n) \cap (I + Rb)$; then we can write

$$r = g + ca^n = h + db$$

for some $g, h \in I$ and $c, d \in R$. Thus $ra = ga + ca^{n+1} = ha + dab$, so that, since $ab, g, h \in I$, we have

$$ca^{n+1} = ha + dab - ga \in I.$$

Hence $c \in (I : a^{n+1}) = (I : a^n)$ (by choice of $n$), so that $r = g + ca^n \in I$. It follows that

$$I = (I + Ra^n) \cap (I + Rb),$$

as claimed.

Now $I$ is irreducible, and $I \subset I + Rb$ because $b \notin I$. Hence $I = I + Ra^n$ and $a^n \in I$. We have proved that $I$ is a primary ideal of $R$. □

**4.35** COROLLARY. *Let $I$ be a proper ideal in the commutative Noetherian ring $R$. Then $I$ has a primary decomposition, and so, by 4.16, it also has a minimal primary decomposition.*

*Proof.* This is now immediate from the last two results: by 4.33, $I$ can be expressed as an intersection of finitely many irreducible ideals of $R$, and an irreducible ideal of $R$ is primary by 4.34. □

Thus all our theory of associated primes of decomposable ideals (see 4.19) applies in particular to arbitrary proper ideals in a commutative Noetherian ring. This is, in fact, a very powerful tool for us to have available when studying such a ring. However, illustrations of this will have to wait until Chapter 8, devoted to the development of the basic theory of such rings.

**4.36** EXERCISE. Let $R$ be a commutative ring and let $X$ be an indeterminate; use the extension and contraction notation of 2.41 in conjunction with the natural ring homomorphism $f : R \to R[X]$. Let $Q$ and $I$ be ideals of $R$.

(i) Show that $Q$ is a primary ideal of $R$ if and only if $Q^e$ is a primary ideal of $R[X]$.

(ii) Show that, if $I$ is a decomposable ideal of $R$ and

$$I = Q_1 \cap \ldots \cap Q_n \quad \text{with } \sqrt{Q_i} = P_i \text{ for } i = 1, \ldots, n$$

is a primary decomposition of $I$, then

$$I^e = Q_1^e \cap \ldots \cap Q_n^e \quad \text{with } \sqrt{Q_i^e} = P_i^e \text{ for } i = 1, \ldots, n$$

is a primary decomposition of the ideal $I^e$ of $R[X]$.

(iii) Show that, if $I$ is a decomposable ideal of $R$, then

$$\mathrm{ass}_{R[X]} I^e = \{ P^e : P \in \mathrm{ass}_R I \} .$$

**4.37** EXERCISE. Let $R$ be a commutative Noetherian ring, and let $Q$ be a $P$-primary ideal of $R$. By 4.33, $Q$ can be expressed as an intersection of finitely many irreducible ideals of $R$. One can refine such an expression to obtain

$$Q = \bigcap_{i=1}^{n} J_i,$$

where each $J_i$ (for $1 \leq i \leq n$) is irreducible and irredundant in the intersection, so that, for all $i = 1, \ldots, n$,

$$\bigcap_{\substack{j=1 \\ j \neq i}}^{n} J_j \not\subseteq J_i.$$

By 4.34, the ideals $J_1, \ldots, J_n$ are all primary.

Prove that $J_i$ is $P$-primary for all $i = 1, \ldots, n$.

**4.38** EXERCISE. Let $R$ be the polynomial ring $K[X_1, \ldots, X_n]$ over the field $K$ in the indeterminates $X_1, \ldots, X_n$, and let $\alpha_1, \ldots, \alpha_n \in K$. Let $r \in \mathbf{N}$ with $1 \leq r \leq n$. Show that, for all choices of $t_1, \ldots, t_r \in \mathbf{N}$, the ideal

$$((X_1 - \alpha_1)^{t_1}, \ldots, (X_r - \alpha_r)^{t_r})$$

of $R$ is primary.

# Chapter 5

# Rings of fractions

This chapter is concerned with a far-reaching generalization of the construction of the field of fractions of an integral domain, which was reviewed in 1.31. Recall the construction: if $R$ is an integral domain, then $S := R \setminus \{0\}$ is a multiplicatively closed subset of $R$ in the sense of 3.43 (that is $1 \in S$ and $S$ is closed under multiplication); an equivalence relation $\sim$ on $R \times S$ given by, for $(a, s), (b, t) \in R \times S$,

$$(a, s) \sim (b, t) \iff at - bs = 0$$

is considered; the equivalence class which contains $(a, s)$ (where $(a, s) \in R \times S$) is denoted by $a/s$; and the set of all the equivalence classes of $\sim$ can be given the structure of a field in such a way that the rules for addition and multiplication resemble exactly the familiar high school rules for addition and multiplication of fractions.

The generalization which concerns us in this chapter applies to any multiplicatively closed subset $S$ of an arbitrary commutative ring $R$: once again, we consider an equivalence relation on the set $R \times S$, but in this case the definition of the relation is more complicated in order to overcome problems created by the possible presence of zerodivisors. Apart from this added complication, the construction is remarkably similar to that of the field of fractions of an integral domain, although the end product does not have quite such good properties: we do not often get a field, and, in fact, the general construction yields what is known as the ring of fractions $S^{-1}R$ of $R$ with respect to the multiplicatively closed subset $S$; this ring of fractions may have non-zero zerodivisors; and, although there is a natural ring homomorphism $f : R \to S^{-1}R$, this map is not automatically injective.

However, on the credit side, we should point out right at the beginning that one of the absolutely fundamental examples of this construction arises when we take for the multiplicatively closed subset $S$ of $R$ the complement $R \setminus P$ of a prime ideal $P$ of $R$: in this case, the new ring of fractions $S^{-1}R$ turns out to be a quasi-local ring, denoted by $R_P$; furthermore, the passage from $R$ to $R_P$ for appropriate $P$, referred to as 'localization at $P$', is often a powerful tool in commutative algebra.

**5.1 LEMMA.** *Let $S$ be a multiplicatively closed subset of the commutative ring $R$. Define a relation $\sim$ on $R \times S$ as follows: for $(a,s),(b,t) \in R \times S$, we write*

$$(a,s) \sim (b,t) \quad \Longleftrightarrow \quad \exists\, u \in S \text{ with } u(ta - sb) = 0.$$

*Then $\sim$ is an equivalence relation on $R \times S$.*

*Proof.* It is clear that $\sim$ is both reflexive and symmetric: recall that $1 \in S$. Suppose that $(a,s)$, $(b,t)$, $(c,u) \in R \times S$ are such that $(a,s) \sim (b,t)$ and $(b,t) \sim (c,u)$. Thus there exist $v, w \in S$ such that $v(ta - sb) = 0 = w(ub - tc)$. The first of these equations yields $wuvta = wuvsb$, and the second yields $vswub = vswtc$. Therefore

$$wtv(ua - sc) = 0 \quad \text{and} \quad wtv \in S.$$

Hence $(a,s) \sim (c,u)$. It follows that $\sim$ is transitive and is therefore an equivalence relation. $\square$

**5.2 PROPOSITION, TERMINOLOGY and NOTATION.** *Let the situation be as in 5.1, so that $S$ is a multiplicatively closed subset of the commutative ring $R$. For $(a,s) \in R \times S$, denote the equivalence class of $\sim$ which contains $(a,s)$ by $a/s$ or*

$$\frac{a}{s},$$

*and denote the set of all equivalence classes of $\sim$ by $S^{-1}R$. Then $S^{-1}R$ can be given the structure of a commutative ring under operations for which*

$$\frac{a}{s} + \frac{b}{t} = \frac{ta + sb}{st}, \qquad \frac{a}{s}\frac{b}{t} = \frac{ab}{st}$$

*for all $a, b \in R$ and $s, t \in S$. This new ring $S^{-1}R$ is called the* ring of fractions *of $R$ with respect to $S$; its zero element is $0/1$ and its multiplicative identity element is $1/1$.*

*There is a ring homomorphism $f : R \to S^{-1}R$ given by $f(r) = r/1$ for all $r \in R$; this is referred to as the* natural ring homomorphism.

*Proof.* This proof consists entirely of a large amount of routine and tedious checking: one must verify that the formulas given in the statement of the proposition for the operations of addition and multiplication are unambiguous, and verify that the axioms for a commutative ring are satisfied. The fact that $f$ is a ring homomorphism is then obvious.

We shall just verify that the formula given for addition is unambiguous, and leave the remaining checking as an exercise for the reader: it is certainly good for the soul of each student of the subject that he or she should carry out all this checking at least once in his or her life! So suppose that $a, a', b, b' \in R$ and $s, s', t, t' \in S$ are such that, in $S^{-1}R$,

$$\frac{a}{s} = \frac{a'}{s'} \quad \text{and} \quad \frac{b}{t} = \frac{b'}{t'}.$$

Thus there exist $u, v \in S$ such that $u(s'a - sa') = 0 = v(t'b - tb')$. The first of these equations yields that

$$uv(s't'ta - stt'a') = 0,$$

while we deduce from the second that

$$uv(s't'sb - sts'b') = 0.$$

Add these two equations to obtain that

$$uv\left(s't'(ta + sb) - st(t'a' + s'b')\right) = 0,$$

so that, since $uv \in S$, we finally deduce that

$$\frac{ta + sb}{st} = \frac{t'a' + s'b'}{s't'}.$$

The rest of the proof is left as an exercise. □

**5.3** ♯EXERCISE. Complete the proof of 5.2. (It is perhaps worth pointing out that, once it has been checked that the formulas given for the rules for addition and multiplication are unambiguous, then this information can be used to simplify considerably the work involved in the checking of the ring axioms. The point is that, given, say, $a, b \in R$ and $s, t \in S$, we can write $a/s \in S^{-1}R$ as

$$\frac{a}{s} = \frac{ta}{ts}$$

(because $1\,((ts)a - s(ta)) = 0$), so that $a/s$ and $b/t$ can be put on a 'common denominator'. Note also that

$$\frac{ta}{ts} + \frac{sb}{ts} = \frac{tsta + tssb}{(ts)^2} = \frac{ta + sb}{ts}.$$

Thus the amount of work in this exercise is perhaps not as great as the reader might first have feared.)

**5.4 REMARKS.** Let the situation be as in 5.1 and 5.2.

(i) Note that $0_{S^{-1}R} = 0/1 = 0/s$ for all $s \in S$.

(ii) Let $a \in R$ and $s \in S$. Then $a/s = 0_{S^{-1}R}$ if and only if there exists $t \in S$ such that $t(1a - s0) = 0$, that is, if and only if there exists $t \in S$ such that $ta = 0$.

(iii) Thus the ring $S^{-1}R$ is trivial, that is, $1/1 = 0/1$, if and only if there exists $t \in S$ such that $t1 = 0$, that is, if and only if $0 \in S$.

(iv) In general, even if $0 \notin S$ so that $S^{-1}R$ is not trivial, the natural ring homomorphism $f : R \to S^{-1}R$ need not be injective: it follows from (ii) above that $\operatorname{Ker} f = \{a \in R : \text{ there exists } t \in S \text{ such that } ta = 0\}$.

(v) We should perhaps reinforce some of the comments which were included as hints for Exercise 5.3. For $a \in R$ and $s, t \in S$, we have $a/s = ta/ts$, so that we can change the denominator in $a/s$ by multiplying numerator and denominator by $t$; this means that a finite number of formal fractions in $S^{-1}R$ can be put on a common denominator.

(vi) Note also that addition of formal fractions in $S^{-1}R$ that already have the same denominator is easy: for $b, c \in R$ and $s \in S$, we have

$$\frac{b}{s} + \frac{c}{s} = \frac{b+c}{s}.$$

How does this new construction of ring of fractions relate to the construction of the field of fractions of an integral domain? Some readers will probably have already realised that the latter is a particular case of the former.

**5.5 REMARK.** Let $R$ be an integral domain, and set $S := R \setminus \{0\}$, a multiplicatively closed subset of $R$. If we construct the ring of fractions $S^{-1}R$ as in 5.2, then we obtain precisely the field of fractions of $R$. This is because $S$ consists of non-zerodivisors on $R$, so that, for $a, b \in R$ and $s, t \in S$, it is the case that there exists $u \in S$ with $u(ta - sb) = 0$ if and only if $ta - sb = 0$. Thus the equivalence relation used in 5.2 (for this special case) is the same as that used in 1.31, and, furthermore, the ring operations on the set of equivalence classes are the same in the two situations.

Note that in this case the natural ring homomorphism $f : R \to S^{-1}R$ is injective and embeds $R$ as a subring of its field of fractions.

**5.6 EXERCISE.** Let $I$ be a proper ideal of the commutative ring $R$, and let $\Phi$ denote the set of all multiplicatively closed subsets of $R$ which are disjoint from $I$. Show that $\Phi$ has at least one maximal member with respect to inclusion, and that $S$ is a maximal member of $\Phi$ if and only if $R \setminus S$ is a minimal prime ideal of $I$ (see 3.52).

**5.7** EXERCISE. Let $S$ be a multiplicatively closed subset of the commutative ring $R$. We say that $S$ is *saturated* precisely when the following condition is satisfied: whenever $a, b \in R$ are such that $ab \in S$, then both $a$ and $b$ belong to $S$.

(i) Show that $S$ is saturated if and only if $R \setminus S$, the complement of $S$ in $R$, is the union of some (possibly empty) family of prime ideals of $R$.

(ii) Let $T$ be an arbitrary multiplicatively closed subset of $R$. Let $\overline{T}$ denote the intersection of all saturated multiplicatively closed subsets of $R$ which contain $T$. Show that $\overline{T}$ is a saturated multiplicatively closed subset of $R$ which contains $T$, so that $\overline{T}$ is the smallest saturated multiplicatively closed subset of $R$ which contains $T$ in the sense that it is contained in every saturated multiplicatively closed subset of $R$ which contains $T$.

(We call $\overline{T}$ the *saturation* of $T$.)

(iii) Prove that, with the notation of (ii) above,

$$\overline{T} = R \setminus \bigcup_{\substack{P \in \mathrm{Spec}(R) \\ P \cap T = \emptyset}} P.$$

It is time that we had an example to show that the natural ring homomorphism from a commutative ring to one of its non-trivial rings of fractions need not be injective.

**5.8** EXAMPLE. Take $R = \mathbf{Z}/6\mathbf{Z}$ and $S = \{\bar{1}, \bar{3}, \bar{5}\}$, where we are denoting the natural image in $R$ of $n \in \mathbf{Z}$ by $\bar{n}$. Then the natural ring homomorphism $f : R \to S^{-1}R$ has $\mathrm{Ker}\, f = \bar{2}R$.

*Proof.* By 5.4(iv), $\mathrm{Ker}\, f = \{\bar{n} \in R : n \in \mathbf{Z} \text{ and } s\bar{n} = 0 \text{ for some } s \in S\}$. Since $\bar{1}, \bar{5}$ are units of $R$, it follows that $\mathrm{Ker}\, f = (0 : \bar{3})$, and it is easily seen that this ideal is $\bar{2}R$. $\square$

We are now going to analyse the natural ring homomorphism from a commutative ring to one of its rings of fractions in greater detail.

**5.9** REMARKS. Let $S$ be a multiplicatively closed subset of the commutative ring $R$, and let $f : R \to S^{-1}R$ be the natural ring homomorphism. Note that $f$ has the following properties.

(i) For each $s \in S$, the element $f(s) = s/1$ is a unit of $S^{-1}R$, having inverse $1/s$.

(ii) By 5.4(iv), if $a \in \mathrm{Ker}\, f$, then there exists $s \in S$ such that $sa = 0$.

(iii) Each element $a/s$ of $S^{-1}R$ (where $a \in R$, $s \in S$) can be written as $a/s = f(a)(f(s))^{-1}$, since

$$\frac{a}{s} = \frac{a}{1}\frac{1}{s} = \frac{a}{1}\left(\frac{s}{1}\right)^{-1} = f(a)(f(s))^{-1}.$$

In the situation of 5.9, the ring $S^{-1}R$ can be regarded as an $R$-algebra by means of the natural ring homomorphism $f$: see 1.9. We shall see shortly that the properties of $f$ described in 5.9 essentially determine $S^{-1}R$ uniquely as an $R$-algebra: of course, we shall have to make precise exactly what we mean by this phrase.

**5.10** PROPOSITION. *Let $S$ be a multiplicatively closed subset of the commutative ring $R$; also, let $f : R \to S^{-1}R$ denote the natural ring homomorphism. Let $R'$ be a second commutative ring, and let $g : R \to R'$ be a ring homomorphism with the property that $g(s)$ is a unit of $R'$ for all $s \in S$. Then there is a unique ring homomorphism $h : S^{-1}R \to R'$ such that $h \circ f = g$.*
*In fact, $h$ is such that*

$$h(a/s) = g(a)\,(g(s))^{-1} \qquad \text{for all } a \in R, \ s \in S.$$

*Proof.* We first show that the formula given in the second paragraph of the statement is unambiguous. So suppose that $a, a' \in R$, $s, s' \in S$ are such that $a/s = a'/s'$ in $S^{-1}R$. Thus there exists $t \in S$ such that $t\,(s'a - sa') = 0$. Apply the ring homomorphism $g$ to this equation to obtain that

$$g(t)\,(g\,(s')\,g\,(a) - g\,(s)\,g\,(a')) = g\,(0) = 0.$$

But, by hypothesis, each of $g\,(t), g\,(s), g\,(s')$ is a unit of $R'$; we therefore multiply the above equation by the product of their inverses to deduce that

$$g\,(a)\,(g\,(s))^{-1} = g\,(a')\,(g\,(s'))^{-1}.$$

It follows that we can define a map $h : S^{-1}R \to R'$ by the formula given in the second paragraph of the statement of the proposition. It is now an easy exercise to check that this $h$ is a ring homomorphism: remember that two formal fractions in $S^{-1}R$ can be put on a common denominator. Observe also that $h \circ f = g$ because, for all $a \in R$, we have $(h \circ f)(a) = h(a/1) = g(a)\,(g(1))^{-1} = g(a)$.

It remains to show that this $h$ is the only ring homomorphism with the stated properties. So suppose that $h' : S^{-1}R \to R'$ is a ring homomorphism such that $h' \circ f = g$. Then, for all $a \in R$, we have $h'(a/1) = (h' \circ f)\,(a) = g(a)$. In particular, for $s \in S$, we have $h'(s/1) = g(s)$; recall that $g(s)$ is a unit of $R'$, and this enables us to deduce that we must have $h'(1/s) = (g(s))^{-1}$ since

$$h'\left(\frac{1}{s}\right) g\,(s) = h'\left(\frac{1}{s}\right) h'\left(\frac{s}{1}\right) = h'\left(\frac{1}{s}\frac{s}{1}\right) = h'\,(1_{S^{-1}R}) = 1_{R'}.$$

It follows that, for all $a \in R$, $s \in S$, we must have

$$h'\left(\frac{a}{s}\right) = h'\left(\frac{a}{1}\frac{1}{s}\right) = h'\left(\frac{a}{1}\right) h'\left(\frac{1}{s}\right) = g(a)(g(s))^{-1},$$

so that there is exactly one ring homomorphism $h$ with the desired properties. $\square$

**5.11** EXERCISE. Let $S$ and $T$ be multiplicatively closed subsets of the commutative ring $R$ such that $S \subseteq T$. Show that there is a ring homomorphism $h : S^{-1}R \to T^{-1}R$ for which $h(a/s) = a/s \in T^{-1}R$ for all $a \in R$ and $s \in S$.

(We refer to $h$ as the *natural ring homomorphism* in this situation.)

**5.12** EXERCISE. Let $S$ and $T$ be multiplicatively closed subsets of the commutative ring $R$ such that $S \subseteq T$. Let $h : S^{-1}R \to T^{-1}R$ be the natural ring homomorphism of 5.11. Show that the following statements are equivalent.

(i) The homomorphism $h$ is an isomorphism.

(ii) For each $t \in T$, the element $t/1 \in S^{-1}R$ is a unit of $S^{-1}R$.

(iii) For each $t \in T$, there exists $a \in R$ such that $at \in S$.

(iv) We have $T \subseteq \overline{S}$, where $\overline{S}$ denotes the saturation of $S$ (see 5.7).

(v) Whenever $P \in \operatorname{Spec}(R)$ is such that $P \cap S = \emptyset$, then $P \cap T = \emptyset$ too.

In the situation of 5.10, the natural homomorphism $f : R \to S^{-1}R$ and the ring homomorphism $g : R \to R'$ turn $S^{-1}R$ and $R'$ into $R$-algebras. The ring homomorphism $h$ given by 5.10 is actually a homomorphism of $R$-algebras in the sense of the following definition.

**5.13** DEFINITIONS. Let $R$ be a commutative ring, and let $R', R''$ be commutative $R$-algebras having structural ring homomorphisms $f' : R \to R'$ and $f'' : R \to R''$. An *$R$-algebra homomorphism* from $R'$ to $R''$ is a ring homomorphism $\psi : R' \to R''$ such that $\psi \circ f' = f''$.

We say that such an $R$-algebra homomorphism $\psi$ is an *$R$-algebra isomorphism* precisely when $\psi$ is a ring isomorphism. Then, $(\psi)^{-1} : R'' \to R'$ is also an $R$-algebra isomorphism, because (it is a ring isomorphism by 1.7 and)

$$(\psi)^{-1} \circ f'' = (\psi)^{-1} \circ \psi \circ f' = f'.$$

**5.14** ♯EXERCISE. Let $R$ be a commutative ring, and let

$$\psi : R_1 \longrightarrow R_2, \qquad \phi : R_2 \longrightarrow R_3$$

be $R$-algebra homomorphisms of commutative $R$-algebras. Show that

$$\phi \circ \psi : R_1 \to R_3$$

is an $R$-algebra homomorphism. Deduce that, if $\psi, \phi$ are $R$-algebra isomorphisms, then so too is $\phi \circ \psi$.

We can now show that, in the situation of 5.9, the properties of the $R$-algebra $S^{-1}R$ described in that result serve to determine $S^{-1}R$ uniquely up to $R$-algebra isomorphism.

**5.15** PROPOSITION. *Let $S$ be a multiplicatively closed subset of the commutative ring $R$. Suppose that $R'$ is a commutative $R$-algebra with structural ring homomorphism $g : R \to R'$, and assume that*
   (i) *$g(s)$ is a unit of $R'$ for all $s \in S$;*
   (ii) *if $a \in \operatorname{Ker} g$, then there exists $s \in S$ such that $sa = 0$;*
   (iii) *each element of $R'$ can be written in the form $g(a)\,(g(s))^{-1}$ for some*
$a \in R$ and $s \in S$.
   *Then there is a unique isomorphism of $R$-algebras $h : S^{-1}R \to R'$; in other words, there is a unique ring isomorphism $h : S^{-1}R \to R'$ such that $h \circ f = g$, where $f : R \to S^{-1}R$ denotes the natural ring homomorphism.*

*Proof.* By 5.10, there is a unique ring homomorphism $h : S^{-1}R \to R'$ such that $h \circ f = g$, and, moreover, $h$ is given by

$$h\left(\frac{a}{s}\right) = g\,(a)\,(g\,(s))^{-1} \qquad \text{for all } a \in R,\ s \in S.$$

It therefore remains only for us to show that $h$ is bijective.

It is clear from condition (iii) of the hypotheses that $h$ is surjective. Suppose that $a \in R$, $s \in S$ are such that $a/s \in \operatorname{Ker} h$. Then $g(a)\,(g(s))^{-1} = 0$, so that $g(a) = 0$ and $a \in \operatorname{Ker} g$. Hence, by condition (ii) of the hypotheses, there exists $t \in S$ such that $ta = 0$, so that $a/s = 0$ in $S^{-1}R$. Hence $h$ is injective too. $\square$

We next give an illustration of the use of the universal mapping property described in 5.10.

**5.16** EXAMPLE. Let $R$ be an integral domain, and let $S$ be a multiplicatively closed subset of $R$ such that $0 \notin S$. Let $K$ denote the field of fractions of $R$, and let $\theta : R \to K$ denote the natural ring homomorphism. Now for each $s \in S$, the element $s/1$ of $K$ is a unit of $K$, because it has inverse $1/s$. It therefore follows from 5.10 that there is a unique $R$-algebra homomorphism $h : S^{-1}R \to K$ (when $S^{-1}R$ and $K$ are regarded as $R$-algebras by means of the natural ring homomorphisms); moreover, we have

$$h\left(\frac{a}{s}\right) = \theta\left(a\right)\left(\theta\left(s\right)\right)^{-1} = \frac{a}{1}\frac{1}{s} = \frac{a}{s}$$

for all $a \in R$, $s \in S$. (The reader should note that there are two uses of the formal symbol $a/s$ here, one to denote an element of $S^{-1}R$ and the other to denote an element of $K$: the objects concerned are formed using different equivalence relations and should not be confused.)

It is clear that both $h$ and the natural homomorphism $f : R \to S^{-1}R$ are injective: we usually use $f$ and $h$ to identify $R$ as a subring of $S^{-1}R$ and $S^{-1}R$ as a subring of $K$.

**5.17** EXAMPLES. Let $R$ denote a general commutative ring.

(i) For a fixed $t \in R$, the set $S := \{t^n : n \in \mathsf{N}_0\}$ is a multiplicatively closed subset of $R$: we used a similar set in the proof of 3.48. In this case, the ring of fractions $S^{-1}R$ is often denoted by $R_t$. Note that, by 5.4(iii), $R_t$ is trivial if and only if $0 \in S$, that is, if and only if $t$ is nilpotent.

The notation just introduced gives a new meaning for the symbol $\mathsf{Z}_t$, for $t \in \mathsf{N}$: by 5.16, we can identify this ring of fractions of $\mathsf{Z}$ with the subring of $\mathsf{Q}$ consisting of all rational numbers which can be written in the form $a/t^n$ for some $a \in \mathsf{Z}$, $n \in \mathsf{N}_0$. For this reason, we shall no longer use the notation of 1.2(iii) to denote the ring of residue classes of integers modulo $t$, and we shall use $\mathsf{Z}/\mathsf{Z}t$ or $\mathsf{Z}/t\mathsf{Z}$ instead.

(ii) Let $J$ be an ideal of $R$. Then the set $1+J = \{1 + c : c \in J\}$ (which is nothing more than the coset of $J$ in $R$ which contains 1) is a multiplicatively closed subset of $R$, since $1 = 1 + 0$ and

$$(1 + c_1)(1 + c_2) = 1 + (c_1 + c_2 + c_1 c_2)$$

for all $c_1, c_2 \in J$. By 5.4(iii), $(1 + J)^{-1}R$ is trivial if and only if $0 \in 1 + J$, and it is easy to see that this occurs if and only if $J = R$.

**5.18** EXERCISE. Let $t \in \mathsf{N}$ and let $p_1, \ldots, p_t$ be $t$ distinct prime numbers. Show that the ring

$$R = \{\alpha \in \mathsf{Q} : \alpha = m/n \text{ for some } m \in \mathsf{Z} \text{ and } n \in \mathsf{N} \text{ such that}$$
$$n \text{ is divisible by none of } p_1, \ldots, p_t\}$$

of Exercise 3.67 is isomorphic to a ring of fractions of $\mathsf{Z}$.

**5.19** EXERCISE. Let $R$ be a commutative ring and let $X$ be an indeterminate. By 5.17(ii), the set $1 + XR[X]$ is a multiplicatively closed subset of $R[X]$. Note that $R[[X]]$ can be regarded as an $R[X]$-algebra by means of the inclusion homomorphism. Show that there is an injective $R[X]$-algebra homomorphism

$$(1 + XR[X])^{-1}R[X] \longrightarrow R[[X]].$$

Probably the most important example of a ring of fractions of a commutative ring $R$ is that where the multiplicatively closed subset concerned is $R \setminus P$ for some prime ideal $P$ of $R$. (In fact, given an ideal $I$ of $R$, the condition that $I \in \operatorname{Spec}(R)$ is equivalent to the condition that $R \setminus I$ is multiplicatively closed.) We give this example a lemma all of its own!

**5.20** LEMMA and DEFINITION. *Let $R$ be a commutative ring and let $P \in \operatorname{Spec}(R)$; let $S := R \setminus P$, a multiplicatively closed subset of $R$. The ring of fractions $S^{-1}R$ is denoted by $R_P$; it is a quasi-local ring, called the localization of $R$ at $P$, with maximal ideal*

$$\left\{ \lambda \in R_P : \lambda = \frac{a}{s} \text{ for some } a \in P,\ s \in S \right\}.$$

*Proof.* Let

$$I = \left\{ \lambda \in R_P : \lambda = \frac{a}{s} \text{ for some } a \in P,\ s \in S \right\}.$$

By 3.13 and 3.14, it is enough for us to show that $I$ is an ideal of $R_P$ and that $I$ is exactly the set of non-units of $R$.

It is easy to see that $I$ is an ideal of $R_P$: in fact (and this is a point to which we shall return in some detail later in the chapter) $I$ is the extension of $P$ to $R_P$ under the natural ring homomorphism. Let $\lambda \in R_P \setminus I$, and take any representation $\lambda = a/s$ with $a \in R,\ s \in S$. We must have $a \notin P$, so that $a/s$ is a unit of $R_P$ with inverse $s/a$. On the other hand, if $\mu$ is a unit of $R_P$, and $\mu = b/t$ for some $b \in R,\ t \in S$, then there exist $c \in R,\ v \in S$ such that

$$\frac{b}{t}\frac{c}{v} = \frac{1}{1}$$

in $R_P$. Therefore, there exists $w \in S$ such that $w(bc - tv) = 0$, so that $wbc = wtv \in R \setminus P$. Hence $b \notin P$, and since this reasoning applies to every representation $\mu = b/t$, with $b \in R,\ t \in S$, of $\mu$ as a formal fraction, it follows that $\mu \notin I$.

We have now proved that the ideal $I$ of $R_P$ is equal to the set of non-units of $R_P$, and so the proof is complete. $\square$

**5.21** EXAMPLE. By 3.34, $2\mathbb{Z}$ is a prime ideal of $\mathbb{Z}$, and so we can form the localization $\mathbb{Z}_{2\mathbb{Z}}$; by 5.16, this localization can be identified with the subring of $\mathbb{Q}$ consisting of all rational numbers which can be expressed in the form $m/n$ with $m, n \in \mathbb{Z}$ and $n$ odd.

Similarly, for a prime number $p$, the ideal $p\mathbb{Z} \in \operatorname{Spec}(\mathbb{Z})$, and the localization $\mathbb{Z}_{p\mathbb{Z}}$ can be identified with

$$\{\gamma \in \mathbb{Q} : \gamma = m/n \text{ for some } m, n \in \mathbb{Z} \text{ with } n \neq 0 \text{ and } \operatorname{GCD}(n, p) = 1\}.$$

**5.22** EXERCISE. Let $K$ be a field and let $a_1, \ldots, a_n \in K$. Let $F$ denote the field of fractions of the integral domain $K[X_1, \ldots, X_n]$ (of polynomials with coefficients in $K$ in indeterminates $X_1, \ldots, X_n$). Show that

$$R = \{\alpha \in F : \alpha = f/g \text{ with } f, g \in K[X_1, \ldots, X_n] \text{ and } g(a_1, \ldots, a_n) \neq 0\}$$

is a subring of $F$ which is isomorphic to a ring of fractions of $K[X_1, \ldots, X_n]$. Is $R$ quasi-local? If so, what can you say about its residue field? Justify your responses.

Now that we have a good fund of examples of rings of fractions, it is time for us to examine the ideal theory of such a ring. In the discussion it will be very convenient for us to use the extension and contraction notation of 2.41 in relation to the natural ring homomorphism from a commutative ring to one of its rings of fractions. To this end, we shall now introduce some notation which will be employed for several results.

**5.23** NOTATION. Until further notice, let $S$ be a multiplicatively closed subset of the commutative ring $R$; let $f : R \to S^{-1}R$ denote the natural ring homomorphism. Use the extension and contraction notation and terminology of 2.41 and 2.45 for $f$. In particular, $\mathcal{I}_R$ denotes the set of all ideals of $R$, $\mathcal{C}_R$ denotes the set of all ideals of $R$ which are contracted from $S^{-1}R$ under $f$, and $\mathcal{E}_{S^{-1}R}$ denotes the subset of $\mathcal{I}_{S^{-1}R}$ consisting of all ideals of $S^{-1}R$ which are extended from $R$ under $f$.

Our first result on the ideals of $S^{-1}R$ is that every ideal of this ring is actually extended from $R$, so that, with the notation of 5.23, $\mathcal{E}_{S^{-1}R} = \mathcal{I}_{S^{-1}R}$, the set of all ideals of $S^{-1}R$.

**5.24** LEMMA. *Let the situation be as in 5.23. Let $\mathcal{J}$ be an ideal of $S^{-1}R$. Then $\mathcal{J} = \mathcal{J}^{ce}$, so that each ideal of $S^{-1}R$ is extended from $R$ and*

$$\mathcal{E}_{S^{-1}R} = \mathcal{I}_{S^{-1}R}.$$

*Proof.* Let $\lambda \in \mathcal{J}$, and consider a representation $\lambda = a/s$, where $a \in R$, $s \in S$. Then

$$f(a) = \frac{a}{1} = \frac{s}{1}\frac{a}{s} \in \mathcal{J},$$

and so $a \in \mathcal{J}^c$. Hence $\lambda = a/s = (1/s) f(a) \in \mathcal{J}^{ce}$. Thus

$$\mathcal{J} \subseteq \mathcal{J}^{ce},$$

and the reverse inclusion is automatic, by 2.44(ii). $\square$

Thus, to describe a typical ideal of $S^{-1}R$, we have only to describe the extension from $R$ of a typical ideal of $R$. We now do this rather carefully.

**5.25** LEMMA. *Let the situation be as in 5.23, and let $I$ be an ideal of $R$. Then*

$$I^e = \left\{ \lambda \in S^{-1}R : \lambda = \frac{a}{s} \text{ for some } a \in I, \ s \in S \right\}.$$

*Proof.* It is clear that, for all $a \in I$ and $s \in S$, we have $a/s = (1/s)f(a) \in I^e$. For the reverse inclusion, let $\lambda \in I^e$. Now $I^e$ is the ideal of $S^{-1}R$ generated by $f(I)$. Thus, by 2.18, there exist $n \in \mathbb{N}$, $h_1, \ldots, h_n \in I$ and $\mu_1, \ldots, \mu_n \in S^{-1}R$ such that $\lambda = \sum_{i=1}^n \mu_i f(h_i)$. But there exist $a_1, \ldots, a_n \in R$ and $s_1, \ldots, s_n \in S$ such that $\mu_i = a_i/s_i$ $(1 \leq i \leq n)$, and so

$$\lambda = \sum_{i=1}^n \frac{a_i}{s_i} \frac{h_i}{1} = \sum_{i=1}^n \frac{a_i h_i}{s_i}.$$

When we put the right-hand side of this equation on a common denominator we see that $\lambda$ can be written as $\lambda = a/s$ with $a \in I$ and $s \in S$. $\square$

**5.26** ♯EXERCISE. Let the situation be as in 5.23. Show that if the ring $R$ is Noetherian, then so too is the ring $S^{-1}R$.

**5.27** REMARK. An important point is involved in 5.25: that result does *not* say that, for $\lambda \in I^e$, every representation for $\lambda$ as a formal fraction $b/t$ with $b \in R$, $t \in S$ must have its numerator $b$ in $I$; all that is claimed in 5.25 on this point is that $\lambda$ has at least one such representation $a/s$ with numerator $a \in I$ (and $s \in S$, of course).

Let us illustrate the point with an example. Consider the ring $\mathbb{Z}_3$ of fractions of $\mathbb{Z}$ with respect to the multiplicatively closed subset $\left\{ 3^i : i \in \mathbb{N}_0 \right\}$ of $\mathbb{Z}$: see 5.17(i). Set $J = 6\mathbb{Z}$, and use the extension and contraction notation as indicated in 5.23 for this example. The element $2/3 \in \mathbb{Z}_3$ clearly has one representation as a formal fraction in which the numerator does not belong to $J = 6\mathbb{Z}$, and yet

$$\frac{2}{3} = \frac{6}{3^2} \in J^e = (6\mathbb{Z})^e$$

by 5.25.

However, in the general case, the situation in this respect is much simpler for prime ideals, and even primary ideals, of $R$ which do not meet $S$, as we shall see in 5.29 below.

**5.28** NOTE. Let the situation be as in 5.23. Let $Q$ be a $P$-primary ideal of $R$. Then $Q \cap S = \emptyset$ if and only if $P \cap S = \emptyset$. One implication here is clear, while the other follows from the fact that, if $a \in P \cap S$ then there exists $n \in \mathbb{N}$ such that $a^n \in Q$, and $a^n \in S$ because $S$ is multiplicatively closed.

**5.29** LEMMA. *Let the situation be as in 5.23, and let $Q$ be a primary ideal of $R$ such that $Q \cap S = \emptyset$. Let $\lambda \in Q^e$. Then every representation $\lambda = a/s$ of $\lambda$ as a formal fraction (with $a \in R$, $s \in S$) must have its numerator $a \in Q$.*

*Furthermore, $Q^{ec} = Q$.*

*Proof.* Consider an arbitrary representation $\lambda = a/s$, with $a \in R$, $s \in S$. Since $\lambda \in Q^e$, there exist $b \in Q$, $t \in S$ such that $\lambda = b/t = a/s$. Therefore there exists $u \in S$ such that $u(sb - ta) = 0$. Hence $(ut)a = usb \in Q$. Now $ut \in S$, and since $Q \cap S = \emptyset$ it follows that every positive power of $ut$ lies outside $Q$. But $Q$ is a primary ideal, and so $a \in Q$, as required.

It is automatic that $Q \subseteq Q^{ec}$: see 2.44(i). To establish the reverse inclusion, let $a \in Q^{ec}$. Thus $a/1 \in Q^e$, and so, by what we have just proved, $a \in Q$. Hence $Q^{ec} \subseteq Q$ and the proof is complete. $\square$

As every prime ideal of $R$ is primary, 5.29 applies in particular to prime ideals of $R$ which are disjoint from $S$. This is such an important point that it is worth our while to record it separately.

**5.30** COROLLARY. *Let the situation be as in 5.23, and let $P \in \operatorname{Spec}(R)$ be such that $P \cap S = \emptyset$. Then every formal fraction representation of every element of $P^e$ must have its numerator in $P$, and, furthermore, $P^{ec} = P$.* $\square$

This corollary will enable us to give a complete description of the prime ideals of $S^{-1}R$ in terms of the prime ideals of $R$; however, before we deal with this, it is desirable for us to record some properties of the operation of extension of ideals from $R$ to $S^{-1}R$.

**5.31** LEMMA. *Let the situation be as in 5.23, and let $I, J$ be ideals of $R$. Then*

(i) $(I + J)^e = I^e + J^e$;
(ii) $(IJ)^e = I^e J^e$;
(iii) $(I \cap J)^e = I^e \cap J^e$;
(iv) $(\sqrt{I})^e = \sqrt{(I^e)}$;
(v) $I^e = S^{-1}R$ *if and only if* $I \cap S \neq \emptyset$.

*Proof.* (i) This is immediate from 2.43(i).
(ii) This is immediate from 2.43(ii).

(iii) Since $I \cap J \subseteq I$, it is clear that $(I \cap J)^e \subseteq I^e$; similarly, $(I \cap J)^e \subseteq J^e$, and so
$$(I \cap J)^e \subseteq I^e \cap J^e.$$
To establish the reverse inclusion, let $\lambda \in I^e \cap J^e$; by 5.25, $\lambda$ can be written as
$$\lambda = \frac{a}{s} = \frac{b}{t} \quad \text{with } a \in I, \ b \in J, \ s,t \in S.$$
(Do not fall into the trap described in 5.27: it is not automatic that, say, $a \in J$ as well as $I$!) It follows that there exists $u \in S$ such that $u(ta - sb) = 0$, so that $uta = usb \in I \cap J$ because $a \in I$ and $b \in J$. We can now write
$$\lambda = \frac{a}{s} = \frac{uta}{uts},$$
and this shows that $\lambda \in (I \cap J)^e$, by 5.25.

(iv) Let $\lambda \in (\sqrt{I})^e$. By 5.25, there exist $a \in \sqrt{I}$ and $s \in S$ such that $\lambda = a/s$. Now there exists $n \in \mathbf{N}$ such that $a^n \in I$. Hence
$$(\lambda)^n = \frac{a^n}{s^n} \in I^e$$
by 5.25, so that $\lambda \in \sqrt{(I^e)}$. This shows that
$$(\sqrt{I})^e \subseteq \sqrt{(I^e)}.$$

The reverse inclusion is not quite so straightforward. Let $\mu \in \sqrt{(I^e)}$, and take a representation $\mu = a/s$ with $a \in R$, $s \in S$. Now there exists $n \in \mathbf{N}$ such that $(\mu)^n = (a/s)^n \in I^e$, and so, by 5.25 once again, there exist $b \in I$, $t \in S$ such that
$$(\mu)^n = \frac{a^n}{s^n} = \frac{b}{t}.$$
Therefore there exists $v \in S$ such that $v(ta^n - s^n b) = 0$, so that $vta^n = vs^n b \in I$. Hence
$$(vta)^n = \left(v^{n-1}t^{n-1}\right)(vta^n) \in I$$
and $vta \in \sqrt{I}$. Thus $\mu = a/s = vta/vts \in (\sqrt{I})^e$, and we have shown that
$$(\sqrt{I})^e \supseteq \sqrt{(I^e)}.$$

(v) ($\Rightarrow$) Assume that $I^e = S^{-1}R$, so that $1/1 \in I^e$. By 5.25, this means that there exist $a \in I$, $s,t \in S$ such that $t(s1 - 1a) = 0$, so that $ts = ta \in I \cap S$.

($\Leftarrow$) Assume that $s \in I \cap S$; then, in $S^{-1}R$, we have $1/1 = s/s \in I^e$ (by 5.25), so that $I^e = S^{-1}R$. $\square$

We can now give a complete description of the prime ideals of $S^{-1}R$.

**5.32** THEOREM. *Let the situation be as in 5.23.*

(i) *If $P \in \operatorname{Spec}(R)$ and $P \cap S \neq \emptyset$, then $P^e = S^{-1}R$.*

(ii) *If $P \in \operatorname{Spec}(R)$ and $P \cap S = \emptyset$, then $P^e \in \operatorname{Spec}(S^{-1}R)$.*

(iii) *If $\mathcal{P} \in \operatorname{Spec}(S^{-1}R)$, then $\mathcal{P}^c \in \operatorname{Spec}(R)$ and $\mathcal{P}^c \cap S = \emptyset$. Also $\mathcal{P}^{ce} = \mathcal{P}$.*

(iv) *The prime ideals of $S^{-1}R$ are precisely the ideals of the form $P^e$, where $P$ is a prime ideal of $R$ such that $P \cap S = \emptyset$. In fact, each prime ideal of $S^{-1}R$ has the form $P^e$ for exactly one $P \in \operatorname{Spec}(R)$ such that $P \cap S = \emptyset$.*

*Proof.* (i) See 5.31(v).

(ii) For $P \in \operatorname{Spec}(R)$ with $P \cap S = \emptyset$, we have $P^{ec} = P$ by 5.30; consequently, $P^e \subset S^{-1}R$ since otherwise we should have $P^{ec} = \left(S^{-1}R\right)^c = R \neq P$.

Let $\lambda = a/s$, $\mu = b/t \in S^{-1}R$, where $a, b \in R$, $s, t \in S$, be such that $\lambda\mu \in P^e$, that is, such that $ab/st \in P^e$. By 5.30, $ab \in P$, so that, since $P$ is prime, either $a \in P$ or $b \in P$. Thus, by 5.25, we have either $\lambda = a/s \in P^e$ or $\mu = b/t \in P^e$. Hence $P^e \in \operatorname{Spec}(S^{-1}R)$.

(iii) For $\mathcal{P} \in \operatorname{Spec}(S^{-1}R)$, it is automatic from 3.27(ii) that $\mathcal{P}^c \in \operatorname{Spec}(R)$. Also, by 5.24, $\mathcal{P}$ is an extended ideal and $\mathcal{P}^{ce} = \mathcal{P}$. Hence we must have $\mathcal{P}^c \cap S = \emptyset$, for otherwise part (i) above would show that $\mathcal{P} = \mathcal{P}^{ce} = S^{-1}R$, a contradiction.

(iv) We have just proved in part (iii) that each prime ideal of $S^{-1}R$ has the form $P^e$ for some prime ideal $P$ of $R$ which is disjoint from $S$. Also, if $P, P'$ are prime ideals of $R$ with $P \cap S = P' \cap S = \emptyset$ and $P^e = P'^e$, then it follows from 5.30 that $P = P^{ec} = P'^{ec} = P'$. $\square$

**5.33** REMARKS. (i) The above Theorem 5.32 is important and will be used many times in the sequel. Most of its results can be summarized in the statement that extension gives us a bijective mapping

$$\{P \in \operatorname{Spec}(R) : P \cap S = \emptyset\} \quad \longrightarrow \quad \operatorname{Spec}(S^{-1}R)$$
$$P \qquad\qquad\qquad \longmapsto \qquad\qquad P^e$$

which preserves inclusion relations. The inverse of this bijection is given by contraction, and that too preserves inclusion relations.

(ii) Let us consider the implications of Theorem 5.32 for the localization of the commutative ring $R$ at a prime ideal $P$ of $R$. In this case, the multiplicatively closed subset concerned is $R \setminus P$, and, for $P' \in \operatorname{Spec}(R)$, we have $P' \cap (R \setminus P) = \emptyset$ if and only if $P' \subseteq P$. Thus, by part (i) above, there is a bijective inclusion-preserving mapping

$$\{P' \in \operatorname{Spec}(R) : P' \subseteq P\} \quad \longrightarrow \quad \operatorname{Spec}(R_P)$$
$$P' \qquad\qquad\qquad \longmapsto \qquad\qquad P'^e$$

whose inverse is also inclusion-preserving and is given by contraction. Since $\{P' \in \operatorname{Spec}(R) : P' \subseteq P\}$ clearly has $P$ as unique maximal element with respect to inclusion, it follows that $\operatorname{Spec}(R_P)$ has $P^e$ as unique maximal element with respect to inclusion. In view of 3.45, this gives another proof that $R_P$ is a quasi-local ring, a fact already established in a more down-to-earth manner in 5.20.

**5.34** EXERCISE. Let $R$ be a non-trivial commutative ring, and assume that, for each $P \in \operatorname{Spec}(R)$, the localization $R_P$ has no non-zero nilpotent element. Show that $R$ has no non-zero nilpotent element.

**5.35** DEFINITION and EXERCISE. We say that a non-trivial commutative ring is *quasi-semi-local* precisely when it has only finitely many maximal ideals.

Let $R$ be a commutative ring, let $n \in \mathbf{N}$, and let $P_1, \ldots, P_n$ be prime ideals of $R$. Show that $S := \bigcap_{i=1}^n (R \setminus P_i)$ is a multiplicatively closed subset of $R$ and that the ring $S^{-1}R$ is quasi-semi-local. Determine the maximal ideals of $S^{-1}R$.

**5.36** NOTATION. In the situation of 5.23, for an ideal $I$ of $R$, the extension $I^e$ of $I$ to $S^{-1}R$ under the natural ring homomorphism $f$ is often denoted by $IS^{-1}R$ instead of the more correct but also more cumbersome $f(I)S^{-1}R$. This notation is used particularly often in the case of a localization at a prime ideal $P$ of $R$: thus we shall frequently denote the unique maximal ideal of $R_P$ by $PR_P$.

It was mentioned in Chapter 4 that the theory of ideals in rings of fractions provides insight into the Second Uniqueness Theorem for Primary Decomposition. We therefore now analyse the behaviour of primary ideals in connection with rings of fractions, and present the results in a theorem which is very similar to 5.32.

**5.37** THEOREM. *Let the situation be as in 5.23.*

(i) *If $Q$ is a primary ideal of $R$ for which $Q \cap S \neq \emptyset$, then $Q^e = S^{-1}R$.*

(ii) *If $Q$ is a $P$-primary ideal of $R$ such that $Q \cap S = \emptyset$, then $Q^e$ is a $P^e$-primary ideal of $S^{-1}R$.*

(iii) *If $Q$ is a $\mathcal{P}$-primary ideal of $S^{-1}R$, then $Q^c$ is a $\mathcal{P}^c$-primary ideal of $R$ such that $Q^c \cap S = \emptyset$. Also $Q^{ce} = Q$.*

(iv) *The primary ideals of $S^{-1}R$ are precisely the ideals of the form $Q^e$, where $Q$ is a primary ideal of $R$ which is disjoint from $S$. In fact, each primary ideal of $S^{-1}R$ has the form $Q^e$ for exactly one primary ideal $Q$ of $R$ for which $Q \cap S = \emptyset$.*

*Proof.* (i) This is immediate from 5.31(v).

(ii) By 5.31(v), we have $Q^e \neq S^{-1}R$; also $\sqrt{(Q^e)} = P^e$ by 5.31(iv). Suppose that $\lambda, \mu \in S^{-1}R$ are such that $\lambda\mu \in Q^e$ but $\mu \notin P^e$. Take representations $\lambda = a/s$, $\mu = b/t$, with $a, b \in R$, $s, t \in S$. Observe that $P \cap S = \emptyset$, by 5.28. It follows from 5.29 that $ab \in Q$ but $b \notin P$. Since $Q$ is $P$-primary, we must therefore have $a \in Q$, so that $\lambda = a/s \in Q^e$. Hence $Q^e$ is $P^e$-primary.

(iii) It is immediate from 4.6 that $Q^c$ is $P^c$-primary. Now $Q$ is an extended ideal of $S^{-1}R$ by 5.24, and we have $Q^{ce} = Q$; hence $Q^c \cap S = \emptyset$ in view of 5.31(v).

(iv) By part (iii) above, each primary ideal of $S^{-1}R$ has the form $Q^e$ for some primary ideal $Q$ of $R$ such that $Q \cap S = \emptyset$. Suppose that $Q$ and $Q'$ are primary ideals of $R$ with $Q \cap S = Q' \cap S = \emptyset$ such that $Q^e = Q'^e$. Then it follows from 5.29 that $Q = Q^{ec} = Q'^{ec} = Q'$. $\square$

**5.38** REMARKS. Once again, it is worth our while to spend some time taking stock of what we have proved, for the results of Theorem 5.37 are very important.

(i) Most of the results of 5.37 can be summarized by the statements that extension of ideals gives us an (inclusion-preserving) bijection from the set of all primary ideals of $R$ which are disjoint from $S$ to the set of all primary ideals of $S^{-1}R$, and, moreover, that the inverse of this bijection is given by contraction of ideals, and this also preserves inclusion relations.

(ii) Sometimes it is necessary to be more precise and to specify the radicals of the primary ideals under consideration. Note that, by 5.32, each prime ideal of $S^{-1}R$ has the form $P^e$ for exactly one prime ideal $P$ of $R$ which is disjoint from $S$. So let $P \in \mathrm{Spec}(R)$ with $P \cap S = \emptyset$. It follows from 5.37 that there is an inclusion-preserving bijection

$$\{Q \in \mathcal{I}_R : Q \text{ is } P\text{-primary}\} \longrightarrow \{Q \in \mathcal{I}_{S^{-1}R} : Q \text{ is } P^e\text{-primary}\}$$
$$Q \longmapsto Q^e$$

given by extension of ideals, whose inverse (also inclusion-preserving) is given by contraction.

**5.39** EXERCISE. Let the situation be as in 5.23.

(i) Let $\mathcal{I}$ be an irreducible ideal (see 4.31) of $S^{-1}R$. Show that $\mathcal{I}^c$ is an irreducible ideal of $R$.

(ii) Let $I$ be an irreducible ideal of $R$ for which $S \cap I = \emptyset$. Suppose that $R$ is Noetherian. Show that $I^e$ is an irreducible ideal of $S^{-1}R$.

Let us illustrate the rather technical results of 5.37 and 5.38 by using them to describe the behaviour of primary decompositions under fraction formation.

**5.40** PROPOSITION. *Let the situation be as in 5.23, and let $I$ be a decomposable ideal of $R$. Let*

$$I = Q_1 \cap \ldots \cap Q_n \qquad \text{with } \sqrt{Q_i} = P_i \text{ for } i = 1, \ldots, n$$

*be a primary decomposition of $I$, and suppose that the terms have been indexed so that, for a suitable $m \in \mathbb{N}_0$ with $0 \leq m \leq n$, we have*

$$P_i \cap S = \emptyset \qquad \text{for } 1 \leq i \leq m,$$

*but*

$$P_j \cap S \neq \emptyset \qquad \text{for } m < j \leq n.$$

*(Both the extreme values $0$ and $n$ are permitted for $m$.) If $m = 0$, then $I^e = S^{-1}R$ and $I^{ec} = R$. However, if $1 \leq m \leq n$, then $I^e$ and $I^{ec}$ are both decomposable ideals, and*

$$I^e = Q_1^e \cap \ldots \cap Q_m^e \qquad \text{with } \sqrt{Q_i^e} = P_i^e \text{ for } i = 1, \ldots, m$$

*and*

$$I^{ec} = Q_1 \cap \ldots \cap Q_m \qquad \text{with } \sqrt{Q_i} = P_i \text{ for } i = 1, \ldots, m$$

*are primary decompositions. Finally, if the initial primary decomposition of $I$ is minimal (and $1 \leq m \leq n$), then these last two primary decompositions of $I^e$ and $I^{ec}$ are also minimal.*

*Proof.* It is clear from 5.31(iii) that $I^e = \bigcap_{i=1}^{n} Q_i^e$; but, by 5.37(i) and (ii), we have that $Q_j^e = S^{-1}R$ for $m < j \leq n$, while $Q_i^e$ is $P_i^e$-primary for $1 \leq i \leq m$. In particular, we see that, if $m = 0$, then $I^e = S^{-1}R$ and so $I^{ec} = R$. We now assume for the remainder of the proof that $1 \leq m \leq n$.

Thus, in these circumstances,

$$I^e = Q_1^e \cap \ldots \cap Q_m^e \qquad \text{with } \sqrt{Q_i^e} = P_i^e \text{ for } i = 1, \ldots, m$$

is a primary decomposition of $I^e$. Contract back to $R$ and use 2.43(iii) to see that $I^{ec} = \bigcap_{i=1}^{m} Q_i^{ec}$. But, by 5.29, we have $Q_i^{ec} = Q_i$ for all $i = 1, \ldots, m$, and so

$$I^{ec} = Q_1 \cap \ldots \cap Q_m \qquad \text{with } \sqrt{Q_i} = P_i \text{ for } i = 1, \ldots, m$$

is a primary decomposition of $I^{ec}$.

Finally, suppose that the initial primary decomposition of $I$ is minimal. Then it is immediate from 5.32(iv) that $P_1^e, \ldots, P_m^e$ are distinct prime ideals of $S^{-1}R$, and, of course, $P_1, \ldots, P_m$ are distinct prime ideals of $R$. Furthermore, we cannot have

$$Q_j \supseteq \bigcap_{\substack{i=1 \\ i \neq j}}^{m} Q_i$$

for some $j$ with $1 \leq j \leq m$, simply because that would imply that

$$Q_j \supseteq \bigcap_{\substack{i=1 \\ i \neq j}}^{n} Q_i,$$

contrary to the minimality of the initial primary decomposition. It now follows that we cannot have

$$Q_j^e \supseteq \bigcap_{\substack{i=1 \\ i \neq j}}^{m} Q_i^e$$

for some $j$ with $1 \leq j \leq m$, because we would be able to deduce from that, on contraction back to $R$ and use of 2.43(iii) and 5.29, that

$$Q_j \supseteq \bigcap_{\substack{i=1 \\ i \neq j}}^{m} Q_i,$$

and we have just seen that this cannot be. $\square$

**5.41** COROLLARY. *Let the situation be as in 5.23, and let $I$ be a decomposable ideal of $R$. If $I^e \neq S^{-1}R$, then both $I^e$ and $I^{ec}$ are decomposable ideals, and*

$$\text{ass } I^e = \{P^e : P \in \text{ass } I \text{ and } P \cap S = \emptyset\},$$

$$\text{ass } I^{ec} = \{P : P \in \text{ass } I \text{ and } P \cap S = \emptyset\}.$$

*Proof.* This is immediate from 5.40. $\square$

**5.42** REMARK. We are now in a position to show how the theory of ideals in rings of fractions can be used to provide a proof of the Second Uniqueness Theorem for Primary Decomposition 4.29.

We use the notation of the statement in 4.29. Set $S = R \setminus P_i$, where $P_i$ is a minimal prime ideal belonging to $I$. Note that, for all $j \in \mathbf{N}$ with $1 \leq j \leq n$ and $j \neq i$, we cannot have $P_j \subseteq P_i$, and so $P_j \cap S \neq \emptyset$. On the other hand, $P_i \cap S = \emptyset$. Now apply the results of 5.40 to this particular choice of $S$: we obtain that

$$Q_i = I^{ec} = Q'_i$$

and the result is proved. $\square$

**5.43** EXERCISE. Let $I$ be a decomposable ideal of the commutative ring $R$. Let

$$I = Q_1 \cap \ldots \cap Q_n \quad \text{with } \sqrt{Q_i} = P_i \text{ for } i = 1, \ldots, n$$

be a minimal primary decomposition of $I$. Let $\mathcal{P}$ be a non-empty subset of ass $I$ with the property that whenever $P \in \mathcal{P}$ and $P' \in$ ass $I$ are such that $P' \subseteq P$, then $P' \in \mathcal{P}$ too (such a subset of ass $I$ is called an *isolated subset of* ass $I$). Show that

$$\bigcap_{\substack{i=1 \\ P_i \in \mathcal{P}}}^{n} Q_i$$

depends only on $I$ and not on the choice of minimal primary decomposition of $I$.

**5.44** EXERCISE. Let $I$ be an ideal of the commutative ring $R$, and use $\bar{r}$ to denote $r + I$ (for each $r \in R$). Let $S$ be a multiplicatively closed subset of $R$, and set $\overline{S} := \{\bar{s} : s \in S\}$, which is clearly a multiplicatively closed subset of $R/I$. Prove that

$$(\overline{S})^{-1}(R/I) \cong S^{-1}R/IS^{-1}R.$$

Deduce that, if $P \in \operatorname{Spec}(R)$ is such that $I \subseteq P$, so that $P/I \in \operatorname{Spec}(R/I)$ by 3.28, then
   (i) $(R/I)_{P/I} \cong R_P/IR_P$, and
   (ii) the residue field of the local ring $R_P$ is isomorphic to the field of fractions of the integral domain $R/P$.

**5.45** EXERCISE. Let $S$ be a multiplicatively closed subset of the commutative ring $R$, and let $P \in \operatorname{Spec}(R)$ be such that $P \cap S = \emptyset$. Hence, by 5.32(ii), we have $PS^{-1}R \in \operatorname{Spec}(S^{-1}R)$. Prove that

$$\left(S^{-1}R\right)_{PS^{-1}R} \cong R_P.$$

**5.46** LEMMA and DEFINITION. *Let $P$ be a prime ideal of the commutative ring $R$, and let $n \in \mathbb{N}$. Use the notation of 5.23 in the particular case in which the multiplicatively closed subset $S$ is $R \setminus P$; thus we are going to* *employ the extension and contraction notation and terminology of 2.41 with reference to the natural ring homomorphism $f : R \to R_P$.*

With this notation, $(P^n)^{ec}$ is a $P$-primary ideal of $R$, called the $n$-th symbolic power of $P$, and denoted by $P^{(n)}$.

*Proof.* By 5.33(ii), $R_P$ is a quasi-local ring with unique maximal ideal $PR_P = P^e$. Now by 5.31(ii), $(P^n)^{ec} = ((P^e)^n)^c$. Furthermore, by 4.9, $(P^e)^n$ is a $P^e$-primary ideal of $R_P$. Since $P^{ec} = P$ by 5.30, it now follows from 4.6 that $((P^e)^n)^c$ is a $P$-primary ideal of $R$, as claimed. $\square$

**5.47** EXERCISE.  Let $P$ be a prime ideal of the commutative ring $R$. This exercise is concerned with the symbolic powers $P^{(n)}$ $(n \in \mathbb{N})$ of $P$ introduced in 5.46. Let $m, n \in \mathbb{N}$. Show that

(i) if $P^n$ has a primary decomposition, then $P$ is its unique isolated prime ideal, and $P^{(n)}$ is the (uniquely determined) $P$-primary term in any minimal primary decomposition of $P^n$;

(ii) if $P^{(m)}P^{(n)}$ has a primary decomposition, then $P$ is its unique isolated prime ideal and $P^{(m+n)}$ is the $P$-primary term in any minimal primary decomposition of $P^{(m)}P^{(n)}$; and

(iii) $P^{(n)} = P^n$ if and only if $P^n$ is $P$-primary.

**5.48** EXERCISE.  If $R$ is a non-trivial commutative ring with the property that $R_P$ is an integral domain for every $P \in \mathrm{Spec}(R)$, must $R$ necessarily be an integral domain? Justify your response.

# Chapter 6

# Modules

At the beginning of Chapter 2 the comment was made that some experienced readers will have found it amazing that a whole first chapter of this book contained no mention of the concept of ideal in a commutative ring. The same experienced readers will have found it equally amazing that there has been no discussion prior to this point in the book of the concept of module over a commutative ring. Experience has indeed shown that the study of the modules over a commutative ring $R$ can provide a great deal of information about $R$ itself. Perhaps one reason for the value of the concept of module is that it can be viewed as putting an ideal $I$ of $R$ and the residue class ring $R/I$ on the same footing. Up to now we have regarded $I$ as a substructure of $R$, while $R/I$ is a factor or 'quotient' structure of $R$: in fact, both can be regarded as $R$-modules.

Modules are to commutative rings what vector spaces are to fields. However, because the underlying structure of the commutative ring can be considerably more complicated and unpleasant than the structure of a field, the theory of modules is much more complicated than the theory of vector spaces: to give one example, the fact that some non-zero elements of a commutative ring may not have inverses means that we cannot expect the ideas of linear independence and linear dependence to play such a significant rôle in module theory as they do in the theory of vector spaces.

It is time we became precise and introduced the formal definition of module.

**6.1** Definition. Let $R$ be a commutative ring. A *module over $R$*, or an *$R$-module*, is an additively written Abelian group $M$ furnished with a 'scalar multiplication' of its elements by elements of $R$, that is, a mapping

$$. : R \times M \to M,$$

such that

  (i) $r.(m + m') = r.m + r.m'$ for all $r \in R$, $m, m' \in M$,

  (ii) $(r + r').m = r.m + r'.m$ for all $r, r' \in R$, $m \in M$,

  (iii) $(rr').m = r.(r'.m)$ for all $r, r' \in R$, $m \in M$, and

  (iv) $1_R.m = m$ for all $m \in M$.

**6.2** REMARKS. (i) In practice, the '.' denoting scalar multiplication of a module element by a ring element is usually omitted.

(ii) The axioms in 6.1 should be familiar to the reader from his undergraduate studies of vector spaces. Indeed, a module over a field $K$ is just a vector space over $K$. In our study of module theory, certain fundamental facts about vector spaces will play a crucial rôle: it will be convenient for us to introduce the abbreviation $K$-*space* for the more cumbersome 'vector space over $K$'.

(iii) The axioms in 6.1 have various easy consequences regarding the manipulation of expressions involving addition, subtraction and scalar multiplication, such as, for example, the fact that

$$(r - r')m = rm - r'm \qquad \text{for all } r, r' \in R \text{ and } m \in M.$$

We shall not dwell on such points.

**6.3** EXAMPLES. Let $R$ be a commutative ring, and let $I$ be an ideal of $R$.

(i) A very important example of an $R$-module is $R$ itself: $R$ is, of course, an Abelian group, the multiplication in $R$ gives us a mapping

$$. : R \times R \longrightarrow R,$$

and the ring axioms ensure that this 'scalar multiplication' turns $R$ into an $R$-module.

(ii) Let $I$ be an ideal of $R$. Since $I$ is closed under addition and under multiplication by arbitrary elements of $R$, it follows that $I$ too is an $R$-module under the addition and multiplication of $R$.

(iii) We show next that the residue class ring $R/I$ can be viewed as an $R$-module. Of course, $R/I$ has a natural Abelian group structure; we need to provide it with a scalar multiplication by elements of $R$. To this end, let $s, s' \in R$ be such that $s + I = s' + I$ in $R/I$, and let $r \in R$. Thus $s - s' \in I$, and so $rs - rs' = r(s - s') \in I$; hence $rs + I = rs' + I$. It follows that we can unambiguously define a mapping

$$\begin{array}{ccc} R \times R/I & \longrightarrow & R/I \\ (r, s + I) & \longmapsto & rs + I \, , \end{array}$$

and it is routine to check that $R/I$ becomes an $R$-module with respect to this 'scalar multiplication'.

Example 6.3(iii) prompts the following word of warning. Regard $\mathbf{Z}/6\mathbf{Z}$ as a $\mathbf{Z}$-module in the manner of that example. Then $3 + 6\mathbf{Z} \neq 0_{\mathbf{Z}/6\mathbf{Z}}$ and, of course, $2 \neq 0$ in $\mathbf{Z}$; however, $2(3 + 6\mathbf{Z}) = 0_{\mathbf{Z}/6\mathbf{Z}}$. Thus the result of multiplication of a non-zero element of the module by a non-zero scalar is, in this particular case, zero. This state of affairs could not happen in a vector space over a field, and the reader is warned to be suitably cautious.

**6.4** ♯EXERCISE. Let $R$ be a commutative ring. Let $S$ be an $R$-algebra with structural ring homomorphism $f : R \to S$. (See 1.9.) Show that $S$ is an $R$-module with respect to its own addition and scalar multiplication given by

$$
\begin{array}{rcl}
R \times S & \longrightarrow & S \\
(r, s) & \longmapsto & f(r)s \, .
\end{array}
$$

Show also that $S$ can be viewed as an $R$-module using its own addition and scalar multiplication given by

$$
\begin{array}{rcl}
R \times S & \longrightarrow & S \\
(r, s) & \longmapsto & sf(r) \, .
\end{array}
$$

(Note that these two $R$-module structures on $S$ are identical in the case in which $S$ is a *commutative* $R$-algebra. In the sequel, we shall only use these ideas in such a situation.)

**6.5** ♯EXERCISE. Let $G$ be an (additively written) Abelian group. Show that there is exactly one way of turning $G$ into a $\mathbf{Z}$-module, and that in this $\mathbf{Z}$-module structure, the 'scalar multiplication' is given by

$$
ng = \left\{
\begin{array}{lll}
g \; + \; \cdots \; + \; g & (n \text{ terms}) & \text{for } n > 0, \\
0_G & & \text{for } n = 0, \\
(-g) \; + \; \cdots \; + \; (-g) & (-n \text{ terms}) & \text{for } n < 0
\end{array}
\right.
$$

for all $g \in G$ and $n \in \mathbf{Z}$.

Deduce that the concept of Abelian group is exactly the same as the concept of $\mathbf{Z}$-module.

**6.6** REMARK. Let $R$ and $S$ be commutative rings, and let $f : R \to S$ be a ring homomorphism. Let $G$ be an $S$-module. Then it is easy to check that $G$ has a structure as $R$-module with respect to (the same addition and) scalar multiplication given by

$$
\begin{array}{rcl}
R \times G & \longrightarrow & G \\
(r, g) & \longmapsto & f(r)g \, .
\end{array}
$$

In these circumstances, we say that $G$ is regarded as an $R$-module *by means of $f$*, or *by restriction of scalars* when there is no ambiguity about which ring homomorphism is being used.

We have, in fact, already come across two situations in this chapter which could be described in terms of restriction of scalars. Firstly, in 6.3(iii), we described how the residue class ring $R/I$, where $I$ is an ideal of the commutative ring $R$, can be regarded as an $R$-module. Another way of arriving at the same $R$-module structure on $R/I$ is to regard $R/I$ as a module over itself in the natural way (see 6.3(i)), and then regard it as an $R$-module by restriction of scalars using the natural surjective ring homomorphism from $R$ to $R/I$.

Secondly, let $R$ be a commutative ring and let $S$ be a commutative $R$-algebra, with structural ring homomorphism $f : R \to S$. In 6.4, we saw that $S$ can be regarded as an $R$-module: in fact, that $R$-module structure can be achieved by regarding $S$ as a module over itself in the natural way, and then 'restricting scalars' using $f$ to regard $S$ as an $R$-module.

The examples in 6.3 give some clear hints about some natural developments of the theory of modules. The fact that an ideal $I$ of a commutative ring $R$ is itself an $R$-module with respect to the operations in $R$ suggests some concept of 'submodule', while the $R$-module structure on $R/I$ hints at a 'factor module' structure. These ideas are absolutely fundamental, and we develop them in the next few results.

**6.7** DEFINITION. Let $M$ be a module over the commutative ring $R$, and let $G$ be a subset of $M$. We say that $G$ is a *submodule of $M$*, or an *$R$-submodule of $M$*, precisely when $G$ is itself an $R$-module with respect to the operations for $M$.

Note that, in the situation of 6.7, a submodule of $M$ is, in particular, an Abelian subgroup of the additive group of $M$, and so must have the same zero element $0_M$ as $M$. Furthermore, $M$ itself is a submodule of $M$, as also is the singleton set $\{0_M\}$; the latter is called the *zero submodule of $M$* and denoted simply by 0.

It will come as no surprise that there is a 'Submodule Criterion'.

**6.8** THE SUBMODULE CRITERION. *Let $R$ be a commutative ring and let $G$ be a subset of the $R$-module $M$. Then $G$ is a submodule of $M$ if and only if the following conditions hold:*
   (i) $G \neq \emptyset$;
   (ii) *whenever $g, g' \in G$ and $r, r' \in R$, then $rg + r'g' \in G$.*

*Proof.* ($\Rightarrow$) This is clear, since $G$ must contain the zero element 0 of $M$, and $G$ must be closed under the addition of $M$ and under scalar multiplication by arbitrary elements of $R$.

($\Leftarrow$) By the Subgroup Criterion, $G$ is an additive subgroup of $M$; also, by (ii), $G$ is closed under scalar multiplication by arbitrary elements of $R$.

It is now clear that properties 6.1(i),(ii),(iii),(iv) are automatically inherited from $M$. □

**6.9** REMARK. It follows from 6.8 that, when a commutative ring $R$ is regarded as a module over itself in the natural way, as described in 6.3(i), then its submodules are precisely its ideals.

The Submodule Criterion 6.8 enables us to develop the theory of generating sets for modules and submodules. Much of the theory is very similar to work on generation of ideals in 2.17, 2.18 and 2.19, and so will be covered by means of exercises.

**6.10** GENERATION OF SUBMODULES. Let $M$ be a module over the commutative ring $R$. By 6.8, the intersection of any non-empty family of submodules of $M$ is again a submodule; we adopt the convention whereby the intersection of the empty family of submodules of $M$ is interpreted as $M$ itself.

Let $J \subseteq M$. We define the *submodule of $M$ generated by $J$* to be the intersection of the (certainly non-empty) family of all submodules of $M$ which contain $J$. Note that this is the *smallest* submodule of $M$ which contains $J$ in the sense that (it *is* one and) it is contained in every other submodule of $M$ which contains $J$.

**6.11** ♯EXERCISE. Let $M$ be a module over the commutative ring $R$, and let $J \subseteq M$; let $G$ be the submodule of $M$ generated by $J$.
 (i) Show that, if $J = \emptyset$, then $G = 0$.
 (ii) Show that, if $J \neq \emptyset$, then

$$G = \left\{ \sum_{i=1}^{n} r_i j_i : n \in \mathbb{N},\ r_1, \ldots, r_n \in R,\ j_1, \ldots, j_n \in J \right\}.$$

 (iii) Show that, if $\emptyset \neq J = \{l_1, \ldots, l_t\}$, then

$$G = \left\{ \sum_{i=1}^{t} r_i l_i : r_1, \ldots, r_t \in R \right\}.$$

(In this case, we say that $G$ is *generated by* $l_1, \ldots, l_t$.)

**6.12** NOTATION and TERMINOLOGY. If, in the situation of 6.11, $J = \{j\}$, then it follows from 6.11(iii) that the submodule $G$ of $M$ generated by $J$ is $\{rj : r \in R\}$: this submodule is often denoted by $Rj$.

We shall say that a submodule $N$ of (the above) $R$-module $M$ is *finitely generated* precisely when it is generated by a finite subset of $M$ (in fact, of $N$, necessarily). An $R$-module is said to be *cyclic* precisely when it can be generated by one element.

**6.13** SUMS OF SUBMODULES. Let $M$ be a module over the commutative ring $R$. Let $(G_\lambda)_{\lambda \in \Lambda}$ be a family of submodules of $M$. We define the *sum* $\sum_{\lambda \in \Lambda} G_\lambda$ to be the submodule of $M$ generated by $\bigcup_{\lambda \in \Lambda} G_\lambda$. In particular, this sum is zero when $\Lambda = \emptyset$.

**6.14** ♯EXERCISE. Let $M$ be a module over the commutative ring $R$.
   (i) Show that the binary operation on the set of all submodules of $M$ given by submodule sum is both commutative and associative.
   (ii) Let $G_1, \ldots, G_n$ be submodules of $M$. Show that

$$\sum_{i=1}^{n} G_i = \left\{ \sum_{i=1}^{n} g_i : g_i \in G_i \text{ for } i = 1, \ldots, n \right\}.$$

We often denote $\sum_{i=1}^{n} G_i$ by $G_1 + \cdots + G_n$.
   (iii) Let $j_1, \ldots, j_n \in M$. Show that the submodule of $M$ generated by $j_1, \ldots, j_n$ is $Rj_1 + \cdots + Rj_n$.

**6.15** DEFINITION and REMARKS. Let $M$ be a module over the commutative ring $R$. Let $I, I'$ be ideals of $R$. We denote by $IM$ the submodule of $M$ generated by $\{rg : r \in I, g \in M\}$. Thus

$$IM = \left\{ \sum_{i=1}^{n} r_i g_i : n \in \mathbb{N}, \ r_1, \ldots, r_n \in I, \ g_1, \ldots, g_n \in M \right\}.$$

   (i) Note that $I(I'M) = (II')M$.
   (ii) For $a \in R$, we write $aM$ instead of $(Ra)M$: in fact,

$$(Ra)M = \{am : m \in M\}.$$

**6.16** DEFINITION. Let $M$ be a module over the commutative ring $R$. Let $G$ be a submodule of $M$, and let $J \subseteq M$, with $J \neq \emptyset$. We denote the ideal

$$\{r \in R : rj \in G \text{ for all } j \in J\}$$

of $R$ by $(G : J)$ (or by $(G :_R J)$ when it is desirable to emphasize the ring concerned). Observe that, if $N$ is the submodule of $M$ generated by $J$, then $(G : J) = (G : N)$. For $m \in M$, we write $(G : m)$ instead of $(G : \{m\})$.
   In the special case in which $G = 0$, the ideal

$$(0 : J) = \{r \in R : rj = 0 \text{ for all } j \in J\}$$

is called the *annihilator of $J$*, and denoted by $\text{Ann}(J)$ or $\text{Ann}_R(J)$. Also, for $m \in M$, we call $(0 : m)$ the *annihilator of $m$*.

**6.17** ♯EXERCISE. Let $I$ be an ideal of the commutative ring $R$. Show that
$I = \operatorname{Ann}_R(R/I) = (0 :_R 1 + I)$.

**6.18** ♯EXERCISE. Let $M$ be a module over the commutative ring $R$, let
$N$, $N'$, $G$ be submodules of $M$, and let

$$(G_\lambda)_{\lambda \in \Lambda} \quad \text{and} \quad (N_\theta)_{\theta \in \Theta}$$

be two families of submodules of $M$. Show that
  (i) $\left( \bigcap_{\lambda \in \Lambda} G_\lambda : N \right) = \bigcap_{\lambda \in \Lambda} (G_\lambda : N)$;
  (ii) $\left( G : \sum_{\theta \in \Theta} N_\theta \right) = \bigcap_{\theta \in \Theta} (G : N_\theta)$.
Deduce that $\operatorname{Ann}(N + N') = \operatorname{Ann}(N) \cap \operatorname{Ann}(N')$.

**6.19** CHANGE OF RINGS. Let $M$ be a module over the commutative ring
$R$. Let $I$ be an ideal of $R$ such that $I \subseteq \operatorname{Ann}(M)$. We show now how $M$
can be given a natural structure as a module over $R/I$.

Let $r, r' \in R$ be such that $r + I = r' + I$, and let $m \in M$. Then
$r - r' \in I \subseteq \operatorname{Ann}(M)$, and so $(r - r')m = 0$ and $rm = r'm$. Hence we can
unambiguously define a mapping

$$\begin{array}{ccc} R/I \times M & \longrightarrow & M \\ (r + I, m) & \longmapsto & rm \end{array}$$

and it is routine to check that this turns the Abelian group $M$ into an $R/I$-
module. Note that the $R$-module and $R/I$-module structures on $M$ are
related in the following way: $(r + I)m = rm$ for all $r \in R$ and all $m \in M$.

It should be noted that *a subset of $M$ is an $R$-submodule if and only if
it is an $R/I$-submodule.*

There is another 'colon' construction in the theory of modules in addition to that introduced in 6.16.

**6.20** DEFINITION. Let $M$ be a module over the commutative ring $R$,
let $G$ be a submodule of $M$, and let $I$ be an ideal of $R$. Then $(G :_M I)$
denotes the submodule $\{m \in M : rm \in G \text{ for all } r \in I\}$ of $M$. Observe that
$G \subseteq (G :_M I)$.

The particular case of this notation in which $G = 0$ is often used: the
submodule $(0 :_M I) = \{m \in M : rm = 0 \text{ for all } r \in I\}$ can be regarded as
the 'annihilator of $I$ in $M$'.

**6.21** ♯EXERCISE. Let $M$ be a module over the commutative ring $R$, let
$G$ be a submodule of $M$, and let $(G_\theta)_{\theta \in \Theta}$ be a family of submodules of $M$;
also let $I$, $J$ be ideals of $R$, and let $(I_\lambda)_{\lambda \in \Lambda}$ be a family of ideals of $R$. Show
that

(i) $((G :_M J) :_M K) = (G :_M JK) = ((G :_M K) :_M J)$;

(ii) $\left(\bigcap_{\theta \in \Theta} G_\theta :_M I\right) = \bigcap_{\theta \in \Theta} (G_\theta :_M I)$;

(iii) $\left(G :_M \sum_{\lambda \in \Lambda} I_\lambda\right) = \bigcap_{\lambda \in \Lambda} (G :_M I_\lambda)$.

We have now given a fairly comprehensive account of the elementary theory of submodules. We also promised earlier a discussion of the theory of residue class modules or factor modules.

**6.22** THE CONSTRUCTION OF RESIDUE CLASS MODULES. Let $M$ be a module over the commutative ring $R$, and let $G$ be a submodule of $M$. Of course, $G$ is a subgroup of the additive Abelian group $M$, and we can form the residue class group $M/G$:

$$M/G = \{m + G : m \in M\}$$

consists of the cosets of $G$ in $M$; two such cosets $m + G$ and $m' + G$ (where $m, m' \in M$) are equal if and only if $m - m' \in G$; and the addition in $M/G$ is such that

$$(m + G) + (x + G) = (m + x) + G \qquad \text{for all } m, x \in M.$$

(Compare the discussion in 2.8.)

Now let $r \in R$ and suppose that $m, m' \in M$ are such that $m+G = m'+G$ in $M/G$. Then $m - m' \in G$, so that, since $G$ is a submodule of $M$, we have $r(m-m') \in G$ and $rm+G = rm'+G$. Hence we can unambiguously define a mapping

$$\begin{aligned} R \times M/G &\longrightarrow M/G \\ (r, m + G) &\longmapsto rm + G \end{aligned}$$

and it is routine to check that this turns the Abelian group $M/G$ into an $R$-module. We call this $R$-module the *residue class module* or *factor module* of $M$ modulo $G$.

**6.23** ♯EXERCISE. Let $M$ be a module over the commutative ring $R$, and let $I$ be an ideal of $R$. Show that $I \subseteq \text{Ann}_R(M/IM)$ and deduce that $M/IM$ has a structure as an $(R/I)$-module under which $(r+I)(m+IM) = rm+IM$ for all $r \in R$ and $m \in M$.

It is important for the reader to have a good grasp of the form of the submodules of a factor module: the next exercise establishes results which are reminiscent of 2.37.

**6.24** ♯EXERCISE: THE SUBMODULES OF A FACTOR MODULE. Let $M$ be a module over the commutative ring $R$, and let $G$ be a submodule of $M$.

(i) Let $G'$ be a submodule of $M$ such that $G' \supseteq G$. Show that $G'/G$ is a submodule of $M/G$.

(ii) Show that each submodule of $M/G$ has the form $G''/G$ for exactly one submodule $G''$ of $M$ such that $G'' \supseteq G$.

(iii) Parts (i) and (ii) establish the existence of a bijective mapping from the set of all submodules of $M$ which contain $G$ to the set of all submodules of $M/G$. Let $G_1, G_2$ be submodules of $M$ which contain $G$. Show that $G_1 \subseteq G_2$ if and only if $G_1/G \subseteq G_2/G$ (so that both the above-mentioned bijective mapping and its inverse preserve inclusion relations).

**6.25** ♮EXERCISE. Let $M$ be a module over the commutative ring $R$. Let $G, G_1, G_2$ be submodules of $M$ with $G_i \supseteq G$ for $i = 1, 2$. Let $I$ be an ideal of $R$. For each of the following choices of the submodule $\mathcal{H}$ of $M/G$, determine the unique submodule $H$ of $M$ which has the properties that $H \supseteq G$ and $H/G = \mathcal{H}$.

(i) $\mathcal{H} = (G_1/G) + (G_2/G)$.

(ii) $\mathcal{H} = I(G_1/G)$.

(iii) $\mathcal{H} = (G_1/G) \cap (G_2/G)$.

(iv) $\mathcal{H} = 0$.

**6.26** EXERCISE. Let $M$ be a module over the commutative ring $R$. Let $G_1, G_2$ be submodules of $M$. Show that $\mathrm{Ann}((G_1 + G_2)/G_1) = (G_1 : G_2)$.

It is high time that we introduced the concept of module homomorphism. This is just the module-theoretic analogue of the concept of linear mapping in vector space theory.

**6.27** DEFINITIONS. Let $M$ and $N$ be modules over the commutative ring $R$. A mapping $f : M \to N$ is said to be a *homomorphism of R-modules* or an *R-module homomorphism*, or just simply an *R-homomorphism*, precisely when

$$f(rm + r'm') = rf(m) + r'f(m') \quad \text{for all } m, m' \in M \text{ and } r, r' \in R.$$

Such an $R$-module homomorphism is said to be a *monomorphism* precisely when it is injective; it is an *epimorphism* if and only if it is surjective. An *R-module isomorphism* is a bijective $R$-module homomorphism.

The mapping $z : M \to N$ defined by $z(m) = 0_N$ for all $m \in M$ is an $R$-homomorphism, called the *zero homomorphism* and denoted by $0$. If $f_i : M \to N$ (for $i = 1, 2$) are $R$-homomorphisms, then the mapping $f_1 + f_2 : M \to N$ defined by $(f_1 + f_2)(m) = f_1(m) + f_2(m)$ for all $m \in M$ is also an $R$-homomorphism, called the *sum* of $f_1$ and $f_2$.

**6.28** ♯EXERCISE. Let $M$ and $N$ be modules over the commutative ring $R$, and supppose that $f : M \to N$ is an isomorphism. Show that the inverse mapping $f^{-1} : N \to M$ is also an isomorphism. In these circumstances, we say that $M$ and $N$ are *isomorphic* $R$-modules, and we write $M \cong N$.

Note that $M \cong M$: the identity mapping $\mathrm{Id}_M$ of $M$ onto itself is an isomorphism.

Show that isomorphic $R$-modules have equal annihilators.

**6.29** REMARK. It should be clear to the reader that if $f : M \to N$ and $g : N \to G$ are homomorphisms of modules over the commutative ring $R$, then the composition $g \circ f : M \to G$ is also an $R$-module homomorphism.

Further, if $f$ and $g$ are isomorphisms, then so too is $g \circ f$.

**6.30** EXERCISE. Let $R$ be a commutative ring, and let $R'$ and $R''$ be commutative $R$-algebras; let $\psi : R' \to R''$ be a ring homomorphism. Show that $\psi$ is an $R$-algebra homomorphism (see 5.13) if and only if $\psi$ is a homomorphism of $R$-modules when $R'$ and $R''$ are regarded as $R$-modules by means of their structural ring homomorphisms.

There is an interrelation between the concepts of submodule and module homomorphism which is reminiscent of the interrelation between ideals and homomorphisms of commutative rings described in 2.12.

**6.31** DEFINITION. Let $M$ be a module over the commutative ring $R$. Let $G$ be a submodule of $M$. Then the mapping $f : M \to M/G$ defined by $f(m) = m + G$ for all $m \in M$ is an $R$-module homomorphism (as is easily checked), called the *natural* or *canonical* homomorphism. Note that $f$ is actually surjective, and so is an epimorphism.

**6.32** ♯EXERCISE and DEFINITIONS. Let $M$ be a module over the commutative ring $R$.

(i) Suppose that $N$ is a second $R$-module, and that $f : M \to N$ is a homomorphism of $R$-modules. The *kernel of* $f$, denoted by $\mathrm{Ker}\, f$, is the set $\{m \in M : f(m) = 0_N\}$. Show that $\mathrm{Ker}\, f$ is a submodule of $M$. Show also that $\mathrm{Ker}\, f = 0$ if and only if $f$ is a monomorphism.

The *image of* $f$, denoted $\mathrm{Im}\, f$, is the subset $f(M) = \{f(m) : m \in M\}$ of $N$. Show that $\mathrm{Im}\, f$ is a submodule of $N$.

(ii) Let $G$ be a submodule of $M$. Show that the kernel of the natural epimorphism from $M$ to $M/G$ of 6.31 is just $G$. Deduce that the natural epimorphism from $M$ to $M/0$ is an isomorphism.

(iii) Deduce that a subset $H$ of $M$ is a submodule of $M$ if and only if there exists an $R$-module homomorphism from $M$ to some $R$-module $M'$ which has kernel equal to $H$.

The reader should note the similarity of the result in 6.32(iii) to that of 2.12, in which we saw that a subset $I$ of a commutative ring $R$ is an ideal of $R$ if and only if $I$ is the kernel of a ring homomorphism from $R$ to some commutative ring $S$. There are further such similarities: for example, we come now to the Isomorphism Theorems for modules.

**6.33** THE FIRST ISOMORPHISM THEOREM FOR MODULES. *Let $M$ and $N$ be modules over the commutative ring $R$, and let $f : M \to N$ be an $R$-homomorphism. Then $f$ induces an isomorphism $\bar{f} : M/\operatorname{Ker} f \to \operatorname{Im} f$ for which $\bar{f}(m + \operatorname{Ker} f) = f(m)$ for all $m \in M$.*

*Proof.* This is so similar to the proof of 2.13 that it will be left as an exercise. $\square$

**6.34** ♯EXERCISE. Prove 6.33, the First Isomorphism Theorem for modules.

The argument which is needed to construct $\bar{f}$ in the First Isomorphism Theorem for modules can be applied to more general situations. For example, the results of the following exercise will be used frequently.

**6.35** ♯EXERCISE. Let $M$ and $N$ be modules over the commutative ring $R$, and let $f : M \to N$ be a homomorphism. Let $G$ be a submodule of $M$ such that $G \subseteq \operatorname{Ker} f$. Show that $f$ induces a homomorphism $g : M/G \to N$ for which $g(m + G) = f(m)$ for all $m \in M$.

Deduce that, if $M'$ is a submodule of $M$ and $N'$ is a submodule of $N$ such that $f(M') \subseteq N'$, then $f$ induces an $R$-homomorphism $\tilde{f} : M/M' \to N/N'$ with the property that $\tilde{f}(m + M') = f(m) + N'$ for all $m \in M$.

**6.36** ♯EXERCISE. Let $R$ be a commutative ring. For an $R$-module $M$, denote by $\mathcal{S}_M$ the set of all submodules of $M$. Let $f : M \to M'$ be an epimorphism of $R$-modules. Show that the mapping

$$\theta \ : \ \{G \in \mathcal{S}_M : G \supseteq \operatorname{Ker} f\} \ \longrightarrow \ \mathcal{S}_{M'}$$
$$G \ \longmapsto \ f(G)$$

is bijective, and that $\theta^{-1}(G') = f^{-1}(G')$ for all $G' \in \mathcal{S}_{M'}$. Show also that $\theta$ and $\theta^{-1}$ preserve inclusion relations.

The name of the last theorem suggests that there is at least a Second Isomorphism Theorem for modules; in fact, there are a Second and a Third such Theorems.

**6.37** THE SECOND ISOMORPHISM THEOREM FOR MODULES. *Let $M$ be a module over the commutative ring $R$. Let $G, G'$ be submodules of $M$ such* *that $G' \supseteq G$, so that, by 6.24, $G'/G$ is a submodule of the $R$-module $M/G$. Then there is an isomorphism*

$$\eta : (M/G)/(G'/G) \longrightarrow M/G'$$

*such that $\eta((m + G) + G'/G) = m + G'$ for all $m \in M$.*

*Proof.* Since $G \subseteq G'$, we can define a mapping $f : M/G \to M/G'$ for which $f(m + G) = m + G'$ for all $m \in M$: if $m, m' \in M$ are such that $m + G = m' + G$, then $m - m' \in G \subseteq G'$, and so $m + G' = m' + G'$. It is routine to check that $f$ is an epimorphism. Furthermore,

$$\begin{aligned}
\text{Ker } f &= \{m + G : m \in M \text{ and } m + G' = 0_{M/G'}\} \\
&= \{m + G : m \in G'\} = G'/G,
\end{aligned}$$

and so the result follows immediately from the First Isomorphism Theorem for modules 6.33. $\square$

**6.38** THE THIRD ISOMORPHISM THEOREM FOR MODULES. *Let $M$ be a module over the commutative ring $R$. Let $G$ and $H$ be submodules of $M$. Then there is an isomorphism*

$$\xi : G/(G \cap H) \longrightarrow (G + H)/H$$

*such that $\xi(g + G \cap H) = g + H$ for all $g \in G$.*

*Proof.* The mapping $f : G \to (G+H)/H$ defined by $f(g) = g+H$ for all $g \in G$ is an $R$-module homomorphism which is surjective, since an arbitrary element of $(G + H)/H$ has the form $g' + h' + H$ for some $g' \in G$, $h' \in H$, and $g' + h' + H = g' + H$. Furthermore,

$$\text{Ker } f = \{g \in G : g + H = 0_{(G+H)/H}\} = \{g \in G : g \in H\} = G \cap H,$$

and so the result follows immediately from the First Isomorphism Theorem for modules 6.33. $\square$

**6.39** DEFINITION. Let $R$ be a commutative ring, let $G, M$ and $N$ be $R$-modules, and let $g : G \to M$ and $f : M \to N$ be $R$-homomorphisms. We say that the sequence

$$G \overset{g}{\longrightarrow} M \overset{f}{\longrightarrow} N$$

is *exact* precisely when $\text{Im } g = \text{Ker } f$.

More generally, we say that a sequence

$$\cdots \longrightarrow M^{n-1} \overset{d^{n-1}}{\longrightarrow} M^n \overset{d^n}{\longrightarrow} M^{n+1} \overset{d^{n+1}}{\longrightarrow} M^{n+2} \longrightarrow \cdots$$

of $R$-modules and $R$-homomorphisms (which, incidentally, can be finite, infinite in both directions or infinite in just one direction) is *exact at* a term $M^r$ in the sequence for which both $d^{r-1}$ and $d^r$ are defined (that is, which is the range of one homomorphism in the sequence and the domain of another) precisely when

$$M^{r-1} \xrightarrow{d^{r-1}} M^r \xrightarrow{d^r} M^{r+1}$$

is an exact sequence; and we say that the whole sequence is *exact* if and only if it is exact at every term $M^r$ for which the concept makes sense, that is, for which both $d^{r-1}$ and $d^r$ are defined.

**6.40** REMARK. Let $M$ be a module over the commutative ring $R$. Observe that there is exactly one $R$-homomorphism $f : 0 \to M$: its image is 0. Also, there is exactly one $R$-homomorphism $g : M \to 0$: its kernel is $M$.

Suppose that $N$ is a second $R$-module and that $h : M \to N$ is an $R$-homomorphism.

(i) The sequence

$$0 \longrightarrow M \xrightarrow{h} N$$

is exact if and only if Ker $h = 0$, that is, if and only if $h$ is a monomorphism.

(ii) Similarly, the sequence

$$M \xrightarrow{h} N \longrightarrow 0$$

is exact if and only if $h$ is an epimorphism.

(iii) Let $G$ be a submodule of $M$. Then there is an exact sequence

$$0 \longrightarrow G \xrightarrow{i} M \xrightarrow{\pi} M/G \longrightarrow 0$$

in which $i$ is the inclusion homomorphism and $\pi$ is the natural canonical epimorphism of 6.31.

The remainder of this chapter is concerned mainly with methods of construction of new $R$-modules from given families of modules over a commutative ring $R$: we are going to discuss the direct sum and direct product of such a family.

**6.41** DEFINITIONS. Let $R$ be a commutative ring and let $(M_\lambda)_{\lambda \in \Lambda}$ be a non-empty family of $R$-modules. Then the Cartesian product set $\prod_{\lambda \in \Lambda} M_\lambda$ is an $R$-module under componentwise operations of addition and scalar multiplication. (This just means that these operations are given by

$$(g_\lambda)_{\lambda \in \Lambda} + (g'_\lambda)_{\lambda \in \Lambda} = (g_\lambda + g'_\lambda)_{\lambda \in \Lambda}$$

and
$$r(g_\lambda)_{\lambda \in \Lambda} = (rg_\lambda)_{\lambda \in \Lambda}$$

for all $(g_\lambda)_{\lambda \in \Lambda}$, $(g'_\lambda)_{\lambda \in \Lambda} \in \prod_{\lambda \in \Lambda} M_\lambda$ and $r \in R$.) This $R$-module is called the *direct product* of the family $(M_\lambda)_{\lambda \in \Lambda}$.

The subset of $\prod_{\lambda \in \Lambda} M_\lambda$ consisting of all families $(g_\lambda)_{\lambda \in \Lambda}$ (with $g_\lambda \in M_\lambda$ for all $\lambda \in \Lambda$, of course) with the property that only at most finitely many of the components $g_\lambda$ are non-zero, is an $R$-submodule of $\prod_{\lambda \in \Lambda} M_\lambda$. We denote this submodule by $\bigoplus_{\lambda \in \Lambda} M_\lambda$, and refer to it as the *direct sum*, or sometimes the *external direct sum*, of the family $(M_\lambda)_{\lambda \in \Lambda}$.

In the case when $\Lambda' = \emptyset$, we interpret both $\bigoplus_{\lambda' \in \Lambda'} M_{\lambda'}$ and $\prod_{\lambda' \in \Lambda'} M_{\lambda'}$ as the zero $R$-module.

Note that when $\Lambda$ is finite, we have $\bigoplus_{\lambda \in \Lambda} M_\lambda = \prod_{\lambda \in \Lambda} M_\lambda$.

**6.42** ♯Exercise. Let the situation be as in 6.41. For each $\mu \in \Lambda$, let $M'_\mu$ denote the subset of $\bigoplus_{\lambda \in \Lambda} M_\lambda$ given by

$$M'_\mu = \left\{ (g_\lambda)_{\lambda \in \Lambda} \in \bigoplus_{\lambda \in \Lambda} M_\lambda : g_\lambda = 0 \text{ for all } \lambda \in \Lambda \text{ with } \lambda \neq \mu \right\}.$$

Show that

(i) $M'_\mu$ is a submodule of $\bigoplus_{\lambda \in \Lambda} M_\lambda$ and $M'_\mu \cong M_\mu$, for all $\mu \in \Lambda$;
(ii) $\sum_{\lambda \in \Lambda} M'_\lambda = \bigoplus_{\lambda \in \Lambda} M_\lambda$; and
(iii) for each $\nu \in \Lambda$, we have

$$M'_\nu \bigcap \sum_{\substack{\lambda \in \Lambda \\ \lambda \neq \nu}} M'_\lambda = 0.$$

It is important for us to be able to recognize when a module is isomorphic to the direct sum of a family of submodules of itself. The concept of 'internal direct sum' helps us to do this.

**6.43** Definition. Let $M$ be a module over the commutative ring $R$. Let $(G_\lambda)_{\lambda \in \Lambda}$ be a non-empty family of submodules of $M$. If $M = \sum_{\lambda \in \Lambda} G_\lambda$, then each element $m \in M$ can be expressed in the form $m = \sum_{i=1}^{n} g_{\lambda_i}$, where $\{\lambda_1, \ldots, \lambda_n\}$ is a finite subset of $\Lambda$ and $g_{\lambda_i} \in G_{\lambda_i}$ for all $i = 1, \ldots, n$. We can actually write this as

$$m = \sum_{\lambda \in \Lambda} g_\lambda,$$

where it is understood that $g_\lambda \in G_\lambda$ for all $\lambda \in \Lambda$ but only finitely many of the $g_\lambda$ are non-zero, so that the summation does make sense. Of course, the phrase 'only finitely many of the $g_\lambda$ are non-zero' is to be interpreted as allowing the possibility that all the $g_\lambda$ are zero.

We say that $M$ is the *direct sum*, or sometimes the *internal direct sum*, of its family of submodules $(G_\lambda)_{\lambda \in \Lambda}$ precisely when each element $m \in M$ can be *uniquely* written in the form

$$m = \sum_{\lambda \in \Lambda} g_\lambda,$$

where $g_\lambda \in G_\lambda$ for all $\lambda \in \Lambda$ and only finitely many of the $g_\lambda$ are non-zero.

Of course, when this is the case, we must have $M = \sum_{\lambda \in \Lambda} G_\lambda$, but the uniqueness aspect of the definition means that additional conditions are satisfied: we denote that the sum is direct by writing $M = \bigoplus_{\lambda \in \Lambda} G_\lambda$.

It might be thought at first sight that this second use of the symbol '$\bigoplus$' in addition to that of 6.41 could lead to confusion. We shall see in Exercise 6.45 below that this will not be the case in practice, simply because, when, in the situation of 6.43, we have $M = \bigoplus_{\lambda \in \Lambda} G_\lambda$, then it is automatic that $M$ is isomorphic to the external direct sum of the family $(G_\lambda)_{\lambda \in \Lambda}$ described in 6.41. We first give another exercise, which, in conjunction with 6.42, demonstrates some similarities between internal and external direct sums.

**6.44** EXERCISE. Let $M$ be a module over the commutative ring $R$. Let $(G_\lambda)_{\lambda \in \Lambda}$ be a non-empty family of submodules of $M$. Prove that $M = \bigoplus_{\lambda \in \Lambda} G_\lambda$, that is, $M$ is the internal direct sum of the family of submodules $(G_\lambda)_{\lambda \in \Lambda}$, if and only if the following conditions are satisfied:

(i) $\sum_{\lambda \in \Lambda} G_\lambda = M$; and
(ii) for each $\nu \in \Lambda$, we have

$$G_\nu \bigcap \sum_{\substack{\lambda \in \Lambda \\ \lambda \neq \nu}} G_\lambda = 0.$$

**6.45** ‡EXERCISE. Let the situation be as in 6.44, and suppose that $M = \bigoplus_{\lambda \in \Lambda} G_\lambda$. Show that $M$ is isomorphic to the external direct sum of the family $(G_\lambda)_{\lambda \in \Lambda}$, as introduced in 6.41.

**6.46** REMARK. Consider again the situation of 6.41 and 6.42, so that $R$ is a commutative ring and $(M_\lambda)_{\lambda \in \Lambda}$ is a non-empty family of $R$-modules. For each $\mu \in \Lambda$, let $M'_\mu$ denote the subset of $\bigoplus_{\lambda \in \Lambda} M_\lambda$ given by

$$M'_\mu = \left\{ (g_\lambda)_{\lambda \in \Lambda} \in \bigoplus_{\lambda \in \Lambda} M_\lambda : g_\lambda = 0 \text{ for all } \lambda \in \Lambda \text{ with } \lambda \neq \mu \right\}.$$

It follows from 6.42 that $M'_\mu$ is a submodule of $\bigoplus_{\lambda \in \Lambda} M_\lambda$ which is isomorphic to $M_\mu$.

The results of Exercises 6.42 and 6.44 show that $\bigoplus_{\lambda \in \Lambda} M_\lambda$ is the internal direct sum of the family $(M'_\lambda)_{\lambda \in \Lambda}$ of its submodules.

There are some very important natural monomorphisms, epimorphisms and exact sequences connected with direct sums. We discuss these next.

**6.47** DEFINITION. Let $(M_\lambda)_{\lambda \in \Lambda}$ be a non-empty family of modules over the commutative ring $R$; let $\mu \in \Lambda$. Set $M = \bigoplus_{\lambda \in \Lambda} M_\lambda$.

It will be convenient to use *almost all* as an abbreviation for 'all except finitely many'.

The *canonical projection of $M = \bigoplus_{\lambda \in \Lambda} M_\lambda$ onto $M_\mu$* is the mapping $p_\mu : M \to M_\mu$ defined by

$$p_\mu \left( (g_\lambda)_{\lambda \in \Lambda} \right) = g_\mu$$

for all $(g_\lambda)_{\lambda \in \Lambda} \in M$ (so that, as usual, it is to be understood that almost all the $g_\lambda$ are zero).

The *canonical injection of $M_\mu$ into $M = \bigoplus_{\lambda \in \Lambda} M_\lambda$* is the mapping $q_\mu : M_\mu \to M$ defined by (for all $z \in M_\mu$)

$$q_\mu(z) = (g_\lambda)_{\lambda \in \Lambda},$$

where $g_\lambda = 0$ for all $\lambda \in \Lambda$ with $\lambda \neq \mu$ and $g_\mu = z$.

Both $p_\mu$ and $q_\mu$ are $R$-homomorphisms; in fact, $p_\mu$ is an epimorphism and $q_\mu$ is a monomorphism. Observe also that

(i) $p_\mu \circ q_\mu = \mathrm{Id}_{M_\mu}$;

(ii) $p_\mu \circ q_\nu = 0$ for all $\nu \in \Lambda$ with $\nu \neq \mu$; and

(iii) when $\Lambda$ is finite, $\sum_{\lambda \in \Lambda} q_\lambda \circ p_\lambda = \mathrm{Id}_M$. (The sum of homomorphisms from one $R$-module to another was defined in 6.27.)

**6.48** ♯EXERCISE. Let $M, M_1, \ldots, M_n$ (where $n \in \mathbf{N}$ with $n \geq 2$) be modules over the commutative ring $R$.

(i) Show that there is an exact sequence

$$0 \longrightarrow M_1 \xrightarrow{q_1} \bigoplus_{i=1}^{n} M_i \xrightarrow{p'_1} \bigoplus_{i=2}^{n} M_i \longrightarrow 0$$

(of $R$-modules and $R$-homomorphisms) in which $q_1$ is the canonical injection and

$$p'_1((m_1, \ldots, m_n)) = (m_2, \ldots, m_n)$$

for all $(m_1, \ldots, m_n) \in \bigoplus_{i=1}^{n} M_i$.

(ii) Suppose that there exist, for each $i = 1, \ldots, n$, homomorphisms $\tilde{p}_i : M \to M_i$ and $\tilde{q}_i : M_i \to M$ such that, for $1 \leq i, j \leq n$,

$$\tilde{p}_i \circ \tilde{q}_i = \text{Id}_{M_i} \qquad \text{and} \qquad \tilde{p}_i \circ \tilde{q}_j = 0 \text{ for } i \neq j,$$

and $\sum_{i=1}^n \tilde{q}_i \circ \tilde{p}_i = \text{Id}_M$. Show that the mapping $f : M \to \bigoplus_{i=1}^n M_i$ defined by

$$f(m) = (\tilde{p}_1(m), \ldots, \tilde{p}_n(m)) \qquad \text{for all } m \in M$$

is an isomorphism.

**6.49** DEFINITIONS. Let $R$ be a commutative ring. An exact sequence of $R$-modules and $R$-homomorphisms of the form

$$0 \longrightarrow L \xrightarrow{f} M \xrightarrow{g} N \longrightarrow 0$$

is called a *short exact sequence*. It is said to *split* precisely when $\text{Im} f = \text{Ker} g$ is a direct summand of $M$, that is, if and only if there is a submodule $G$ of $M$ such that $M = \text{Ker} g \oplus G$.

Thus an example of a short exact sequence is

$$0 \longrightarrow H \xrightarrow{i} M \xrightarrow{\pi} M/H \longrightarrow 0,$$

where $H$ is a submodule of $M$, $i$ is the inclusion homomorphism and $\pi$ is the canonical epimorphism.

An example of a split short exact sequence is

$$0 \longrightarrow M_1 \xrightarrow{q_1} M_1 \oplus M_2 \xrightarrow{p_2} M_2 \longrightarrow 0,$$

where $M_1, M_2$ are $R$-modules, $q_1$ is the canonical injection (see 6.47) and $p_2$ is the canonical projection.

**6.50** EXERCISE. Let $R$ be a commutative ring, and let

$$0 \longrightarrow L \xrightarrow{f} M \xrightarrow{g} N \longrightarrow 0$$

be a short exact sequence of $R$-modules and $R$-homomorphisms. Show that this sequence splits if and only if there exist $R$-homomorphisms $h : N \to M$ and $e : M \to L$ such that

$$e \circ f = \text{Id}_L, \quad g \circ h = \text{Id}_N, \quad e \circ h = 0, \quad f \circ e + h \circ g = \text{Id}_M.$$

The concept of direct sum is intimately related to the idea of 'free' module; a free module is, roughly speaking, one which has the module-theoretic analogue of a basis in vector space theory. As motivation, consider a module $M$ over a commutative ring $R$, and suppose that $M$ is generated by its subset $\{g_\lambda : \lambda \in \Lambda\}$, for some family $(g_\lambda)_{\lambda \in \Lambda}$ of elements of $M$. Then each element $m \in M$ can, by 6.11, be expressed in the form $m = \sum_{\lambda \in \Lambda} r_\lambda g_\lambda$, where $r_\lambda \in R$ for all $\lambda \in \Lambda$ and only finitely many of the $r_\lambda$ are non-zero. (In the case in which $\Lambda = \emptyset$, an empty sum is to be regarded as 0.) However, it is not necessarily the case that such an $m$ can be *uniquely* expressed in this way. To give an example, consider the $\mathbf{Z}$-module $\mathbf{Z}/2\mathbf{Z} \oplus \mathbf{Z}/5\mathbf{Z}$: this is generated by $(1 + 2\mathbf{Z}, 0 + 5\mathbf{Z})$ and $(0 + 2\mathbf{Z}, 1 + 5\mathbf{Z})$ (and not by just $(1 + 2\mathbf{Z}, 0 + 5\mathbf{Z})$, nor by just $(0 + 2\mathbf{Z}, 1 + 5\mathbf{Z})$), but

$$3(1+2\mathbf{Z},0+5\mathbf{Z}) + 6(0+2\mathbf{Z},1+5\mathbf{Z}) = 1(1+2\mathbf{Z},0+5\mathbf{Z}) + 1(0+2\mathbf{Z},1+5\mathbf{Z}).$$

Indeed, if, in the above general situation, each $m \in M$ can be uniquely written in the above form $m = \sum_{\lambda \in \Lambda} r_\lambda g_\lambda$, then we say that $(g_\lambda)_{\lambda \in \Lambda}$ is a 'base' for $M$ and that $M$ is a 'free' $R$-module. Let us give the formal definitions.

**6.51** DEFINITIONS. Let $M$ be a module over the commutative ring $R$. A *base* for $M$ is a family $(e_\lambda)_{\lambda \in \Lambda}$ of elements of $M$ such that

(i) $\{e_\lambda : \lambda \in \Lambda\}$ generates $M$; and

(ii) each $m \in M$ can be uniquely written in the form $m = \sum_{\lambda \in \Lambda} r_\lambda e_\lambda$, where $r_\lambda \in R$ for all $\lambda \in \Lambda$ and only finitely many of the $r_\lambda$ are non-zero.

An $R$-module is said to be *free* precisely when it has a base.

Note that $R$ itself is a free $R$-module, having a base formed by the element $1_R$. The zero $R$-module 0 is a free $R$-module with an empty base.

**6.52** ‡EXERCISE. Let $M$ be a module over the commutative ring $R$, and let $(e_\lambda)_{\lambda \in \Lambda}$ be a family of elements of $M$. Show that $(e_\lambda)_{\lambda \in \Lambda}$ is a base for $\sum_{\lambda \in \Lambda} Re_\lambda$ if and only if the following condition is satisfied: whenever $(r_\lambda)_{\lambda \in \Lambda}$ is a family of elements of $R$, with almost all the $r_\lambda$ zero, such that

$$\sum_{\lambda \in \Lambda} r_\lambda e_\lambda = 0,$$

then $r_\lambda = 0$ for all $\lambda \in \Lambda$.

A connection between free modules and direct sums is given by the following.

**6.53** PROPOSITION. *Let $R$ be a commutative ring.*

(i) *Let* $(R_\lambda)_{\lambda \in \Lambda}$ *be a family of R-modules with* $R_\lambda = R$ *for all* $\lambda \in \Lambda$. *Then* $\bigoplus_{\lambda \in \Lambda} R_\lambda$ *is a free R-module, with (when* $\Lambda \neq \emptyset$*) base* $(e_\lambda)_{\lambda \in \Lambda}$, *where, for each* $\mu \in \Lambda$, *the element* $e_\mu \in \bigoplus_{\lambda \in \Lambda} R_\lambda$ *has its component in* $R_\mu$ *equal to 1 and all its other components zero.*

(ii) *Let* $M$ *be an R-module. Then* $M$ *is free if and only if* $M$ *is isomorphic to an R-module of the type described in part* (i) *above (that is, loosely, if and only if* $M$ *is isomorphic to a direct sum of copies of* $R$*).*

*In fact, if* $M$ *has a base* $(e_\lambda)_{\lambda \in \Lambda}$, *then* $M \cong \bigoplus_{\lambda \in \Lambda} R_\lambda$, *where* $R_\lambda = R$ *for all* $\lambda \in \Lambda$.

*Proof.* (i) This is straightforward, and left as an exercise.

(ii) ($\Rightarrow$) Let $M$ be a free $R$-module having base $(e_\lambda)_{\lambda \in \Lambda}$. The claim is clear when $\Lambda = \emptyset$; so suppose $\Lambda \neq \emptyset$. For each $\lambda \in \Lambda$, let $R_\lambda = R$; define

$$f : \bigoplus_{\lambda \in \Lambda} R_\lambda \longrightarrow M$$

by the rule that $f((r_\lambda)_{\lambda \in \Lambda}) = \sum_{\lambda \in \Lambda} r_\lambda e_\lambda$ for all $(r_\lambda)_{\lambda \in \Lambda} \in \bigoplus_{\lambda \in \Lambda} R_\lambda$. (This makes sense because an element of $\bigoplus_{\lambda \in \Lambda} R_\lambda$ has only finitely many non-zero components.) Of course, $f$ is an $R$-homomorphism. Since $\{e_\lambda : \lambda \in \Lambda\}$ is a generating set for $M$, the mapping $f$ is surjective; since $(e_\lambda)_{\lambda \in \Lambda}$ is a base for $M$, it follows that $f$ is injective.

($\Leftarrow$) Once it has been realised that, if $M'$ and $M''$ are $R$-modules which are isomorphic, then $M'$ is free if and only if $M''$ is free, this part follows immediately from (i) above. $\square$

**6.54** ♯EXERCISE. Provide a proof for 6.53(i).

Just as, in vector space theory, bases permit easy descriptions of linear mappings between vector spaces, so a base for a free module $F$ over a commutative ring $R$ permits easy descriptions of $R$-homomorphisms from $F$ to other $R$-modules.

**6.55** REMARK. Let $F$ be a free module over the commutative ring $R$, and let $(e_\lambda)_{\lambda \in \Lambda}$ be a base for $F$. Let $M$ be an $R$-module, and let $(x_\lambda)_{\lambda \in \Lambda}$ be a family of elements of $M$. Then there is exactly one $R$-module homomorphism $f : F \to M$ for which $f(e_\lambda) = x_\lambda$ for all $\lambda \in \Lambda$.

Indeed, a homomorphism $f$ as described in the preceding paragraph must satisfy $f(\sum_{\lambda \in \Lambda} r_\lambda e_\lambda) = \sum_{\lambda \in \Lambda} r_\lambda x_\lambda$ for each family $(r_\lambda)_{\lambda \in \Lambda}$ of elements of $R$ with only finitely many non-zero members. Furthermore, it is easy to use the fact that $(e_\lambda)_{\lambda \in \Lambda}$ is a base for $F$ to see that the above formula unambiguously defines an $R$-homomorphism from $F$ to $M$.

It is convenient to have available the construction of a free module having a previously specified family of symbols as a base.

**6.56** REMARK. Let $(e_\lambda)_{\lambda \in \Lambda}$ be a family of symbols, indexed by a non-empty set $\Lambda$. Let $R$ be a commutative ring. We show how to construct a free $R$-module $F$ which essentially has $(e_\lambda)_{\lambda \in \Lambda}$ as a base.

Let $F$ be the set of all formal expressions $\sum_{\lambda \in \Lambda} r_\lambda e_\lambda$, where $r_\lambda \in R$ for all $\lambda \in \Lambda$ and only finitely many of the $r_\lambda$ are non-zero. We can define a law of composition $+$ on $F$ and a scalar multiplication by

$$\sum_{\lambda \in \Lambda} r_\lambda e_\lambda + \sum_{\lambda \in \Lambda} s_\lambda e_\lambda = \sum_{\lambda \in \Lambda} (r_\lambda + s_\lambda) e_\lambda$$

and

$$r \left( \sum_{\lambda \in \Lambda} r_\lambda e_\lambda \right) = \sum_{\lambda \in \Lambda} (r r_\lambda) e_\lambda$$

for all $\sum_{\lambda \in \Lambda} r_\lambda e_\lambda$, $\sum_{\lambda \in \Lambda} s_\lambda e_\lambda \in F$ and all $r \in R$. Then it is easy to check that these operations provide $F$ with a structure as $R$-module. Furthermore, if, for each $\mu \in \Lambda$, we identify $e_\mu$ with the formal expression $\sum_{\lambda \in \Lambda} r_\lambda e_\lambda \in F$ which has $r_\mu = 1$ and $r_\lambda = 0$ for all $\lambda \in \Lambda$ with $\lambda \neq \mu$, then $(e_\lambda)_{\lambda \in \Lambda}$ is actually a base for $F$, and this again is easy to check.

One application of 6.56 is to prove the often-needed fact that an arbitrary module over a commutative ring $R$ is an $R$-homomorphic image of some free $R$-module. We shall make use of this fact later in the book. The proof is an easy consequence of 6.55 and 6.56.

**6.57** PROPOSITION. *Let $M$ be a module over the commutative ring $R$. There exist a free $R$-module $F$ and an $R$-module epimorphism $f : F \to M$.*

*Also, if $M$ is finitely generated by $n$ elements, then $F$ can be taken to be a free $R$-module with a finite base having $n$ members.*

*Proof.* All the claims are clear when $M = 0$ and $n = 0$. So we suppose that $M \neq 0$. Let $(x_\lambda)_{\lambda \in \Lambda}$ be a (non-empty) family of elements of $M$ such that $\{x_\lambda : \lambda \in \Lambda\}$ generates $M$: if the worst comes to the worst, we can always use the fact that $M$ generates itself!

Now let $(e_\lambda)_{\lambda \in \Lambda}$ be a family of symbols indexed by $\Lambda$, and use 6.56 to construct a free $R$-module $F$ which has $(e_\lambda)_{\lambda \in \Lambda}$ as a base. Now use 6.55 to construct an $R$-homomorphism $f : F \to M$ for which $f(e_\lambda) = x_\lambda$ for all $\lambda \in \Lambda$. It is easy to see that $f$ is surjective. $\square$

Our last result in this chapter introduces the important concept of the 'rank' of a free module with a finite base over a non-trivial commutative ring.

**6.58** PROPOSITION and DEFINITION. *Let $R$ be a non-trivial commutative ring, and let $F$ be a free $R$-module with a finite base. Then every base for $F$ is finite, and any two bases for $F$ have the same number of members. The number of members in a base for $F$ is called the* rank *of $F$.*

*Proof.* Clearly we can assume that $F \neq 0$. Let $(e_\lambda)_{\lambda \in \Lambda}$ be a base for $F$. Since $R$ is not trivial, it has a maximal ideal, $M$ say, by 3.9. Now, by 6.23, the $R$-module $F/MF$ is annihilated by $M$ and can be given the structure of an $R/M$-module, that is, a vector space over $R/M$, in which $(r + M)(y + MF) = ry + MF$ for all $r \in R$ and $y \in F$. We show next that $(e_\lambda + MF)_{\lambda \in \Lambda}$ is a basis for this $R/M$-space.

It is easy to see that $\{e_\lambda + MF : \lambda \in \Lambda\}$ is a generating set for the $R/M$-space $F/MF$. Let $(\rho_\lambda)_{\lambda \in \Lambda}$ be a family of elements of $R/M$ with only finitely many non-zero members for which

$$\sum_{\lambda \in \Lambda} \rho_\lambda (e_\lambda + MF) = 0_{F/MF}.$$

Now $0_{R/M} = 0_R + M$, and so there is a family $(r_\lambda)_{\lambda \in \Lambda}$ of elements of $R$ with only finitely many non-zero members such that $\rho_\lambda = r_\lambda + M$ for all $\lambda \in \Lambda$. Hence

$$\sum_{\lambda \in \Lambda} (r_\lambda e_\lambda + MF) = 0_{F/MF},$$

so that $\sum_{\lambda \in \Lambda} r_\lambda e_\lambda \in MF$. Since $\{e_\lambda : \lambda \in \Lambda\}$ is a generating set for $F$, it follows easily from 6.15 that there is a family $(a_\lambda)_{\lambda \in \Lambda}$ of elements of $M$ with only finitely many non-zero members such that

$$\sum_{\lambda \in \Lambda} r_\lambda e_\lambda = \sum_{\lambda \in \Lambda} a_\lambda e_\lambda.$$

Since $(e_\lambda)_{\lambda \in \Lambda}$ is a base for $F$, it follows that $r_\lambda = a_\lambda \in M$ for all $\lambda \in \Lambda$. Thus $\rho_\lambda = r_\lambda + M = 0_{R/M}$ for all $\lambda \in \Lambda$. Hence $(e_\lambda + MF)_{\lambda \in \Lambda}$ is a basis for the $R/M$-space $F/MF$, as claimed.

The results of the proposition now follow immediately from standard facts from the theory of bases for vector spaces. $\square$

The general principal, employed in the above proof of 6.58, of manipulating certain situations in module theory so that familiar properties of vector spaces can be used to good effect is one that is useful in the theory of finitely generated modules over quasi-local rings: we shall explore this in our work in Chapter 9.

**6.59** EXERCISE. Suppose that $F$ is a free module over the non-trivial commutative ring $R$, and that $F$ is finitely generated. Show that every base for $F$ is finite.

**6.60** FURTHER STEPS. There are some fairly basic topics in module theory, such as tensor products, modules of homomorphisms, and projective modules, which have been omitted from this chapter because their use will be avoided in this book. However, the reader should be warned that, if he or she wishes to continue with the study of commutative algebra much beyond the scope of this book, then these topics are ones to which attention will need to be paid. They are all central to homological algebra, and the author found Northcott's book [11] a helpful introduction to them. Also, there are accounts of the basic ideas about tensor products in [8, Appendix A] and [1, Chapter 2].

In 6.27, we were close to defining the module $\text{Hom}_R(M, N)$, where $M$ and $N$ are modules over the commutative ring $R$: in fact, $\text{Hom}_R(M, N)$ is just the set of all $R$-homomorphisms from $M$ to $N$ with an $R$-module structure based on the addition defined in 6.27 and a scalar multiplication for which $(rf)(m) = rf(m)$ for $r \in R$, $f \in \text{Hom}_R(M, N)$ and $m \in M$. Also, it is not a great step from free modules to projective modules, because an $R$-module is projective if and only if it is a direct summand of a free $R$-module. However, we shall have to leave the interested reader to explore these topics from other sources.

# Chapter 7

# Chain conditions on modules

The introductory Chapter 6 on modules was concerned with the very basic principles of the theory of modules over commutative rings, and, indeed, one could take the view that Chapter 6, although it contains important and fundamental mathematics for our purposes, does not contain much exciting mathematics. In this chapter, we shall see that certain 'finiteness conditions' on modules over commutative rings can lead to information about the structures of the modules. Whether or not the reader finds the results of this chapter more interesting than those of Chapter 6 is obviously a matter of personal taste, but the author certainly finds some of the theorems presented in this chapter very attractive.

The first 'finiteness conditions' on modules which we shall consider are the so-called 'chain conditions'. The work in 3.35, 3.36, 3.37 and 3.38 is relevant here. Recall from 3.36 that, if $(V, \preceq)$ is a non-empty partially ordered set, then $(V, \preceq)$ satisfies the ascending chain condition if and only if it satisfies the maximal condition. (The statement that $(V, \preceq)$ satisfies the ascending chain condition means that, whenever $(v_i)_{i \in \mathbb{N}}$ is a family of elements of $V$ such that

$$v_1 \preceq v_2 \preceq \ldots \preceq v_i \preceq v_{i+1} \preceq \ldots,$$

then there exists $k \in \mathbb{N}$ such that $v_k = v_{k+i}$ for all $i \in \mathbb{N}$. Also, $(V, \preceq)$ satisfies the maximal condition precisely when every non-empty subset of $V$ contains a maximal element (with respect to $\preceq$). See 3.35.) We now apply these ideas to the set $\mathcal{S}_M$ of all submodules of a module $M$ over a commutative ring $R$. We say that $M$ is 'Noetherian' precisely when $\mathcal{S}_M$

123

satisfies the ascending chain condition with respect to inclusion, that is, precisely when the partially ordered set $(\mathcal{S}_M, \preceq)$ satisfies the equivalent conditions of 3.36 when $\preceq$ is given by, for $G_1, G_2 \in \mathcal{S}_M$,

$$G_1 \preceq G_2 \quad \Longleftrightarrow \quad G_1 \subseteq G_2.$$

Also, we say that $M$ is 'Artinian' precisely when $\mathcal{S}_M$ satisfies the ascending chain condition with respect to reverse inclusion, that is, precisely when the partially ordered set $(\mathcal{S}_M, \preceq_1)$ satisfies the equivalent conditions of 3.36 when $\preceq_1$ is given by, for $G_1, G_2 \in \mathcal{S}_M$,

$$G_1 \preceq_1 G_2 \quad \Longleftrightarrow \quad G_1 \supseteq G_2.$$

The names 'Noetherian' and 'Artinian' are in honour of Emmy Noether and Emil Artin, both of whom made fundamental contributions to the subject.

**7.1 DEFINITION.** Let $M$ be a module over the commutative ring $R$. We say that $M$ is a *Noetherian* $R$-module precisely when it satisfies the following conditions (which are equivalent, by 3.36).

(i) Whenever $(G_i)_{i \in \mathbb{N}}$ is a family of submodules of $M$ such that

$$G_1 \subseteq G_2 \subseteq \ldots \subseteq G_i \subseteq G_{i+1} \subseteq \ldots,$$

then there exists $k \in \mathbb{N}$ such that $G_k = G_{k+i}$ for all $i \in \mathbb{N}$.

(ii) Every non-empty set of submodules of $M$ contains a maximal element with respect to inclusion.

(Condition (i) above is called *the ascending chain condition* for submodules of $M$, while condition (ii) is called the *maximal condition* for submodules of $M$.)

**7.2 EXERCISE.** Let $M$ be a Noetherian module over the commutative ring $R$. Let $u : M \to M$ be an $R$-epimorphism of $M$ onto itself. Prove that $u$ is an isomorphism. (Here is a hint: $\operatorname{Ker} u \subseteq \operatorname{Ker}(u \circ u)$.)

**7.3 DEFINITION.** Let $M$ be a module over the commutative ring $R$. We say that $M$ is an *Artinian* $R$-module precisely when it satisfies the following conditions (which are equivalent, by 3.36).

(i) Whenever $(G_i)_{i \in \mathbb{N}}$ is a family of submodules of $M$ such that

$$G_1 \supseteq G_2 \supseteq \ldots \supseteq G_i \supseteq G_{i+1} \supseteq \ldots,$$

then there exists $k \in \mathbb{N}$ such that $G_k = G_{k+i}$ for all $i \in \mathbb{N}$.

(ii) Every non-empty set of submodules of $M$ contains a minimal element with respect to inclusion.

(Condition (i) above is called *the descending chain condition* for submodules of $M$, while condition (ii) is called the *minimal condition* for submodules of $M$.)

**7.4** EXERCISE. Let $M$ be an Artinian module over the commutative ring $R$. Let $v : M \to M$ be an $R$-monomorphism of $M$ into itself. Prove that $v$ is an isomorphism.

**7.5** REMARK. Let $R$ be a commutative ring. It should be clear to the reader from 6.9 that, when $R$ is regarded as a module over itself in the natural way, then $R$ is a Noetherian $R$-module if and only if $R$ is a Noetherian ring as defined in 3.37, simply because the $R$-submodules of $R$ are precisely the ideals of $R$. Of course, there is a concept of 'Artinian' commutative ring, and we define this next.

**7.6** DEFINITION. Let $R$ be a commutative ring. We say that $R$ is an *Artinian ring* precisely when it satisfies the following conditions (which are equivalent, by 3.36).

(i) Whenever $(I_i)_{i \in \mathbb{N}}$ is a family of ideals of $R$ such that

$$I_1 \supseteq I_2 \supseteq \ldots \supseteq I_i \supseteq I_{i+1} \supseteq \ldots,$$

then there exists $k \in \mathbb{N}$ such that $I_k = I_{k+i}$ for all $i \in \mathbb{N}$.

(ii) Every non-empty set of ideals of $R$ contains a minimal element with respect to inclusion.

We next consider some examples to show that the concepts of Noetherian module and Artinian module are different.

**7.7** EXAMPLE. Since $\mathbb{Z}$ is a principal ideal domain, it is a Noetherian ring, by 3.38. However, $\mathbb{Z}$ is not an Artinian ring because

$$2\mathbb{Z} \supset 2^2\mathbb{Z} \supset \ldots \supset 2^i\mathbb{Z} \supset 2^{i+1}\mathbb{Z} \supset \ldots$$

is a strictly descending chain of ideals of $\mathbb{Z}$ (since, for all $i \in \mathbb{N}$, we have $2^{i+1} = 2^i 2 \in 2^i\mathbb{Z}$, and also $2^i \notin 2^{i+1}\mathbb{Z}$ since an equation $2^i = 2^{i+1}r$ for some $r \in \mathbb{Z}$ would contradict the fact that $\mathbb{Z}$ is a UFD).

Note that this example also provides an instance of a module (over a commutative ring) which is Noetherian but not Artinian.

**7.8** ♯EXERCISE. (i) Show that a field is both an Artinian and a Noetherian ring.

(ii) Show that an Artinian PID must be a field.

Although we have seen in 7.7 an example of a commutative Noetherian ring which is not Artinian, we shall prove in Chapter 8 that every commutative Artinian ring must in fact be Noetherian. Thus, in order to find an example of an Artinian, non-Noetherian module over a commutative ring, we have to look beyond commutative rings considered as modules over themselves.

**7.9** DEFINITION. Let $M$ be a module over the commutative ring $R$. We say that a submodule $G$ of $M$ is *proper* precisely when $G \neq M$.

The above definition extends the concept of 'proper ideal' to modules over a commutative ring.

**7.10** EXAMPLE. Let $p$ be a fixed prime number. Then

$$E(p) := \left\{ \alpha \in \mathbb{Q}/\mathbb{Z} : \alpha = \frac{r}{p^n} + \mathbb{Z} \text{ for some } r \in \mathbb{Z} \text{ and } n \in \mathbb{N}_0 \right\}$$

is a submodule of the $\mathbb{Z}$-module $\mathbb{Q}/\mathbb{Z}$.

For each $t \in \mathbb{N}_0$ set

$$G_t := \left\{ \alpha \in \mathbb{Q}/\mathbb{Z} : \alpha = \frac{r}{p^t} + \mathbb{Z} \text{ for some } r \in \mathbb{Z} \right\}.$$

Then

(i) $G_t$ is the submodule of $E(p)$ generated by $(1/p^t)+\mathbb{Z}$, for each $t \in \mathbb{N}_0$ (so that $G_0 = 0$);

(ii) each proper submodule of $E(p)$ is equal to $G_i$ for some $i \in \mathbb{N}_0$; and

(iii) we have

$$G_0 \subset G_1 \subset \ldots \subset G_n \subset G_{n+1} \subset \ldots,$$

and $E(p)$ is an Artinian, non-Noetherian $\mathbb{Z}$-module.

*Proof.* It is easy to check that $E(p)$ is a submodule of the $\mathbb{Z}$-module $\mathbb{Q}/\mathbb{Z}$: we leave this checking to the reader.

(i) Let $t \in \mathbb{N}_0$. Then

$$G_t = \left\{ r\left( \frac{1}{p^t} + \mathbb{Z} \right) : r \in \mathbb{Z} \right\}$$

is the submodule of $E(p)$ generated by $(1/p^t) + \mathbb{Z}$: see 6.11(iii).

(ii) Let $H$ be a proper submodule of $E(p)$. If $H = 0$, then $H = G_0$; we therefore assume that $H \neq 0$. Let $0 \neq \alpha \in H$. Now there exist $r \in \mathbb{Z}$ and $t \in \mathbb{N}_0$ such that $\alpha = (r/p^t) + \mathbb{Z}$; moreover, since $\alpha \neq 0$, we must have $r \notin p^t\mathbb{Z}$, so that ($r \neq 0$ and) the highest power of $p$ which is a factor of $r$ is smaller than $p^t$. After cancelling any common powers of $p$, we see that we can write

$$\alpha = \frac{r'}{p^{t'}} + \mathbb{Z} \quad \text{with } t' \in \mathbb{N}, \text{ GCD}(r',p) = 1.$$

The next step is to show that, if $0 \neq \alpha_1 \in H$ and

$$\alpha_1 = \frac{r_1}{p^{t_1}} + \mathbb{Z} \quad \text{with } t_1 \in \mathbb{N}, \text{ GCD}(r_1,p) = 1,$$

then $(1/p^{t_1}) + \mathbf{Z} \in H$, so that $G_{t_1} \subseteq H$ in view of part (i). To establish this, note that $\mathrm{GCD}(r_1, p^{t_1}) = 1$, so that, by [15, Theorem 2.4.2], there exist $a, b \in \mathbf{Z}$ such that $ar_1 + bp^{t_1} = 1$. Hence, since $1 - ar_1 \in p^{t_1}\mathbf{Z}$, we have

$$\frac{1}{p^{t_1}} + \mathbf{Z} = \frac{ar_1}{p^{t_1}} + \mathbf{Z} = a\alpha_1 \in H,$$

as claimed.

Now note that $E(p) = \bigcup_{i \in \mathbf{N}_0} G_i$ and

$$G_0 \subseteq G_1 \subseteq \ldots \subseteq G_n \subseteq G_{n+1} \subseteq \ldots$$

(since $(1/p^n) + \mathbf{Z} = p((1/p^{n+1}) + \mathbf{Z})$ for each $n \in \mathbf{N}_0$). Since $H$ is a proper submodule of $E(p)$, it thus follows that there is a greatest integer $i \in \mathbf{N}$ such that $G_i \subseteq H$: if this were not so, then, for each $j \in \mathbf{N}$, there would exist $n_j \in \mathbf{N}$ with $n_j \geq j$ and $G_{n_j} \subseteq H$, so that $G_j \subseteq H$, and this would lead to the contradiction that $H = E(p)$. Let $m$ be this greatest integer. We claim that $G_m = H$. Of course, by definition of $m$, we have $G_m \subseteq H$.

Suppose that $G_m \subset H$ and look for a contradiction. Then there exists $\alpha_2 \in H \setminus G_m$. We can write

$$\alpha_2 = \frac{r_2}{p^{t_2}} + \mathbf{Z} \qquad \text{with } t_2 \in \mathbf{N}, \ \mathrm{GCD}(r_2, p) = 1.$$

Now $t_2 > m$ since $\alpha_2 \notin G_m$. It follows from the paragraph before last in this proof that $G_{t_2} \subseteq H$, contrary to the definition of $m$. Hence $H = G_m$, as claimed.

(iii) Let $i \in \mathbf{N}_0$. We show that $(1/p^{i+1}) + \mathbf{Z} \notin G_i$. Indeed, if we had $(1/p^{i+1}) + \mathbf{Z} \in G_i$, then there would exist $r \in \mathbf{Z}$ such that

$$\frac{1}{p^{i+1}} - \frac{r}{p^i} \in \mathbf{Z},$$

so that $1 - rp \in p^{i+1}\mathbf{Z}$, a contradiction. Hence

$$G_0 \subset G_1 \subset \ldots \subset G_n \subset G_{n+1} \subset \ldots.$$

This shows that $E(p)$ is not a Noetherian $\mathbf{Z}$-module.

The fact that $E(p)$ is an Artinian $\mathbf{Z}$-module follows from part (ii): any strictly descending chain of submodules of $E(p)$ would have second term equal to $G_i$ for some $i \in \mathbf{N}_0$, and we have now proved that there are only finitely many different submodules of $E(p)$ which are contained in $G_i$. Thus we see that there does not exist an infinite strictly descending chain of submodules of $E(p)$. $\square$

As vector spaces are particular examples of modules, it is natural to ask what it means for a vector space to be a Noetherian module, and what it means for a vector space to be an Artinian module. We answer these questions in the next proposition, which shows that, for a field $K$, the concepts of Noetherian $K$-module and Artinian $K$-module coincide. In view of the examples we have seen in 7.7 and 7.10, this gives another instance where the theory of $K$-spaces is simpler than the general theory of modules over an arbitrary commutative ring.

**7.11** Notation. For a finite-dimensional vector space $V$ over a field $K$, we shall denote the dimension of $V$ by vdim $V$ (or by $\text{vdim}_K V$ when it is desirable to indicate the ground field under consideration).

**7.12** Proposition. *Let $K$ be a field, and let $V$ be a vector space over $K$. Then the following statements are equivalent:*
 (i) *$V$ is a finite-dimensional $K$-space;*
 (ii) *$V$ is a Noetherian $K$-module;*
 (iii) *$V$ is an Artinian $K$-module.*

*Proof.* This is really just elementary vector space theory, but, nevertheless, we give the details.

(i) $\Rightarrow$ (ii), (i) $\Rightarrow$ (iii) Assume that $V$ is finite-dimensional, with dimension $n$, say. Recall that if $L$ is a subspace of $V$, then $L$ too has finite dimension, and vdim $L \leq n$; recall also that if $M$ is a second subspace of $V$ such that $L \subset M$, then vdim $L <$ vdim $M$. It follows from these facts that any finite chain

$$L_0 \subset L_1 \subset \ldots \subset L_{t-1} \subset L_t$$

of subspaces of $V$ (with *strict* inclusions) with $t + 1$ terms must have $t \leq n$. Hence $V$ is a Noetherian $K$-module and also an Artinian $K$-module.

(ii) $\Rightarrow$ (i), (iii) $\Rightarrow$ (i) Assume that $V$ is not finite-dimensional: we shall deduce that $V$ is not a Noetherian $K$-module and that it is not an Artinian $K$-module. There exists an infinite sequence $(w_i)_{i \in \mathbf{N}}$ of elements of $V$ such that, for every $n \in \mathbf{N}$, the family $(w_i)_{i=1}^n$ is linearly independent. For each $n \in \mathbf{N}$, set

$$L_n = \sum_{i=1}^{n} K w_i \quad \text{and} \quad M_n = \sum_{i=n+1}^{\infty} K w_i,$$

so that, in particular, $L_n$ is finite-dimensional and vdim $L_n = n$. Since

$$L_1 \subset L_2 \subset \ldots \subset L_n \subset L_{n+1} \subset \ldots,$$

we see that $V$ is not a Noetherian $K$-module. Also, since, for each $n \in \mathbf{N}$, we have $M_{n+1} \subseteq M_n$ and $w_{n+1} \notin M_{n+1}$, there is an infinite strictly descending chain

$$M_1 \supset M_2 \supset \ldots \supset M_n \supset M_{n+1} \supset \ldots$$

of subspaces of $V$, and $V$ is not an Artinian $K$-module. $\square$

There is another very important characterization of Noetherian modules: to discuss this, we need the concept of finitely generated module. Recall from 6.12 that a module $M$ over the commutative ring $R$ is said to be finitely generated if it is generated by some finite subset $J$ of itself, and recall from 6.11 that, when this is the case and $\emptyset \neq J = \{j_1, \ldots, j_t\}$, then each element of $M$ can be written (not necessarily uniquely) as an '$R$-linear combination' $\sum_{i=i}^{t} r_i j_i$ for suitable $r_1, \ldots, r_t \in R$. Also the zero submodule 0 of $M$ is finitely generated, by the empty set.

**7.13** PROPOSITION. *Let $M$ be a module over the commutative ring $R$. Then $M$ is Noetherian if and only if every submodule of $M$ is finitely generated.*

*Proof.* ($\Rightarrow$) Let $G$ be a submodule of $M$. Suppose that $G$ is not finitely generated and look for a contradiction. Let $\Gamma$ be the set of all submodules of $G$ which are finitely generated. Then $\Gamma \neq \emptyset$ since $0 \in \Gamma$. Since every submodule of $G$ is also a submodule of $M$, it follows from the maximal condition that $\Gamma$ has a maximal member with respect to inclusion, $N$ say; also, $N \subset G$ because we are assuming that $G$ is not finitely generated. Let $g \in G \setminus N$; then $N + Rg$ is a finitely generated submodule of $G$ and $N \subset N + Rg$ because $g \in (N + Rg) \setminus N$. Thus we have a contradiction to the maximality of $N$ in $\Gamma$.

Thus $G$ must be finitely generated.

($\Leftarrow$) Let

$$L_1 \subseteq L_2 \subseteq \ldots \subseteq L_n \subseteq L_{n+1} \subseteq \ldots$$

be an ascending chain of submodules of $M$. Then $G = \bigcup_{i \in \mathbb{N}} L_i$ is a submodule of $M$: it is clearly closed under scalar multiplication by arbitrary elements of $R$, and, if $g, h \in G$, then there exist $i, j \in \mathbb{N}$ such that $g \in L_i$ and $h \in L_j$, and since either $L_i \subseteq L_j$ or $L_j \subseteq L_i$ it follows that $g + h \in G$.

By hypothesis, $G$ is a finitely generated $R$-module: suppose that it is generated by $g_1, \ldots, g_t$, where $t \in \mathbb{N}$. (Of course, we can include 0 in a generating set for $G$.) For each $i = 1, \ldots, t$, there exists $n_i \in \mathbb{N}$ such that $g_i \in L_{n_i}$. Let $k = \max\{n_1, \ldots, n_t\}$. Then $g_i \in L_k$ for all $i = 1, \ldots, t$, so that

$$G = \sum_{i=1}^{t} Rg_i \subseteq L_k \subseteq L_{k+1} \subseteq \ldots \subseteq L_{k+i} \subseteq \ldots \subseteq G.$$

Hence $L_k = L_{k+i}$ for all $i \in \mathbb{N}$ and the ascending chain with which we started is eventually stationary. It follows that $M$ is Noetherian. $\square$

It follows from 7.13 that a commutative ring $R$ is Noetherian if and only if every ideal of $R$ is finitely generated. This is a very important fact for us, and will be much reinforced in Chapter 8.

**7.14** LEMMA. *Let $M$ be a module over the commutative ring $R$.*

(i) *If $M$ is Noetherian, then so too is every submodule and factor module of $M$.*

(ii) *If $M$ is Artinian, then so too is every submodule and factor module of $M$.*

*Proof.* (i) Assume that $M$ is Noetherian and let $G$ be a submodule of $M$. Since every submodule of $G$ is a submodule of $M$, it is clear from the definition of Noetherian $R$-module in 7.1 that $G$ is Noetherian. Also it follows from 6.24 that an ascending chain of submodules of $M/G$ must have the form

$$G_1/G \subseteq G_2/G \subseteq \ldots \subseteq G_n/G \subseteq G_{n+1}/G \subseteq \ldots,$$

where

$$G_1 \subseteq G_2 \subseteq \ldots \subseteq G_n \subseteq G_{n+1} \subseteq \ldots$$

is an ascending chain of submodules of $M$ all of which contain $G$. Since the latter chain must eventually become stationary, so must the former.

(ii) This can be proved in a very similar manner to the way in which (i) was proved above, and so it is left as an exercise for the reader. $\square$

**7.15** ♯EXERCISE. Prove part (ii) of 7.14.

**7.16** REMARK. It should be clear to the reader from, for example, 6.36 that if $M_1$ and $M_2$ are isomorphic $R$-modules (where $R$ is a commutative ring), then $M_1$ is Noetherian if and only if $M_2$ is, and $M_1$ is Artinian if and only if $M_2$ is.

The concepts of Noetherian module and Artinian module interact nicely with the idea of short exact sequences of modules introduced in 6.49. We explore this next.

**7.17** PROPOSITION. *Let $M$ be a module over the commutative ring $R$, and let $G$ be a submodule of $M$.*

(i) *The $R$-module $M$ is Noetherian if and only if both $G$ and $M/G$ are Noetherian.*

(ii) *The $R$-module $M$ is Artinian if and only if both $G$ and $M/G$ are Artinian.*

*Proof.* (i) ($\Rightarrow$) This was dealt with in 7.14.

($\Leftarrow$) Let

$$L_1 \subseteq L_2 \subseteq \ldots \subseteq L_n \subseteq L_{n+1} \subseteq \ldots$$

be an ascending chain of submodules of $M$. In order to use the hypotheses that $G$ and $M/G$ are Noetherian, we are going to consider an ascending chain of submodules of $G$ and an ascending chain of submodules of $M/G$, both derived from the above chain. First,

$$G \cap L_1 \subseteq G \cap L_2 \subseteq \ldots \subseteq G \cap L_n \subseteq G \cap L_{n+1} \subseteq \ldots$$

is a chain of submodules of $G$, and so there exists $k_1 \in \mathbf{N}$ such that $G \cap L_{k_1} = G \cap L_{k_1+i}$ for all $i \in \mathbf{N}$.

To obtain an ascending chain of submodules of $M/G$, we need, by 6.24, an ascending chain of submodules of $M$ all of whose terms contain $G$. We do not know whether $L_1$ (for example) contains $G$. However,

$$G + L_1 \subseteq G + L_2 \subseteq \ldots \subseteq G + L_n \subseteq G + L_{n+1} \subseteq \ldots$$

is a chain of submodules of $M$ all of whose terms contain $G$, and so

$$(G + L_1)/G \subseteq (G + L_2)/G \subseteq \ldots \subseteq (G + L_n)/G \subseteq (G + L_{n+1})/G \subseteq \ldots$$

is a chain of submodules of $M/G$. Thus there exists $k_2 \in \mathbf{N}$ such that $(G + L_{k_2})/G = (G + L_{k_2+i})/G$ for all $i \in \mathbf{N}$, so that $G + L_{k_2} = G + L_{k_2+i}$ for all $i \in \mathbf{N}$.

Let $k = \max \{k_1, k_2\}$. We show that, for each $i \in \mathbf{N}$, we have $L_k = L_{k+i}$. Of course, $L_k \subseteq L_{k+i}$. Let $g \in L_{k+i}$. Now we know that

$$G \cap L_k = G \cap L_{k+i} \quad \text{and} \quad G + L_k = G + L_{k+i}.$$

Since $g \in L_{k+i} \subseteq G + L_{k+i} = G + L_k$, there exist $a \in G$ and $b \in L_k$ such that $g = a + b$. Hence

$$a = g - b \in G \cap L_{k+i} = G \cap L_k,$$

so that both $a$ and $b$ belong to $L_k$ and $g = a + b \in L_k$. Therefore $L_{k+i} \subseteq L_k$ and the proof is complete.

(ii) This is left as an exercise. $\square$

**7.18** ‡EXERCISE. Prove part (ii) of 7.17.

**7.19** COROLLARY. *Let $R$ be a commutative ring, and let*

$$0 \longrightarrow L \overset{f}{\longrightarrow} M \overset{g}{\longrightarrow} N \longrightarrow 0$$

*be a short exact sequence of R-modules and R-homomorphisms. (See* 6.49.*)*

  (i) *The R-module M is Noetherian if and only if L and N are Noetherian.*

  (ii) *The R-module M is Artinian if and only if L and N are Artinian.*

*Proof.* This is essentially an easy consequence of 7.16 and 7.17. This time, we give the details for part (ii) and leave those for part (i) as an exercise.

(ii) Note that $L \cong \operatorname{Im} f = \operatorname{Ker} g$, and that, by the the First Isomorphism Theorem for modules 6.33, we also have $M/\operatorname{Ker} g \cong N$. Thus, by 7.17, $M$ is Artinian if and only if $\operatorname{Ker} g$ and $M/\operatorname{Ker} g$ are Artinian; and, by 7.16, this is the case if and only if both $L$ and $N$ are Artinian. $\square$

**7.20** ♯EXERCISE. Prove part (i) of 7.19.

**7.21** COROLLARY. *Let $M_1, \ldots, M_n$ (where $n \in \mathbb{N}$) be modules over the commutative ring R.*

  (i) *The direct sum $\bigoplus_{i=1}^{n} M_i$ is Noetherian if and only if $M_1, \ldots, M_n$ are all Noetherian.*

  (ii) *The direct sum $\bigoplus_{i=1}^{n} M_i$ is Artinian if and only if $M_1, \ldots, M_n$ are all Artinian.*

*Proof.* We prove these results by induction on $n$. In the case in which $n = 1$ there is nothing to prove, since $\bigoplus_{i=1}^{1} M_i$ is clearly isomorphic to $M_1$. So we suppose, inductively, that $n > 1$ and the results have both been proved for smaller values of $n$. By 6.48, there is an exact sequence

$$0 \longrightarrow M_1 \longrightarrow \bigoplus_{i=1}^{n} M_i \longrightarrow \bigoplus_{i=2}^{n} M_i \longrightarrow 0,$$

and so it follows from 7.19 that $\bigoplus_{i=1}^{n} M_i$ is Noetherian (respectively Artinian) if and only if both $M_1$ and $\bigoplus_{i=2}^{n} M_i$ are. But, by the inductive hypothesis, $\bigoplus_{i=2}^{n} M_i$ is Noetherian (respectively Artinian) if and only if $M_2, \ldots, M_n$ all are. With this observation, we can complete the inductive step and the proof. $\square$

This result has a very important corollary. One obvious consequence is that, if $R$ is a commutative Noetherian (respectively Artinian) ring, then every free $R$-module $F$ with a finite base is Noetherian (respectively Artinian), simply because, by 6.53, $F$ is isomorphic to a direct sum of finitely many copies of $R$. However, we can do even better.

**7.22** COROLLARY. *Let $R$ be a commutative ring.*

(i) *If R is a Noetherian ring, then every finitely generated R-module is Noetherian.*

(ii) *If R is an Artinian ring, then every finitely generated R-module is Artinian.*

*Proof.* Let $M$ be a finitely generated $R$-module. By 6.57, there exists a free $R$-module $F$ with a finite base and an $R$-epimorphism $f : F \to M$. If $R$ is a Noetherian (respectively Artinian) ring, then, as was explained immediately before the statement of this corollary, $F$ is a Noetherian (respectively Artinian) $R$-module, and so it follows from 7.19 that $M$ is a Noetherian (respectively Artinian) $R$-module too. $\square$

**7.23** EXERCISE. Let $M$ be a module over the commutative ring $R$, and suppose that $G_1$ and $G_2$ are submodules of $M$ such that $M/G_1$ and $M/G_2$ are both Noetherian. Show that $M/(G_1 \cap G_2)$ is Noetherian.

**7.24** LEMMA. *Let M be a module over the commutative ring R, and let $m \in M$. Then there is an isomorphism of R-modules*

$$f : R/(0 : m) \xrightarrow{\cong} Rm$$

*such that $f(r + (0 : m)) = rm$ for all $r \in R$.*

*Proof.* The mapping $g : R \to Rm$ defined by $g(r) = rm$ for all $r \in R$ is clearly an $R$-epimorphism from $R$ onto the submodule $Rm$ of $M$, and since

$$\operatorname{Ker} g = \{r \in R : rm = 0\} = (0 : m),$$

the claim follows immediately from the First Isomorphism Theorem for modules 6.33. $\square$

**7.25** ♯EXERCISE. Let $M$ be a module over the commutative ring $R$. Show that $M$ is cyclic (see 6.12) if and only if $M$ is isomorphic to an $R$-module of the form $R/I$ for some ideal $I$ of $R$.

**7.26** REMARK. Let $M$ be a module over the commutative ring $R$, and let $I$ be an ideal of $R$ such that $I \subseteq \operatorname{Ann}(M)$, so that, as was explained in 6.19, $M$ can be regarded as an $R/I$-module in a natural way. Recall also from 6.19 that a subset of $M$ is an $R$-submodule if and only if it is an $R/I$-submodule. It therefore follows that $M$ is Noetherian (respectively Artinian) as $R$-module if and only if it is Noetherian (respectively Artinian) as $R/I$-module.

In particular, if $J$ is an ideal of $R$, then it follows from the above that $R/J$ is Noetherian (respectively Artinian) as $R$-module if and only if it is a Noetherian (respectively Artinian) ring.

**7.27** EXERCISE. Let $M$ be a Noetherian module over the commutative ring $R$. Show that $R/\mathrm{Ann}(M)$ is a Noetherian ring.

The above exercise means that, for many purposes, the study of a Noetherian module $M$ over a commutative ring $R$ can be reduced to the situation in which the underlying ring is actually a commutative Noetherian ring, because, by 6.19, $M$ can be regarded as a module over $R/\mathrm{Ann}(M)$ in a natural way, and when this is done, a subset of $M$ is an $R$-submodule if and only if it is an $R/\mathrm{Ann}(M)$-submodule. Thus it is not unreasonable to take the view that if one is going to study Noetherian modules over commutative rings, then one might as well just study finitely generated modules over commutative Noetherian rings: observe that, by 7.22, a finitely generated module over a commutative Noetherian ring $R$ is a Noetherian $R$-module.

**7.28** EXERCISE. Let $M$ be a finitely generated Artinian module over the commutative ring $R$. Show that $R/\mathrm{Ann}(M)$ is an Artinian ring.

**7.29** EXERCISE. Consider $\mathbf{Q}$ as a $\mathbf{Z}$-module. Is $\mathbf{Q}$ an Artinian $\mathbf{Z}$-module? Is $\mathbf{Q}$ a Noetherian $\mathbf{Z}$-module? Justify your responses.

We saw in 7.12 that a vector space over a field $K$, that is a $K$-module, is a Noetherian $K$-module if and only if it is Artinian. We can now use our 'change of rings' ideas of 6.19 and 7.26 to make a considerable improvement on that vector space result.

**7.30** THEOREM. *Let $G$ be a module over the commutative ring $R$, and assume that $G$ is annihilated by the product of finitely many (not necessarily distinct) maximal ideals of $R$, that is, there exist $n \in \mathbf{N}$ and maximal ideals $M_1, \ldots, M_n$ of $R$ such that*

$$M_1 \ldots M_n G = 0.$$

*Then $G$ is a Noetherian $R$-module if and only if $G$ is an Artinian $R$-module.*

*Proof.* We argue by induction on $n$.

In the case in which $n = 1$, so that $G$ is annihilated by the maximal ideal $M_1$ of $R$, we note that, by 6.19 and 7.26, we can regard $G$ as a module over $R/M_1$, that is, as an $R/M_1$-space, in a natural way, and, when this is done, $G$ is a Noetherian (respectively Artinian) $R/M_1$-space if and only if it is a Noetherian (respectively Artinian) $R$-module. But, by 7.12, $G$ is a Noetherian $R/M_1$-space if and only if it is an Artinian $R/M_1$-space. It therefore follows that $G$ is a Noetherian $R$-module if and only if it is an Artinian $R$-module.

Now suppose, inductively, that $n > 1$ and that the result has been proved for smaller values of $n$. The natural exact sequence

$$0 \longrightarrow M_n G \longrightarrow G \longrightarrow G/M_n G \longrightarrow 0$$

of 6.40(iii), used in conjunction with 7.19, shows that $G$ is Noetherian (respectively Artinian) as $R$-module if and only if both $M_n G$ and $G/M_n G$ are Noetherian (respectively Artinian) as $R$-modules. Now the $R$-module $G/M_n G$ is annihilated by the maximal ideal $M_n$ of $R$, and so it follows from what we have proved in the preceding paragraph that $G/M_n G$ is a Noetherian $R$-module if and only if it is an Artinian $R$-module. Also, the $R$-module $M_n G$ is annihilated by the product $M_1 \ldots M_{n-1}$ of $n-1$ maximal ideals of $R$, and so it follows from the inductive hypothesis that $M_n G$ is a Noetherian $R$-module if and only if it is an Artinian $R$-module. Hence $G$ is a Noetherian $R$-module if and only if it is an Artinian $R$-module, and the inductive step, and therefore the proof, are complete. $\square$

Theorem 7.30 will be very useful in Chapter 8 in our discussion of commutative Artinian rings.

The results of 7.12 and 7.30 show that there are many examples of modules over commutative rings which satisfy both the ascending and descending chain conditions. Indeed, if $G$ is a finitely generated module over a commutative Noetherian ring $R$ and $M_1, \ldots, M_n$ are maximal ideals of $R$, then $G$ and $G/M_1 \ldots M_n G$ are Noetherian $R$-modules by 7.22 and 7.14, and since $G/M_1 \ldots M_n G$ is annihilated by $M_1 \ldots M_n$, it satisfies both chain conditions by 7.30.

This discussion of modules which satisfy both chain conditions leads naturally to the topics of composition series and modules of finite length.

**7.31** DEFINITION. Let $G$ be a module over the commutative ring $R$. We say that $G$ is a *simple* $R$-module precisely when $G \neq 0$ and the only submodules of $G$ are 0 and $G$ itself.

**7.32** LEMMA. *Let $G$ be a module over the commutative ring $R$. Then $G$ is simple if and only if $G$ is isomorphic to an $R$-module of the form $R/M$ for some maximal ideal $M$ of $R$.*

*Proof.* ($\Leftarrow$) Let $M$ be a maximal ideal of $R$. By 3.1, the field $R/M$ has exactly two ideals, namely itself and its zero ideal. Hence, by 6.19, the $R$-module $R/M$ has exactly two submodules, namely itself and its zero submodule.

($\Rightarrow$) Suppose that $G$ is a simple $R$-module. Since $G \neq 0$, there exists $g \in G$ with $g \neq 0$. Hence $0 \neq Rg \subseteq G$, and since $Rg$ is a submodule of $G$ by

6.12, we must have $Rg = G$. Thus $G$ is a cyclic $R$-module, and so, by 7.25, $G \cong R/I$ for some ideal $I$ of $R$. Since $G$ has exactly two submodules, it follows from 6.36 and 6.19 that the ring $R/I$ has exactly two ideals; hence, by 3.1 and 3.3, $I$ is a maximal ideal of $R$. $\square$

**7.33** DEFINITION and REMARKS. Let $G$ be a module over the commutative ring $R$. A *strict-chain* of submodules of $G$ is a finite, strictly increasing chain

$$G_0 \subset G_1 \subset \ldots \subset G_{n-1} \subset G_n$$

of submodules of $G$ such that $G_0 = 0$ and $G_n = G$. The *length* of the strict-chain is the number of links, that is, one less than the number of terms (so that the displayed strict-chain above has length $n$). We consider

$$0$$

to be a strict-chain of length 0 of submodules of the zero $R$-module 0.

A strict-chain of submodules of $G$ given by

$$0 = G_0 \subset G_1 \subset \ldots \subset G_{n-1} \subset G_n = G$$

is said to be a *composition series for* $G$ precisely when $G_i/G_{i-1}$ is a simple $R$-module for each $i = 1, \ldots, n$. Note that, by 6.24, this is the case if and only if it is impossible to extend the strict-chain by the insertion of an extra term to make a strict-chain of length $n + 1$. Thus a strict-chain of submodules of $G$ is a composition series for $G$ if and only if it is a 'maximal' strict-chain (in an obvious sense).

Our next few results establish some absolutely fundamental facts about composition series.

**7.34** THEOREM. *Let $G$ be a module over the commutative ring $R$, and assume that $G$ has a composition series of length $n$. Then*
   (i) *no strict-chain of submodules of $G$ can have length greater than $n$;*
   (ii) *every composition series for $G$ has length exactly $n$; and*
   (iii) *each strict-chain of submodules of $G$ of length $n' \leq n$ can be extended to a composition series for $G$ by the insertion of $n - n'$ additional terms; in particular,*
   (iv) *each strict-chain of submodules of $G$ of length $n$ is a composition series for $G$.*

*Proof.* Clearly, we can assume that $n > 0$. For each $R$-module $M$, let us denote by $\ell(M)$ the smallest length of a composition series for $M$ if $M$ has a composition series, and let us set $\ell(M) = \infty$ if $M$ does not have a composition series. As a first step in the proof, we show that, if $H$ is a proper submodule of $G$, then $\ell(H) < \ell(G)$.

Let $\ell(G) = t$ and let

$$0 = G_0 \subset G_1 \subset \ldots \subset G_{t-1} \subset G_t = G$$

provide a composition series for $G$ of length $t$. For each $i = 0, \ldots, t$, let $H_i = H \cap G_i$. Now, by the First Isomorphism Theorem for modules 6.33, for each $i = 1, \ldots, t$, the composite $R$-homomorphism

$$H_i = H \cap G_i \longrightarrow G_i \longrightarrow G_i/G_{i-1}$$

(in which the first map is the inclusion homomorphism and the second is the canonical epimorphism) has kernel equal to $H \cap G_i \cap G_{i-1} = H \cap G_{i-1} = H_{i-1}$ and so induces an $R$-monomorphism

$$\begin{aligned} \psi_i \ : \ H_i/H_{i-1} &\longrightarrow G_i/G_{i-1} \\ h + H_{i-1} &\longmapsto h + G_{i-1}. \end{aligned}$$

Thus $H_i/H_{i-1}$ is isomorphic to a submodule of $G_i/G_{i-1}$, which is simple, and so $H_i/H_{i-1}$ is either 0 or simple; indeed, $H_i/H_{i-1}$ is simple if and only if $\psi_i$ is an isomorphism. Thus, if we remove any repetitions of terms in

$$0 = H_0 \subseteq H_1 \subseteq \ldots \subseteq H_{t-1} \subseteq H_t = H \cap G_t = H,$$

we shall obtain a composition series for $H$. Thus $\ell(H) \leq \ell(G)$. Furthermore, we must have $\ell(H) < \ell(G)$, for otherwise the above process must lead to

$$H_0 \subset H_1 \subset \ldots \subset H_{t-1} \subset H_t$$

as a composition series for $H$, so that $H_i/H_i \cap G_{i-1} = H_i/H_{i-1} \neq 0$ for all $i = 1, \ldots, t$; since $H_0 = 0 = G_0$, it would then follow successively that

$$H_1 = G_1, \ H_2 = G_2, \ \ldots, \ H_t = G_t,$$

contradicting the fact that $H \subset G$. Thus we have shown that $\ell(H) < \ell(G)$, as claimed. Note also that we have shown that every submodule of $G$ has a composition series.

(i) Now let

$$G_0' \subset G_1' \subset \ldots \subset G_{r-1}' \subset G_r'$$

be a strict-chain of submodules of $G$, so that $G_0' = 0$ and $G_r' = G$. Now $\ell(0) = 0$, and so it follows from the preceding paragraph that

$$0 = \ell(G_0') < \ell(G_1') < \ldots < \ell(G_{r-1}') < \ell(G_r') = \ell(G).$$

Hence $r \leq \ell(G) \leq n$. Therefore, since $G$ has a composition series of length $n$ and a composition series for $G$ is, in particular, a strict-chain of submodules of $G$, it follows that $n \leq \ell(G)$, so that $n = \ell(G)$.

(ii) Now suppose that $G$ has a composition series of length $n_1$. Then $n_1 \leq \ell(G) = n$ by part (i) because a composition series is a strict-chain, while $\ell(G) \leq n_1$ by definition of $\ell(G)$.

(iii), (iv) These are now immediate from parts (i) and (ii) and the remarks in 7.33: a strict-chain of submodules of $G$ of length $n' < n = \ell(G)$ cannot be a composition series for $G$ because, by part (ii), all composition series for $G$ have length $n$, and so it can be extended to a strict-chain of length $n' + 1$ by the insertion of an extra term; on the other hand, a strict-chain of submodules of $G$ of length $n$ must already be a composition series for $G$ because otherwise it could be extended to a strict-chain of submodules of $G$ of length $n + 1$, contrary to part (i). $\square$

**7.35** DEFINITION. Let $G$ be a module over the commutative ring $R$. We say that $G$ *has finite length* precisely when $G$ has a composition series. When this is the case, the *length* of $G$, denoted by $\ell(G)$, (or $\ell_R(G)$ when it is desirable to emphasize the underlying ring) is defined to be the length of any composition series for $G$: we have just seen in 7.34 that all composition series for $G$ have the same length.

When $G$ does not have finite length, that is when $G$ does not have a composition series, we shall sometimes indicate this by writing $\ell(G) = \infty$.

It was hinted earlier that there is a connection between composition series and the ascending and descending chain conditions for submodules of a module. The relevant result comes next.

**7.36** PROPOSITION. *Let $G$ be a module over the commutative ring $R$. Then $G$ has finite length if and only if $G$ is both Noetherian and Artinian, that is, if and only if $G$ satisfies both the ascending and descending chain conditions for submodules.*

*Proof.* ($\Rightarrow$) Assume that $G$ has finite length $\ell(G)$. Then it follows from 7.34 that any ascending chain of submodules of $G$ cannot have more than $\ell(G)$ of its inclusions strict, and so must be eventually stationary. Similarly, any descending chain of submodules of $G$ must be eventually stationary.

($\Leftarrow$) Assume that $G$ is both Noetherian and Artinian, so that, by 7.1 and 7.3, it satisfies both the maximal and minimal conditions for submodules. We suppose that $G$ does not have a composition series and look for a contradiction. Then

$$\Theta := \{M : M \text{ is a submodule of } G \text{ and } \ell(M) = \infty\}$$

is not empty, since $G \in \Theta$. Hence, by the minimal condition, $\Theta$ has a minimal member, $H$ say. Now $H \neq 0$, since the zero submodule of $G$ certainly has a composition series: see 7.33. Thus, by the maximal condition, the set of proper submodules of $H$ has at least one maximal member, $H'$ say. By choice of $H$ and the fact that $H' \subset H$, we see that $H'$ has a composition series. Let

$$H'_0 \subset H'_1 \subset \ldots \subset H'_{t-1} \subset H'_t$$

be a composition series for $H'$, so that $H'_0 = 0$, $H'_t = H'$, and $t = \ell(H')$. Since $H'$ is a maximal proper submodule of $H$, it follows from 6.24 that $H/H'$ has exactly two submodules and so is simple. Hence

$$H'_0 \subset H'_1 \subset \ldots \subset H'_{t-1} \subset H'_t \subset H$$

is a composition series for $H$. This is a contradiction!

Hence $G$ must have a composition series. $\square$

We have seen in 7.34 that any two composition series for a module of finite length over a commutative ring have the same length. In fact, two such composition series have even stronger similarities concerned with their so-called 'composition factors' (which we define next). These stronger similarities are specified in the famous Jordan–Hölder Theorem, which is given below as 7.39.

**7.37** DEFINITIONS. Let $G$ be a module over the commutative ring $R$, and suppose that $G$ has finite length. Let

$$G_0 \subset G_1 \subset \ldots \subset G_{n-1} \subset G_n$$

be a composition series for $G$ (so that $G_0 = 0$, $G_n = G$ and $n = \ell(G)$). Then we call the family of simple $R$-modules $(G_i/G_{i-1})_{i=1}^{n}$ the *family of composition factors* of the above composition series. (Of course, this family is empty when $G = 0$.)

Now suppose that $G \neq 0$ and that

$$G'_0 \subset G'_1 \subset \ldots \subset G'_{n-1} \subset G'_n$$

is a second composition series for $G$. (We have here made use of the fact, proved in 7.34, that any two composition series for $G$ have the same length.) We say that the above two composition series for $G$ are *isomorphic* precisely when there exists a permutation $\phi$ of the set $\{1, \ldots, n\}$ of the first $n$ positive integers such that, for all $i = 1, \ldots, n$,

$$G_i/G_{i-1} \cong G'_{\phi(i)}/G'_{\phi(i)-1}.$$

**7.38** LEMMA. *Let $G$ be a module over the commutative ring $R$, and let $H, H'$ be submodules of $G$ such that $H \neq H'$ and both $G/H$ and $G/H'$ are simple. Then*

$$G/H \cong H'/(H \cap H') \quad \text{and} \quad G/H' \cong H/(H \cap H').$$

*Proof.* We show first that $H \subset H + H'$. If this were not the case, then we should have $H = H + H'$, so that, as $H \neq H'$, it would follow that $H' \subset H \subset G$, contradicting the fact that $G/H'$ is simple (in view of 6.24). Hence

$$H \subset H + H' \subseteq G,$$

so that, as $G/H$ is simple, it again follows from 6.24 that $H + H' = G$. Therefore, by the Third Isomorphism Theorem for modules 6.38,

$$G/H = (H + H')/H \cong H'/(H \cap H').$$

The other isomorphism also follows on reversing the rôles of $H$ and $H'$. □

**7.39** THE JORDAN–HÖLDER THEOREM. *Let $G$ be a non-zero module of finite length over the commutative ring $R$. Then every pair of composition series for $G$ are isomorphic (in the sense of 7.37).*

*Proof.* Since $G \neq 0$, we have $n := \ell(G) \geq 1$: we use induction on $n$. The claim is clear when $n = 1$, and so we assume that $n > 1$ and that the result has been proved for smaller values of $n$. Let

$$G_0 \subset G_1 \subset \ldots \subset G_{n-1} \subset G_n$$

and

$$G'_0 \subset G'_1 \subset \ldots \subset G'_{n-1} \subset G'_n$$

be two composition series for $G$ (so that $G_0 = G'_0 = 0$ and $G_n = G'_n = G$). The argument for the inductive step splits into two cases.

The first case is where $G_{n-1} = G'_{n-1}$. Then we have

$$G_n/G_{n-1} = G'_n/G'_{n-1}$$

and both

$$G_0 \subset G_1 \subset \ldots \subset G_{n-1}$$

and

$$G'_0 \subset G'_1 \subset \ldots \subset G'_{n-1}$$

are composition series for $G_{n-1} = G'_{n-1}$. Since $\ell(G_{n-1}) = n - 1$, we can apply the inductive hypothesis to these two composition series for $G_{n-1}$ and the desired result in this case follows easily.

The second case is where $G_{n-1} \neq G'_{n-1}$. Then we set $H = G_{n-1} \cap G'_{n-1}$. By Lemma 7.38,

$$G_n/G_{n-1} \cong G'_{n-1}/H \quad \text{and} \quad G'_n/G'_{n-1} \cong G_{n-1}/H,$$

so that all four of these modules are simple. Thus, if $H = 0$ (so that both $G_{n-1}$ and $G'_{n-1}$ are simple and $n = 2$), the desired conclusion has been obtained. Thus we assume that $H \neq 0$.

In this case,

$$0 \subset H \subset G_{n-1} \subset G_n$$

is a strict-chain of submodules of $G = G_n$, and both $G_n/G_{n-1}$ and $G_{n-1}/H$ are simple. Now, by Theorem 7.34(iii), the above strict-chain can be extended by the insertion of extra terms to a composition series for $G$; since such a composition series for $G$ must have length $n$, it follows that $\ell(H) = n - 2$. In particular, we obtain a composition series

$$H_0 \subset H_1 \subset \ldots \subset H_{n-3} \subset H_{n-2}$$

for $H$ (so that $H_0 = 0$ and $H_{n-2} = H$). The comments in the preceding paragraph now show that the two composition series

$$H_0 \subset H_1 \subset \ldots \subset H_{n-3} \subset H_{n-2} \subset G_{n-1} \subset G_n$$

and

$$H_0 \subset H_1 \subset \ldots \subset H_{n-3} \subset H_{n-2} \subset G'_{n-1} \subset G'_n$$

for $G$ are isomorphic. But we can now use the inductive hypothesis (on two composition series for $G_{n-1}$) to see that the two composition series

$$G_0 \subset G_1 \subset \ldots \subset G_{n-1} \subset G_n$$

and

$$H_0 \subset H_1 \subset \ldots \subset H_{n-3} \subset H_{n-2} \subset G_{n-1} \subset G_n$$

for $G$ are isomorphic. Similarly, the composition series

$$H_0 \subset H_1 \subset \ldots \subset H_{n-3} \subset H_{n-2} \subset G'_{n-1} \subset G'_n$$

and

$$G'_0 \subset G'_1 \subset \ldots \subset G'_{n-1} \subset G'_n$$

are isomorphic, and so we can complete the inductive step.

The theorem is therefore proved by induction. $\square$

**7.40** REMARK. It should be clear to the reader from 7.16, 7.36 and 6.36 that if $G_1$ and $G_2$ are isomorphic $R$-modules (where $R$ is a commutative ring), then $G_1$ has finite length if and only if $G_2$ has finite length, and, when this is the case, $\ell(G_1) = \ell(G_2)$.

**7.41** THEOREM. *Let $R$ be a commutative ring, and let*

$$0 \longrightarrow L \stackrel{f}{\longrightarrow} M \stackrel{g}{\longrightarrow} N \longrightarrow 0$$

*be a short exact sequence of $R$-modules and $R$-homomorphisms.*

(i) *The $R$-module $M$ has finite length if and only if $L$ and $N$ both have finite length.*

(ii) *When $L, M, N$ all have finite length, then*

$$\ell(M) = \ell(L) + \ell(N).$$

*Proof.* (i) This follows easily from 7.19 and 7.36: by 7.36, the $R$-module $M$ has finite length if and only if it is both Noetherian and Artinian; by 7.19, this is the case if and only if $L$ and $N$ are both Noetherian and Artinian; and, by 7.36 again, this is the case if and only if both $L$ and $N$ have finite length.

(ii) Note that $L \cong \operatorname{Im} f = \operatorname{Ker} g$, and that, by the the First Isomorphism Theorem for modules 6.33, we also have $M/\operatorname{Ker} g \cong N$. Thus, by 7.40, $\operatorname{Ker} g$ and $M/\operatorname{Ker} g$ have finite length, and $\ell(L) = \ell(\operatorname{Ker} g)$ and $\ell(N) = \ell(M/\operatorname{Ker} g)$. It is thus sufficient for us to show that if $G$ is a submodule of the $R$-module $M$ (and $M$ has finite length), then $\ell(M) = \ell(G) + \ell(M/G)$. This we do.

The desired result is clear if either $G = 0$ or $G = M$, and so we suppose that

$$0 \subset G \subset M.$$

By Theorem 7.34, the above strict-chain of submodules of $M$ can be extended, by the insertion of extra terms, into a composition series for $M$, say

$$M_0 \subset M_1 \subset \ldots \subset M_{n-1} \subset M_n,$$

where $M_0 = 0$, $M_n = M$ and $n = \ell(M)$. Suppose that $M_t = G$. Then

$$M_0 \subset M_1 \subset \ldots \subset M_t$$

is a composition series for $G$, and it follows from 6.24 that

$$M_t/G \subset M_{t+1}/G \subset \ldots \subset M_n/G$$

is a composition series for $M/G$. Hence $\ell(G) + \ell(M/G) = t + (n-t) = n = \ell(M)$, as required. $\square$

Let $K$ be a field. In 7.12, we saw that the concepts of Noetherian $K$-module and Artinian $K$-module coincide, and, indeed, that if $V$ is a vector space over $K$, then $V$ is a finite-dimensional $K$-space if and only if it is both a Noetherian $K$-module and an Artinian $K$-module. It thus follows from 7.36 that $V$ is a finite-dimensional $K$-space if and only if it is a $K$-module of finite length. We check now that, when this is the case, $\text{vdim}_K V = \ell(V)$.

**7.42** PROPOSITION. *Let $V$ be a vector space over the field $K$. Then $V$ is a finite-dimensional $K$-space if and only if it is a $K$-module of finite length, and, when this is the case, $\text{vdim}_K V = \ell(V)$.*

*Proof.* The first claim was explained immediately before the statement of the proposition; for the second, we argue by induction on $n := \text{vdim}_K V$. When $n = 0$, we have $V = 0$ and the result is clear; when $n = 1$, the only subspaces of $V$ are 0 and $V$ itself (and these are different), and so $0 \subset V$ is a composition series for the $K$-module $V$, so that $\ell(V) = 1$. We therefore suppose that $n > 1$ and that the result has been proved for smaller values of $n$.

Let $v \in V$ with $v \neq 0$; set $U = Kv$, a 1-dimensional subspace of $V$. Thus there is an exact sequence

$$0 \longrightarrow U \overset{i}{\longrightarrow} V \overset{f}{\longrightarrow} V/U \longrightarrow 0$$

of $K$-spaces and $K$-linear maps, in which $i$ is the inclusion map and $f$ is the canonical epimorphism. Now ($U$ and $V/U$ are finite-dimensional and)

$$\text{vdim}_K V = \text{vdim}_K(\text{Ker } f) + \text{vdim}_K(\text{Im } f) = \text{vdim}_K U + \text{vdim}_K(V/U),$$

and so $\text{vdim}_K(V/U) = n-1$. Thus $\text{vdim}_K(V/U) = \ell(V/U)$ by the inductive hypothesis, while $\text{vdim}_K U = \ell(U)$ by the first paragraph of this proof. Hence

$$\text{vdim}_K V = \text{vdim}_K U + \text{vdim}_K(V/U) = \ell(U) + \ell(V/U) = \ell(V)$$

by 7.41, and so the inductive step is complete.

The result is therefore proved by induction. $\square$

**7.43** EXERCISE. Let

$$0 \longrightarrow G_n \overset{d_n}{\longrightarrow} G_{n-1} \longrightarrow \cdots \longrightarrow G_i \overset{d_i}{\longrightarrow} G_{i-1} \longrightarrow \cdots \longrightarrow G_1 \overset{d_1}{\longrightarrow} G_0 \longrightarrow 0$$

be an exact sequence of modules and homomorphisms over the commutative ring $R$ (where $n \in \mathbb{N}$ and $n > 1$), and suppose that $G_i$ has finite length for all $i = 1, \ldots, n - 1$. Show that $G_0$ and $G_n$ have finite length, and that

$$\sum_{i=0}^{n} (-1)^i \ell(G_i) = 0.$$

**7.44 EXERCISE.** Let

$$0 \xrightarrow{d_{n+1}} G_n \xrightarrow{d_n} G_{n-1} \longrightarrow \cdots \longrightarrow G_i \xrightarrow{d_i} G_{i-1} \longrightarrow \cdots \longrightarrow G_1 \xrightarrow{d_1} G_0 \xrightarrow{d_0} 0$$

be a sequence of modules and homomorphisms over the commutative ring $R$ (where $n \in \mathbb{N}$ and $n > 1$) such that $d_i \circ d_{i+1} = 0$ for all $i = 1, \ldots, n-1$, and suppose that $G_i$ has finite length for all $i = 0, \ldots, n$.

For each $i = 0, \ldots, n$, set $H_i = \operatorname{Ker} d_i / \operatorname{Im} d_{i+1}$. Show that $H_i$ has finite length for all $i = 0, \ldots, n$, and that

$$\sum_{i=0}^{n} (-1)^i \ell(H_i) = \sum_{i=0}^{n} (-1)^i \ell(G_i).$$

**7.45 EXERCISE.** Let $G$ be a module over the non-trivial commutative Noetherian ring $R$. Show that $G$ has finite length if and only if $G$ is finitely generated and there exist $n \in \mathbb{N}$ and maximal ideals $M_1, \ldots, M_n$ of $R$ (not necessarily distinct) such that

$$M_1 \ldots M_n G = 0.$$

**7.46 ♯EXERCISE.** Let $R$ be a principal ideal domain which is not a field. Let $G$ be an $R$-module. Show that $G$ has finite length if and only if $G$ is finitely generated and there exists $r \in R$ with $r \neq 0$ such that $rG = 0$.

**7.47 EXERCISE.** Find $\ell_{\mathbb{Z}}((\mathbb{Z}/\mathbb{Z}20) \oplus (\mathbb{Z}/\mathbb{Z}27))$. Determine a composition series for the $\mathbb{Z}$-module $(\mathbb{Z}/\mathbb{Z}20) \oplus (\mathbb{Z}/\mathbb{Z}27)$.

# Chapter 8

# Commutative Noetherian rings

The first part of this chapter is concerned with some very basic and important results in the theory of commutative Noetherian rings. Of course, a commutative ring $R$ is Noetherian if and only if it is Noetherian when viewed as a module over itself in the natural way. Thus, many of the results on Noetherian modules over commutative rings obtained in Chapter 7 will be relevant in this chapter. Also, we have already encountered earlier in the book some fundamental facts about commutative Noetherian rings: one that particularly comes to mind is 4.35, in which we showed that every proper ideal in a commutative Noetherian ring has a primary decomposition. Thus part of this chapter involves reminders of earlier work; however, some important results which have not yet appeared in the book, such as Hilbert's Basis Theorem and Krull's Intersection Theorem, are presented in this chapter.

Towards the end of this chapter, we shall establish some basic facts about commutative Artinian rings, including the fact that a commutative ring $R$ is Artinian if and only if it is Noetherian and every prime ideal of $R$ is maximal.

**8.1 REMINDER.** (See 3.37, 7.5 and 7.13.) Let $R$ be a commutative ring. Then $R$ is said to be *Noetherian* precisely when it satisfies the following equivalent conditions:

(i) $R$ satisfies the ascending chain condition for ideals; that is, whenever

$$I_1 \subseteq I_2 \subseteq \ldots \subseteq I_i \subseteq I_{i+1} \subseteq \ldots$$

is an ascending chain of ideals of $R$, then there exists $k \in \mathbf{N}$ such that $I_k = I_{k+i}$ for all $i \in \mathbf{N}$;

(ii) every non-empty set of ideals of $R$ has a maximal member with respect to inclusion; and

(iii) every ideal of $R$ is finitely generated.

We point out next that various ring-theoretic operations on commutative Noetherian rings produce again commutative Noetherian rings.

**8.2** LEMMA. *Let $R$ and $R'$ be commutative rings, and let $f : R \to R'$ be a surjective ring homomorphism. If $R$ is Noetherian, then so too is $R'$.*

*In particular, if $I$ is an ideal of $R$ and $R$ is Noetherian, then $R/I$ is also a commutative Noetherian ring.*

*Proof.* By the Isomorphism Theorem for commutative rings 2.13, we have $R/\operatorname{Ker} f \cong R'$. Since it is clear (from, for example, 2.46) that if two commutative rings are isomorphic, then one is Noetherian if and only if the other is Noetherian, it is enough for us to prove the second claim that $R/I$ is Noetherian whenever $I$ is an ideal of $R$ and $R$ is Noetherian. This we do.

By 2.37 and 2.39, an ascending chain of ideals of $R/I$ will have the form

$$I_1/I \subseteq I_2/I \subseteq \ldots \subseteq I_i/I \subseteq I_{i+1}/I \subseteq \ldots,$$

where

$$I_1 \subseteq I_2 \subseteq \ldots \subseteq I_i \subseteq I_{i+1} \subseteq \ldots$$

is an ascending chain of ideals of $R$ all of which contain $I$. Since $R$ is Noetherian, there exists $k \in \mathbf{N}$ such that $I_k = I_{k+i}$ for all $i \in \mathbf{N}$, and so $I_k/I = I_{k+i}/I$ for all $i \in \mathbf{N}$. Therefore $R/I$ is Noetherian. $\square$

The result of the next Lemma 8.3 was actually covered in Exercise 5.26, but in view of the importance of the result we give a solution for that exercise now.

**8.3** LEMMA. *Let $R$ be a commutative Noetherian ring and let $S$ be a multiplicatively closed subset of $R$. Then the ring of fractions $S^{-1}R$ of $R$ with respect to $S$ is again Noetherian.*

*Proof.* Let

$$\mathcal{I}_1 \subseteq \mathcal{I}_2 \subseteq \ldots \subseteq \mathcal{I}_i \subseteq \mathcal{I}_{i+1} \subseteq \ldots$$

be an ascending chain of ideals of $S^{-1}R$. Use the extension and contraction notation of 2.41 and 2.45 with reference to the natural ring homomorphism $f : R \to S^{-1}R$. Then

$$\mathcal{I}_1^c \subseteq \mathcal{I}_2^c \subseteq \ldots \subseteq \mathcal{I}_i^c \subseteq \mathcal{I}_{i+1}^c \subseteq \ldots$$

is an ascending chain of ideals of $R$, and so there exists $k \in \mathbb{N}$ such that $\mathcal{I}_k^c = \mathcal{I}_{k+i}^c$ for all $i \in \mathbb{N}$. Hence $\mathcal{I}_k^{ce} = \mathcal{I}_{k+i}^{ce}$ for all $i \in \mathbb{N}$. But, by 5.24, we have $\mathcal{I}_i^{ce} = \mathcal{I}_i$ for all $i \in \mathbb{N}$, and so $\mathcal{I}_k = \mathcal{I}_{k+i}$ for all $i \in \mathbb{N}$. It follows that $S^{-1}R$ is Noetherian. $\square$

**8.4** Lemma. *Let $R$ and $R'$ be commutative rings, and let $f : R \to R'$ be a ring homomorphism. Assume that $R$ is Noetherian and that $R'$, when viewed as an $R$-module by means of $f$ (see 6.6), is finitely generated. Then $R'$ is a Noetherian ring.*

*Proof.* By 7.22, $R'$ is a Noetherian $R$-module. However, every ideal of $R'$ is automatically an $R$-submodule of $R'$, and so, since $R'$ satisfies the ascending chain condition for $R$-submodules, it automatically satisfies the ascending chain condition for ideals. $\square$

**8.5** Exercise. Show that the subring $\mathbb{Z}[\sqrt{-5}]$ of the field $\mathbb{C}$ is Noetherian.

Two fundamental methods of constructing new commutative rings from a given commutative ring $R$ are the formation of polynomial rings and rings of formal power series over $R$: we shall see during the course of the next few results that, if $R$ is Noetherian, then both the ring $R[X_1, \ldots, X_n]$ of polynomials over $R$ in the $n$ indeterminates $X_1, \ldots, X_n$ and the ring $R[[X_1, \ldots, X_n]]$ of formal power series are again commutative Noetherian rings. We deal with the case of polynomial rings first: this is the subject of Hilbert's famous and celebrated Basis Theorem. We first have a preliminary lemma.

**8.6** Lemma. *Let $R$ be a commutative ring, and let $X$ be an indeterminate. Let $\mathcal{I}, \mathcal{J}$ be ideals of $R[X]$ such that $\mathcal{I} \subseteq \mathcal{J}$. For all $i \in \mathbb{N}_0$, set*

$$L_i(\mathcal{I}) := \left\{ a_i \in R : \text{ there exist } a_{i-1}, \ldots, a_0 \in R \text{ with } \sum_{j=0}^{i} a_j X^j \in \mathcal{I} \right\}.$$

(i) *For all $i \in \mathbb{N}_0$, the set $L_i(\mathcal{I})$ is an ideal of $R$, and $L_i(\mathcal{I}) \subseteq L_i(\mathcal{J})$.*
(ii) *We have*

$$L_0(\mathcal{I}) \subseteq L_1(\mathcal{I}) \subseteq \ldots \subseteq L_n(\mathcal{I}) \subseteq L_{n+1}(\mathcal{I}) \subseteq \ldots .$$

(iii) *(Recall that $\mathcal{I} \subseteq \mathcal{J}$.) If $L_n(\mathcal{I}) = L_n(\mathcal{J})$ for all $n \in \mathbb{N}_0$, then $\mathcal{I} = \mathcal{J}$.*

*Proof.* (i) This is clear.
(ii) For this, we need only observe that, if $a_0, a_1, \ldots, a_i \in R$ are such that $f = \sum_{j=0}^{i} a_j X^j \in \mathcal{I}$, then $Xf \in \mathcal{I}$ also, so that $a_i \in L_{i+1}(\mathcal{I})$.

(iii) We have $\mathcal{I} \subseteq \mathcal{J}$, by hypothesis.  Let us suppose that $\mathcal{I} \subset \mathcal{J}$ and look for a contradiction. Then there must exist at least one non-zero polynomial in $\mathcal{J}\backslash\mathcal{I}$: among such polynomials, choose one of smallest degree, say $g = \sum_{j=0}^{n} b_j X^j$ with $b_n \neq 0$. Then $b_n \in L_n(\mathcal{J}) = L_n(\mathcal{I})$, so that there exists

$$h = b_n X^n + c_{n-1} X^{n-1} + \cdots + c_0 \in \mathcal{I}.$$

But then $g - h \in \mathcal{J}\backslash\mathcal{I}$, and since $\deg(g-h) \leq n-1$, we have a contradiction to the definition of $n$. Hence $\mathcal{I} = \mathcal{J}$. $\square$

**8.7** HILBERT'S BASIS THEOREM. *Let $R$ be a commutative Noetherian ring, and let $X$ be an indeterminate. Then the ring $R[X]$ of polynomials is again a Noetherian ring.*

*Proof.* Let

$$\mathcal{I}_0 \subseteq \mathcal{I}_1 \subseteq \ldots \subseteq \mathcal{I}_j \subseteq \mathcal{I}_{j+1} \subseteq \ldots$$

be an ascending chain of ideals of $R[X]$. By 8.6(i),(ii), we have, for each $i \in \mathbf{N}_0$,

$$L_i(\mathcal{I}_0) \subseteq L_i(\mathcal{I}_1) \subseteq \ldots \subseteq L_i(\mathcal{I}_j) \subseteq L_i(\mathcal{I}_{j+1}) \subseteq \ldots$$

and, for each $j \in \mathbf{N}_0$,

$$L_0(\mathcal{I}_j) \subseteq L_1(\mathcal{I}_j) \subseteq \ldots \subseteq L_i(\mathcal{I}_j) \subseteq L_{i+1}(\mathcal{I}_j) \subseteq \ldots.$$

Since $R$ is Noetherian, there exist $p, q \in \mathbf{N}_0$ such that $L_p(\mathcal{I}_q)$ is a maximal member of $\{L_i(\mathcal{I}_j) : i, j \in \mathbf{N}_0\}$. It therefore follows that, for all $i \in \mathbf{N}_0$ with $i \geq p$, we have

$$L_i(\mathcal{I}_j) = L_i(\mathcal{I}_q) \ (= L_p(\mathcal{I}_q)) \qquad \text{for all } j \geq q.$$

But it also follows from $p$ uses of the ascending chain condition that there exists $q' \in \mathbf{N}_0$ such that, for all $i = 0, \ldots, p-1$,

$$L_i(\mathcal{I}_j) = L_i(\mathcal{I}_{q'}) \qquad \text{for all } j \geq q'.$$

Set $t = \max\{q, q'\}$: we have

$$L_i(\mathcal{I}_j) = L_i(\mathcal{I}_t) \qquad \text{for all } i \in \mathbf{N}_0$$

for each $j \in \mathbf{N}_0$ with $j \geq t$. It therefore follows from 8.6(iii) that $\mathcal{I}_j = \mathcal{I}_t$ for all $j \geq t$, and so our original ascending chain of ideals of $R[X]$ is eventually stationary. Hence $R[X]$ is Noetherian. $\square$

**8.8** COROLLARY. *Let $R$ be a commutative Noetherian ring. Then the polynomial ring $R[X_1, \ldots, X_n]$ over $R$ in $n$ indeterminates $X_1, \ldots, X_n$ is also Noetherian.*

*Proof.* This follows immediately from Hilbert's Basis Theorem (8.7 above) on use of induction, because, if $n > 1$, then $R[X_1, \ldots, X_n] = R[X_1, \ldots, X_{n-1}][X_n]$. □

The next corollary is concerned with (commutative) finitely generated algebras over a commutative Noetherian ring.

**8.9** REMARKS and DEFINITIONS. Let $R$ be a commutative ring and let $S$ be a commutative $R$-algebra with structural ring homomorphism $f : R \to S$; set $R' = \operatorname{Im} f$. By an *$R$-subalgebra of $S$* we mean a subring of $S$ which contains $R' = \operatorname{Im} f$; observe that $f$ provides such a subring $S'$ with a structure as $R$-algebra, and then the inclusion mapping $i : S' \to S$ is an $R$-algebra homomorphism in the sense of 5.13.

It should be clear from 1.5 that the intersection of any non-empty family of $R$-subalgebras of $S$ is again an $R$-subalgebra of $S$. For a subset $\Gamma$ of $S$, we define *the $R$-subalgebra of $S$ generated by* $\Gamma$ to be the intersection of the (non-empty) family of all $R$-subalgebras of $S$ which contain $\Gamma$; in the notation of 1.11, this is just $R'[\Gamma]$. Of course, $R'[\Gamma]$ is the smallest $R$-subalgebra of $S$ which contains $\Gamma$ in the sense that (it is one and) it is contained in every other $R$-subalgebra of $S$ which contains $\Gamma$.

We say that an $R$-subalgebra $S'$ of $S$ is *finitely generated* precisely when $S' = R'[\Delta]$ for some finite subset $\Delta$ of $S$, that is, if and only if there exist $\alpha_1, \ldots, \alpha_n \in S$ such that $S' = R'[\alpha_1, \ldots, \alpha_n]$.

**8.10** ‡EXERCISE. Let $R$ be a commutative ring and let $S$ be a commutative $R$-algebra with structural ring homomorphism $f : R \to S$; set $R' = \operatorname{Im} f$. Let $\alpha_1, \ldots, \alpha_n \in S$. Show that $R'[\alpha_1, \ldots, \alpha_n]$ is equal to

$$\left\{ \sum_{i=(i_1,\ldots,i_n)\in\Lambda} r_i' \alpha_1^{i_1} \ldots \alpha_n^{i_n} : \Lambda \subseteq \mathbf{N}_0{}^n, \ \Lambda \text{ finite}, \ r_i' \in R' \ \forall \, i \in \Lambda \right\}.$$

**8.11** COROLLARY. *Let $R$ be a commutative Noetherian ring. Suppose that the commutative ring $S$ is a finitely generated $R$-algebra. Then $S$ too is a Noetherian ring.*

*Proof.* Let $f : R \to S$ be the structural ring homomorphism, and set $R' = \operatorname{Im} f$. By hypothesis, there exist $\alpha_1, \ldots, \alpha_n \in S$ such that $S = R'[\alpha_1, \ldots, \alpha_n]$. By 8.10, this ring is equal to

$$\left\{ \sum_{i=(i_1,\ldots,i_n)\in\Lambda} r_i' \alpha_1^{i_1} \ldots \alpha_n^{i_n} : \Lambda \subseteq \mathbf{N}_0{}^n, \ \Lambda \text{ finite}, \ r_i' \in R' \ \forall \, i \in \Lambda \right\}.$$

By 8.8, the commutative ring $R[X_1, \ldots, X_n]$ of polynomials over $R$ in the $n$ indeterminates $X_1, \ldots, X_n$ is again Noetherian. Now, by 1.16, there is a ring homomorphism $g : R[X_1, \ldots, X_n] \rightarrow S$ which extends $f$ and which is such that $g(X_i) = \alpha_i$ for all $i = 1, \ldots, n$. The description of $S = R'[\alpha_1, \ldots, \alpha_n]$ given in the above display shows that $g$ is surjective, and so it follows from 8.2 that $S$ is a Noetherian ring. □

We next turn our attention to rings of formal power series over a commutative Noetherian ring $R$. One proof of the fact that the ring $R[[X]]$ of all formal power series in the indeterminate $X$ with coefficients in $R$ is again Noetherian is given by Matsumura in [8, Theorem 3.3]: since this book is intended as a preparation for books like Matsumura's, we do not repeat that proof here. The reader may, however, find the proof presented in 8.13 below, which uses a theorem of I. S. Cohen (see [8, Theorem 3.4]), an interesting alternative.

**8.12** THEOREM (I. S. Cohen). *Let $R$ be a commutative ring with the property that each of its prime ideals is finitely generated. Then $R$ is Noetherian.*

*Proof.* Suppose that $R$ is not Noetherian, and look for a contradiction. Then

$$\Theta := \{I : I \text{ is an ideal of } R \text{ which is not finitely generated}\}$$

is non-empty, by 8.1. Partially order $\Theta$ by inclusion: we shall apply Zorn's Lemma to this partially ordered set. Let $\Phi$ be a non-empty totally ordered subset of $\Theta$. Then $J := \bigcup_{I \in \Phi} I$ is an ideal of $R$ (because if $I', I'' \in \Phi$, then either $I' \subseteq I''$ or $I' \supseteq I''$): we aim to show that $J$ is not finitely generated, so that it belongs to $\Theta$ and is therefore an upper bound for $\Phi$ in $\Theta$.

Suppose that $J$ is finitely generated, say by $a_1, \ldots, a_t$. Then for each $i = 1, \ldots, t$, there exists $I_i \in \Phi$ such that $a_i \in I_i$. Since $\Phi$ is totally ordered, there exists $h \in \mathbb{N}$ with $1 \leq h \leq t$ such that $I_i \subseteq I_h$ for all $i = 1, \ldots, t$. Then we have

$$J = Ra_1 + \cdots + Ra_t \subseteq I_h \subseteq J,$$

so that $I_h$ is finitely generated. This contradiction shows that $J$ is not finitely generated. Hence $J \in \Theta$, and $J$ is an upper bound for $\Phi$ in $\Theta$. We can now apply Zorn's Lemma to see that $\Theta$ has a maximal element. We shall achieve a contradiction by showing that each maximal element $P$ of $\Theta$ is prime.

First of all, $P \subset R$ since $R = R1$ is finitely generated whereas $P$ is not. Let $a, b \in R \setminus P$ and suppose that $ab \in P$. We shall obtain a contradiction. Since $P \subset P + Ra$, it follows from the maximality of $P$ in $\Theta$ that $P + Ra$ is finitely generated, by $p_1 + r_1a, \ldots, p_n + r_na$, where $p_1, \ldots, p_n \in P$ and

$r_1, \ldots, r_n \in R$, say. Let $K = (P : a)$. Since $K \supseteq P + Rb \supset P$, it again follows from the maximality of $P$ in $\Theta$ that $K$ is finitely generated; therefore the ideal $aK$ is also finitely generated.

We claim now that $P = Rp_1 + \cdots + Rp_n + aK$. Of course,

$$P \supseteq Rp_1 + \cdots + Rp_n + aK.$$

Let $r \in P \subset P + aR$. Then there exist $c_1, \ldots, c_n \in R$ such that

$$r = c_1(p_1 + r_1 a) + \cdots + c_n(p_n + r_n a).$$

Now $(\sum_{i=1}^{n} c_i r_i)a = r - \sum_{i=1}^{n} c_i p_i \in P$, so that $\sum_{i=1}^{n} c_i r_i \in (P : a) = K$. Hence

$$r = \sum_{i=1}^{n} c_i p_i + \left( \sum_{i=1}^{n} c_i r_i \right) a \in \sum_{i=1}^{n} Rp_i + aK.$$

Thus $P \subseteq Rp_1 + \cdots + Rp_n + aK$, and so $P = Rp_1 + \cdots + Rp_n + aK$, as claimed. Thus $P$ is finitely generated, a contradiction. Therefore $ab \notin P$. Hence $P$ is prime.

We have therefore found a prime ideal of $R$ which is not finitely generated. This contradiction shows that $R$ must be Noetherian. $\square$

**8.13** THEOREM. *Let $R$ be a commutative Noetherian ring. Then the ring $R[[X]]$ of formal power series in the indeterminate $X$ with coefficients in $R$ is again Noetherian.*

*Proof.* We propose to use 8.12. Thus let $\mathcal{P} \in \mathrm{Spec}(R[[X]])$. The map

$$h \quad : \quad \begin{array}{ccc} R[[X]] & \longrightarrow & R \\ \sum_{j=0}^{\infty} r_j X^j & \longmapsto & r_0 \end{array}$$

is a surjective ring homomorphism, so that $h(\mathcal{P})$ is actually an ideal of $R$. Since $R$ is Noetherian, $h(\mathcal{P})$ is finitely generated, say by $a_0^{(1)}, \ldots, a_0^{(t)}$. For each $i = 1, \ldots, t$, there exists

$$f^{(i)} = a_0^{(i)} + a_1^{(i)} X + \cdots + a_n^{(i)} X^n + \cdots \in \mathcal{P}$$

which has 0-th coefficient $a_0^{(i)}$. We distinguish two cases, according as $X \in \mathcal{P}$ or $X \notin \mathcal{P}$.

First suppose that $X \in \mathcal{P}$. Then, for each $i = 1, \ldots, t$,

$$a_0^{(i)} = f^{(i)} - X \sum_{j=0}^{\infty} a_{j+1}^{(i)} X^j \in \mathcal{P}.$$

Also, given $f = \sum_{j=0}^{\infty} b_j X^j \in \mathcal{P}$, we have $b_0 \in h(\mathcal{P}) = \sum_{i=1}^{t} Ra_0^{(i)}$. Hence

$$f = b_0 + X \sum_{j=0}^{\infty} b_{j+1} X^j \in \sum_{i=1}^{t} R[[X]]a_0^{(i)} + X R[[X]].$$

It follows that $\mathcal{P} = \sum_{i=1}^{t} R[[X]]a_0^{(i)} + X R[[X]]$, so that $\mathcal{P}$ is finitely generated in this case.

Now suppose that $X \notin \mathcal{P}$. In this case, we shall show that

$$\mathcal{P} = \sum_{i=1}^{t} R[[X]]f^{(i)}.$$

Of course, $f^{(i)} \in \mathcal{P}$ for all $i = 1, \ldots, t$. Let $f = \sum_{j=0}^{\infty} b_j X^j \in \mathcal{P}$. Hence

$$b_0 \in h(\mathcal{P}) = \sum_{i=1}^{t} Ra_0^{(i)},$$

and so there exist $b_0^{(1)}, \ldots, b_0^{(t)} \in R$ such that $b_0 = b_0^{(1)}a_0^{(1)} + \cdots + b_0^{(t)}a_0^{(t)}$. Hence $f - \sum_{i=1}^{t} b_0^{(i)}f^{(i)}$ has 0 for its 0-th coefficient, and so

$$f - \sum_{i=1}^{t} b_0^{(i)}f^{(i)} = X g_1$$

for some $g_1 \in R[[X]]$.

Assume, inductively, that $v \in \mathbf{N}$ and we have found

$$b_0^{(1)}, \ldots, b_0^{(t)}, b_1^{(1)}, \ldots, b_1^{(t)}, \ldots, b_{v-1}^{(1)}, \ldots, b_{v-1}^{(t)} \in R$$

such that

$$f - \sum_{i=1}^{t} \left( \sum_{j=0}^{v-1} b_j^{(i)} X^j \right) f^{(i)} = X^v g_v$$

for some $g_v \in R[[X]]$; this is certainly the case when $v = 1$. Since the left-hand side of the last displayed equation belongs to $\mathcal{P}$ and $X \notin \mathcal{P}$, we have $g_v \in \mathcal{P}$ because $\mathcal{P}$ is prime. By the preceding paragraph, there exist $b_v^{(1)}, \ldots, b_v^{(t)} \in R$ such that $g_v - \sum_{i=1}^{t} b_v^{(i)}f^{(i)} = X g_{v+1}$ for some $g_{v+1} \in R[[X]]$. Hence

$$f - \sum_{i=1}^{t} \left( \sum_{j=0}^{v} b_j^{(i)} X^j \right) f^{(i)} = X^v \left( g_v - \sum_{i=1}^{t} b_v^{(i)}f^{(i)} \right) = X^{v+1} g_{v+1}.$$

This completes the inductive step. Therefore, we find, by induction, elements $b_j^{(1)}, \ldots, b_j^{(t)} \in R$ ($j \in \mathbf{N}_0$) such that, for all $v \in \mathbf{N}$, we have

$$f - \sum_{i=1}^{t} \left( \sum_{j=0}^{v-1} b_j^{(i)} X^j \right) f^{(i)} = X^v g_v$$

with $g_v \in R[[X]]$. For each $i = 1, \ldots, t$, let $e^{(i)} = \sum_{j=0}^{\infty} b_j^{(i)} X^j \in R[[X]]$. Then $f = \sum_{i=1}^{t} e^{(i)} f^{(i)}$ because, for each $v \in \mathbf{N}$, we have

$$f - \sum_{i=1}^{t} e^{(i)} f^{(i)} = X^v g_v - \sum_{i=1}^{t} \left( \sum_{j=v}^{\infty} b_j^{(i)} X^j \right) f^{(i)} \in X^v R[[X]]$$

and it is easy to see that $\bigcap_{v \in \mathbf{N}} X^v R[[X]] = 0$.

We have therefore proved that, in this case, $\mathcal{P} = \sum_{i=1}^{t} f^{(i)} R[[X]]$, so that $\mathcal{P}$ is finitely generated.

Thus every prime ideal of $R[[X]]$ is finitely generated, and so $R[[X]]$ is a Noetherian ring by 8.12. $\square$

**8.14** COROLLARY. *Let $R$ be a commutative Noetherian ring. Then the ring $R[[X_1, \ldots, X_n]]$ of formal power series in $n$ indeterminates $X_1, \ldots, X_n$ with coefficients in $R$ is again Noetherian.*

*Proof.* This is an easy consequence of 8.13 and 1.20, where it was shown that $R[[X_1, \ldots, X_n]] \cong R[[X_1, \ldots, X_{n-1}]][[X_n]]$ when $n > 1$. A simple use of induction will therefore prove the corollary. $\square$

**8.15** EXERCISE. Let $R$ be a commutative ring, and $X$ an indeterminate. If $R[X]$ is Noetherian, must $R$ be Noetherian? If $R[[X]]$ is Noetherian, must $R$ be Noetherian? Justify your responses.

So far in this chapter, we have seen that various ring-theoretic operations when performed on a commutative Noetherian ring $R$, such as the formation of residue class rings, the formation of rings of fractions, and the formation of rings of polynomials or rings of formal power series in finitely many indeterminates with coefficients in $R$, always keep us within the class of commutative Noetherian rings. This gives us a vast supply of examples of commutative Noetherian rings, for we can start with a very easy example of such a ring, like a field or $\mathbf{Z}$, and apply in turn a finite sequence of operations of the above types, and the result will be a commutative Noetherian ring.

Before we go on to discuss properties of commutative Noetherian rings, it is perhaps worth pointing out that the property of being Noetherian is *not* automatically inherited by subrings of commutative Noetherian rings. The following example settles this point.

**8.16** EXAMPLE. Consider the ring $R_\infty = K[X_1, \ldots, X_n, \ldots]$ of polynomials with coefficients in a field $K$ in the countably infinite family of indeterminates $(X_i)_{i \in \mathbb{N}}$: see 2.21. For each $n \in \mathbb{N}$, set $R_n = K[X_1, \ldots, X_n]$. Then

$$R_\infty := \bigcup_{n \in \mathbb{N}} R_n = \bigcup_{n \in \mathbb{N}} K[X_1, \ldots, X_n],$$

with the (unique) ring structure such that $R_n$ is a subring of $R_\infty$ for each $n \in \mathbb{N}$. In 2.21, we found an ideal of $R_\infty$ which is not finitely generated, and so it follows from 8.1 that $R_\infty$ is not Noetherian. But $R_\infty$ is an integral domain (because each $R_n$ is, and, for $n, m \in \mathbb{N}$, we have that either $R_n$ is a subring of $R_m$ or $R_m$ is a subring of $R_n$). Hence $R_\infty$ has a field of fractions, $L$ say, and, of course, $L$ is a Noetherian ring. Since $R_\infty$ is isomorphic to a subring of $L$, it follows that a subring of a commutative Noetherian ring need not necessarily be Noetherian.

Now that we have a good supply of examples of commutative Noetherian rings, it is appropriate for us to develop some further properties of these rings. This is a good point at which to remind the reader of some of the things we have already achieved in this direction.

**8.17** REMINDERS. Let $R$ be a commutative Noetherian ring.

Each proper ideal $I$ of $R$ has a primary decomposition: see 4.35. Thus the consequences of the theory of primary decomposition, including the First and Second Uniqueness Theorems, can be used when studying commutative Noetherian rings. In particular, ass $I$ denotes the (finite) set of prime ideals which belong to $I$ for primary decomposition, and the minimal members of ass $I$ are precisely the minimal prime ideals of $I$: see 4.24.

Thus $I$ has only finitely many minimal prime ideals.

We now develop some refinements of the theory of primary decomposition which are particularly relevant in commutative Noetherian rings.

**8.18** DEFINITION. Let $M$ be a module over the commutative ring $R$. A *zerodivisor on $M$* is an element $r \in R$ for which there exists $m \in M$ such that $m \neq 0$ but $rm = 0$. An element of $R$ which is not a zerodivisor on $M$ is often referred to as a *non-zerodivisor on $M$*.

The set of all zerodivisors on $M$ is denoted by $\mathrm{Zdv}(M)$ (or by $\mathrm{Zdv}_R(M)$ when it is desirable to emphasize the underlying ring).

Of course, when the commutative ring $R$ is regarded as a module over itself in the natural way, then a zerodivisor on $R$ in the sense of 8.18 is exactly the same as a zerodivisor in $R$ in the sense of 1.21. Observe that there are no zerodivisors on the zero $R$-module, that is, $\mathrm{Zdv}(0) = \emptyset$, simply because there does not exist a non-zero element in 0. However, for a non-zero $R$-module $M$, we always have $0 \in \mathrm{Zdv}(M)$.

Our next result relates the associated prime ideals of a decomposable ideal $I$ of the commutative ring $R$ to the set of all zerodivisors on the $R$-module $R/I$.

**8.19** PROPOSITION. *Let $I$ be a decomposable (see 4.15) ideal of the commutative ring $R$. Then*

$$\mathrm{Zdv}_R(R/I) = \bigcup_{P \in \mathrm{ass}\, I} P.$$

*Proof.* Let $a \in \mathrm{Zdv}_R(R/I)$, so that there exists $r \in R$ such that $r + I \neq 0_{R/I}$ but $a(r + I) = 0_{R/I}$. This means that $r \notin I$ but $ar \in I$. Let

$$I = Q_1 \cap \ldots \cap Q_n \quad \text{with } \sqrt{Q_i} = P_i \text{ for } i = 1, \ldots, n$$

be a minimal primary decomposition of $I$. Since $r \notin I$, there exists $j \in \mathbb{N}$ with $1 \leq j \leq n$ such that $r \notin Q_j$. Since $ar \in I \subseteq Q_j$, it follows from the fact that $Q_j$ is a $P_j$-primary ideal that $a \in P_j \in \mathrm{ass}\, I$. Hence

$$\mathrm{Zdv}_R(R/I) \subseteq \bigcup_{P \in \mathrm{ass}\, I} P.$$

We now establish the reverse inclusion. Let $P \in \mathrm{ass}\, I$, so that, by 4.17 and 4.19, there exists $a \in R$ such that $(I : a)$ is $P$-primary. Because $(I : a) \subseteq P \subset R$, we have $a \notin I$. Let $r \in P = \sqrt{(I : a)}$. Thus there exists $j \in \mathbb{N}$ such that $r^j \in (I : a)$, so that $r^j a \in I$. Let $t$ be the least positive integer $j$ such that $r^j a \in I$; then $r^{t-1} a \notin I$ (even if $t = 1$, simply because $a \notin I$), whereas $r^t a \in I$. Thus

$$r^{t-1} a + I \neq 0_{R/I} \quad \text{but} \quad r(r^{t-1} a + I) = 0_{R/I}.$$

Thus $r \in \mathrm{Zdv}_R(R/I)$, and so it follows that

$$\mathrm{Zdv}_R(R/I) \supseteq \bigcup_{P \in \mathrm{ass}\, I} P.$$

Hence $\mathrm{Zdv}_R(R/I) = \bigcup_{P \in \mathrm{ass}\, I} P.$ $\square$

**8.20** REMARK. Let $I$ be a proper ideal of the commutative Noetherian ring $R$. We have seen (see 8.17) that $I$ is decomposable, and our results show that the finite set of prime ideals $\mathrm{ass}\, I$ is intimately related to both the radical of $I$ and the set of all zerodivisors on the $R$-module $R/I$: by 3.54, we have $\sqrt{I} = \bigcap_{P \in \mathrm{Min}(I)} P$, and since, by 4.24, each member of $\mathrm{ass}\, I$

contains a member of $\text{Min}(I)$ and $\text{Min}(I)$ is precisely the set of isolated prime ideals belonging to $I$, it follows that

$$\sqrt{I} = \bigcap_{P \in \text{ass } I} P;$$

also, we have just seen in 8.19 that $\text{Zdv}_R(R/I) = \bigcup_{P \in \text{ass } I} P$.

In particular, when $R$ is non-trivial, we can take $I = 0$ in the above: we obtain that the finite set $\text{ass } 0$ of prime ideals of $R$ provides descriptions of both the set of all nilpotent elements in $R$ and the set of all zerodivisors in $R$, because

$$\sqrt{0} = \bigcap_{P \in \text{ass } 0} P \quad \text{and} \quad \text{Zdv}(R) = \bigcup_{P \in \text{ass } 0} P.$$

We are now going to work towards a refinement of 4.17 which is available when the underlying commutative ring $R$ is Noetherian. We shall need the following lemma, which is of interest in its own right.

**8.21** LEMMA. *Let $I$ be an ideal of the commutative ring $R$, and suppose that $\sqrt{I}$ is finitely generated. Then there exists $n \in \mathbb{N}$ such that $(\sqrt{I})^n \subseteq I$, that is, $I$ contains a power of its radical.*

*Consequently, every ideal in a commutative Noetherian ring contains a power of its radical.*

*Proof.* Suppose that $\sqrt{I}$ is generated by $a_1, \ldots, a_k$. Thus, for each $i = 1, \ldots, k$, there exists $n_i \in \mathbb{N}$ such that $a_i^{n_i} \in I$. Set $n = 1 + \sum_{i=1}^{k}(n_i - 1)$. It follows from 2.28(iii) that $(\sqrt{I})^n$ is the ideal of $R$ generated by

$$L := \left\{ a_1^{r_1} a_2^{r_2} \ldots a_k^{r_k} : r_1, \ldots, r_k \in \mathbb{N}_0, \ \sum_{i=1}^{k} r_i = n \right\}.$$

Now, whenever $r_1, \ldots, r_k$ are non-negative integers which sum to $n$, we must have $r_j \geq n_j$ for at least one integer $j$ with $1 \leq j \leq k$ (for otherwise $\sum_{i=1}^{k} r_i \leq \sum_{i=1}^{k}(n_i - 1) < n$, which is a contradiction), so that

$$a_1^{r_1} \ldots a_j^{r_j} \ldots a_k^{r_k} \in I.$$

Hence $L \subseteq I$, so that $(\sqrt{I})^n = RL \subseteq I$.

The final claim of the lemma is now immediate from 8.1, which shows that every ideal in a commutative Noetherian ring is finitely generated. $\square$

**8.22** PROPOSITION. *Let $I$ be a proper ideal of the commutative Noetherian ring $R$, and let $P \in \text{Spec}(R)$. Then $P \in \text{ass } I$ if and only if there exists $a \in R$ such that $(I : a) = P$, that is, if and only if there exists $\lambda \in R/I$ such that $(0 :_R \lambda) = \text{Ann}_R(\lambda) = P$.*

*Proof.* ($\Leftarrow$) Suppose there exists $a \in R$ such that $(I : a) = P$; then, in particular, $(I : a)$ is $P$-primary, and so $P \in \mathrm{ass}\, I$ by 4.17.

($\Rightarrow$) Let

$$I = Q_1 \cap \ldots \cap Q_n \qquad \text{with } \sqrt{Q_i} = P_i \text{ for } i = 1, \ldots, n$$

be a minimal primary decomposition of $I$. Let $j \in \mathbb{N}$ with $1 \leq j \leq n$, and set

$$I_j = \bigcap_{\substack{i=1 \\ i \neq j}}^{n} Q_i,$$

so that $I \subset I_j \not\subseteq Q_j$ by virtue of the minimality of the above primary decomposition. By 8.21, there exists $t \in \mathbb{N}$ such that $P_j^t \subseteq Q_j$; hence

$$P_j^t I_j \subseteq Q_j I_j \subseteq Q_j \cap I_j = I.$$

Let $u$ be the least $t \in \mathbb{N}$ such that $P_j^t I_j \subseteq I$. Thus $P_j^u I_j \subseteq I$ and $P_j^{u-1} I_j \not\subseteq I$ (even if $u = 1$, simply because $I_j \not\subseteq I$).

Choose $a \in P_j^{u-1} I_j \setminus I$, so that $a \in I_j \setminus I$. Thus

$$(I : a) = \left( \bigcap_{i=1}^{n} Q_i : a \right) = \bigcap_{i=1}^{n} (Q_i : a) = (Q_j : a),$$

and this is $P_j$-primary (by 4.14). But, since $aP_j \subseteq P_j^u I_j \subseteq I$, we have

$$P_j \subseteq (I : a) \subseteq P_j,$$

and so $P_j = (I : a)$.

To complete the proof, it is only necessary to observe that, for $b \in R$, we have $(I : b) = (0 :_R b + I)$, the annihilator of the element $b + I$ of the $R$-module $R/I$. $\square$

In Chapter 9, we shall use 8.22 as a starting point for a discussion of the theory of Associated prime ideals of modules over commutative Noetherian rings.

One application of primary decomposition in commutative Noetherian rings which we can develop now concerns an approach to Krull's Intersection Theorem.

**8.23** THEOREM. *Let $I$ be an ideal of the commutative Noetherian ring $R$, and let $J = \bigcap_{n=1}^{\infty} I^n$. Then $J = IJ$.*

*Proof.* If $I = R$, then the claim is clear, and so we assume that $I$ is a proper ideal of $R$. Since $IJ \subseteq J \subseteq I$, we see that $IJ$ is also a proper ideal of $R$, and so, by 8.17, has a primary decomposition. Let

$$IJ = Q_1 \cap \ldots \cap Q_n \quad \text{with } \sqrt{Q_i} = P_i \text{ for } i = 1, \ldots, n$$

be a minimal primary decomposition of $IJ$. We shall show that $J \subseteq IJ$ by showing that $J \subseteq Q_i$ for each $i = 1, \ldots, n$.

Suppose that, for some $i \in \mathbb{N}$ with $1 \le i \le n$ we have $J \not\subseteq Q_i$, so that there exists $a \in J \setminus Q_i$. Since

$$aI \subseteq IJ = Q_1 \cap \ldots \cap Q_n \subseteq Q_i$$

and $Q_i$ is $P_i$-primary, it follows from the fact that $a \notin Q_i$ that $I \subseteq P_i$. But $P_i = \sqrt{Q_i}$, and so, by 8.21, there exists $t \in \mathbb{N}$ such that $P_i^t \subseteq Q_i$. Hence

$$J = \bigcap_{n=1}^{\infty} I^n \subseteq I^t \subseteq P_i^t \subseteq Q_i.$$

This is a contradiction. Hence $J \subseteq Q_i$ for all $i = 1, \ldots, n$, and so $J \subseteq IJ$. The result follows. $\square$

Nakayama's Lemma will help us to make interesting deductions from 8.23. It is perhaps surprising that we have not before now encountered a situation where Nakayama's Lemma is needed. However, it is high time the reader was introduced to it.

**8.24** NAKAYAMA'S LEMMA. *Let $M$ be a finitely generated module over the commutative ring $R$, and let $I$ be an ideal of $R$ such that $I \subseteq \mathrm{Jac}(R)$, the Jacobson radical of $R$ (see 3.16). Assume that $M = IM$. Then $M = 0$.*

*Proof.* We suppose that $M \ne 0$ and look for a contradiction. Let $L = \{g_1, \ldots, g_n\}$ be a minimal generating set (with $n$ elements) for $M$: this means that $M$ is generated by $L$ but by no proper subset of $L$. (We are here using the fact that $M$, being non-zero, is not generated by its empty subset!)

Now $g_1 \in IM$, and so there exist $a_1, \ldots, a_n \in I$ such that $g_1 = \sum_{i=1}^{n} a_i g_i$. Hence

$$(1 - a_1)g_1 = a_2 g_2 + \cdots + a_n g_n.$$

But, since $a_1 \in I \subseteq \mathrm{Jac}(R)$, it follows from 3.17 that $1 - a_1$ is a unit of $R$, with inverse $u$ say. Hence $g_1 = \sum_{i=2}^{n} u a_i g_i$. It follows from this that $M$ is generated by $\{g_2, \ldots, g_n\}$, a proper subset of $L$. This contradiction shows that $M = 0$. $\square$

**8.25** CInfty COROLLARY: KRULL'S INTERSECTION THEOREM. *Let $I$ be an ideal of the commutative Noetherian ring $R$ such that $I \subseteq \mathrm{Jac}(R)$. Then*

$$\bigcap_{n=1}^{\infty} I^n = 0.$$

*Proof.* Set $J = \bigcap_{n=1}^{\infty} I^n$. By 8.23, $J = IJ$. But, since $R$ is Noetherian, $J$ is a finitely generated ideal of $R$, and so is a finitely generated $R$-module. Hence $J = 0$ by Nakayama's Lemma 8.24. □

Probably the most important applications of Krull's Intersection Theorem are to local rings.

**8.26** DEFINITION. By a *local ring* we shall mean a commutative Noetherian ring which is quasi-local (see 3.12). By a *semi-local ring* we shall mean a commutative Noetherian ring which is quasi-semi-local (see 5.35).

Thus, throughout this book, the phrase 'local ring' will always mean 'commutative *Noetherian* ring with exactly one maximal ideal'. We shall use the terminology '$(R, M)$ is a local ring' (and variants thereof) to indicate that $R$ is a local ring and $M$ is its unique maximal ideal. Recall from 3.12 that the residue field of the local ring $(R, M)$ is the field $R/M$.

Of course, the Jacobson radical of a local ring $(R, M)$ is just $M$, and so we deduce the following from 8.25.

**8.27** COROLLARY. *Let $(R, M)$ be a local ring. Then $\bigcap_{n=1}^{\infty} M^n = 0$.* □

**8.28** EXERCISE. Let $(R, M)$ be a local ring, and let $I$ be an ideal of $R$. Show that $\bigcap_{n=1}^{\infty}(I + M^n) = I$.

**8.29** EXERCISE. Let $(R, M)$ be a local ring, and let $Q$ be an $M$-primary ideal of $R$. Note that the $R$-module $(Q : M)/Q$ is annihilated by $M$, and so, by 6.19, can be regarded as a vector space over $R/M$ in a natural way. Show that the following statements are equivalent:

(i) $Q$ is irreducible (see 4.31);
(ii) $\mathrm{vdim}_{R/M}(Q : M)/Q = 1$;
(iii) the set of all ideals of $R$ which strictly contain $Q$ admits $(Q : M)$ as smallest element.

**8.30** EXERCISE. Let $Q$ be a $P$-primary ideal of the commutative Noetherian ring $R$. By 4.37, we can express $Q$ as an irredundant intersection of finitely many irreducible ideals of $R$. Prove that the number of terms in such an irredundant intersection is an invariant of $Q$, that is, is independent of the choice of expression for $Q$ as an irredundant intersection of finitely many irreducible ideals of $R$.

(Here are some hints: localize at $P$ and use 5.39(ii) to reduce to the case where $(R, M)$ is local and $P = M$; then use Exercise 8.29 above to show that the number of terms is equal to $\text{vdim}_{R/M}(Q : M)/Q$.)

Krull's Intersection Theorem can be taken as a starting point for beginnings of a discussion of topological aspects of local rings. Although we shall not go far down this avenue in this book, some relevant ideas are explored in the next exercise.

**8.31** EXERCISE. Let $R$ be a commutative ring and let $I$ be an ideal of $R$ such that $\bigcap_{n=1}^{\infty} I^n = 0$. (Note that Krull's Intersection Theorem gives a good supply of examples of this situation.) Define

$$\rho : R \times R \longrightarrow \mathbf{R}$$

as follows: set $\rho(a, a) = 0$ for all $a \in R$; and, for $a, b \in R$ with $a \neq b$, note that there exists a greatest integer $t \in \mathbf{N}_0$ such that $a - b \in I^t$ (or else the fact that

$$I \supseteq I^2 \supseteq \ldots \supseteq I^i \supseteq I^{i+1} \supseteq \ldots$$

is a descending chain would mean that $a - b \in \bigcap_{n=1}^{\infty} I^n = 0$, a contradiction) and define $\rho(a, b) = 2^{-t}$.

(i) Show that $\rho$ is a metric on $R$.

(ii) Let $a, b, c \in R$ be such that $\rho(a, b) \neq \rho(a, c)$. Show that $\rho(b, c) = \rho(a, b)$ or $\rho(a, c)$. (This result can be interpreted as telling us that every triangle in the metric space $(R, \rho)$ is isosceles!)

(iii) Let $U \subset R$ and let $a \in R$. Show that $U$ contains an open set (in the metric space $(R, \rho)$) which contains $a$ if and only if $a + I^n \subseteq U$ for some $n \in \mathbf{N}_0$.

Show that a subset of $R$ is open if and only if it is a union of cosets of powers of $I$.

(iv) Let $(\hat{R}, \hat{\rho})$ denote the completion of the metric space $(R, \rho)$: see [2, p. 51], for example (but be cautious over the arguments in [2, p. 55]). (Thus $\hat{R}$ is the set of all equivalence classes of the equivalence relation $\sim$ on the set $C$ of all Cauchy sequences of elements of the metric space $(R, \rho)$ given by, for $(a_n)_{n \in \mathbf{N}}, (b_n)_{n \in \mathbf{N}} \in C$,

$$(a_n)_{n \in \mathbf{N}} \sim (b_n)_{n \in \mathbf{N}} \quad \Longleftrightarrow \quad \lim_{n \to \infty} \rho(a_n, b_n) = 0.$$

Also, given $\hat{a}, \hat{b} \in \hat{R}$ represented by Cauchy sequences $(a_n)_{n \in \mathbf{N}}, (b_n)_{n \in \mathbf{N}}$ respectively (so that, to be precise, $(a_n)_{n \in \mathbf{N}} \in \hat{a}$ and $(b_n)_{n \in \mathbf{N}} \in \hat{b}$), we have

$$\hat{\rho}(\hat{a}, \hat{b}) = \lim_{n \to \infty} \rho(a_n, b_n).$$

This part of the exercise is for readers who are familiar with the completion of a metric space!)

Show that $\hat{R}$ can be given the structure of a commutative ring in such a way that the natural injective mapping from $R$ to $\hat{R}$ is a ring homomorphism (so that $R$ can be identified with a subring of $\hat{R}$) under operations for which, for $(a_n)_{n \in \mathbb{N}} \in \hat{a} \in \hat{R}$ and $(b_n)_{n \in \mathbb{N}} \in \hat{b} \in \hat{R}$, we have

$$(a_n + b_n)_{n \in \mathbb{N}} \in \hat{a} + \hat{b} \quad \text{and} \quad (a_n b_n)_{n \in \mathbb{N}} \in \hat{a}\hat{b}.$$

**8.32** FURTHER STEPS. The above Exercise 8.31 is the only excursion into the topological aspects of commutative rings which we shall make in this book. However, the reader should be warned that that is just 'the tip of the iceberg', so to speak, because the theory of completions of commutative rings is a very extensive and important topic. By use of appropriate topologies, one can produce a 'completion' of a commutative ring $R$ based on an arbitrary ideal $I$ of $R$: one does not have to restrict attention to ideals which have the property that the intersection of their powers is zero.

Several of the books cited in the Bibliography have sections on completions and related matters: see, for example, Atiyah and Macdonald [1, Chapter 10], Matsumura [7, Chapter 9] and [8, Section 8], Nagata [9, Chapter II], Northcott [10, Chapter V] and [12, Chapter 9] and Zariski and Samuel [19, Chapter VIII].

It should also be mentioned that the theory of inverse limits (see, for example, [8, p. 271]) has a significant rôle to play in the theory of completions. Also, completions of commutative Noetherian rings provide examples of what are called 'flat' modules: a full understanding of these requires knowledge of tensor products, and so, once again, we find motivation for the student to learn about tensor products!

Another application of Krull's Intersection Theorem is to facts concerning uniqueness, or rather the lack of uniqueness, in certain aspects of minimal primary decompositions of proper ideals in a commutative Noetherian ring $R$. Recall that, for a proper ideal $I$ of $R$, the number of terms and the set of radicals of the primary terms in any minimal primary decomposition of $I$ are independent of the choice of such minimal primary decomposition (by the First Uniqueness Theorem 4.18), and also the primary terms which correspond to the minimal prime ideals of $I$ again depend only on $I$ (by the Second Uniqueness Theorem 4.29). We can use Krull's Intersection Theorem as one means to approach the fact that, if

$$I = Q_1 \cap \ldots \cap Q_n \quad \text{with } \sqrt{Q_i} = P_i \text{ for } i = 1, \ldots, n$$

is a minimal primary decomposition of $I$ and $P_i$ is an embedded prime ideal of $I$, then there are infinitely many different $P_i$-primary ideals which could replace $Q_i$ in the above minimal primary decomposition so that the result is still a minimal primary decomposition of $I$. These ideas are covered by the next sequence of exercises.

**8.33 EXERCISE.** Let $(R, M)$ be a local ring.

(i) Show that, if there exists a non-maximal prime ideal of $R$, then $M^n \supset M^{n+1}$ for every $n \in \mathbf{N}$.

(ii) Show that, if $I$ is a proper ideal of $R$ and $\sqrt{I} \neq M$, then $I + M^n \supset I + M^{n+1}$ for every $n \in \mathbf{N}$.

**8.34 EXERCISE.** Let $P$ be a prime ideal of the commutative Noetherian ring $R$. This exercise is partly concerned with the symbolic powers $P^{(n)}$ $(n \in \mathbf{N})$ of $P$ introduced in 5.46.

(i) Show that, if $P$ is not a minimal prime ideal of $0$, then $P^{(n)} \supset P^{(n+1)}$ for every $n \in \mathbf{N}$.

(ii) Let $I$ be an ideal of $R$ such that $I \subset P$ but $P$ is not a minimal prime ideal of $I$. Use the extension and contraction notation of 2.41 in conjunction with the natural ring homomorphism $R \to R_P$. Show that

$$(I + P^n)^{ec} \supset (I + P^{n+1})^{ec} \qquad \text{for every } n \in \mathbf{N}.$$

Show also that $(I + P^n)^{ec}$ is $P$-primary for every $n \in \mathbf{N}$.

**8.35 EXERCISE.** Let $Q$ be a $P$-primary ideal of the commutative Noetherian ring $R$. Show that there exists $n \in \mathbf{N}$ such that $P^{(n)} \subseteq Q$.

**8.36 EXERCISE.** Let $I$ be a proper ideal of the commutative Noetherian ring $R$, and let

$$I = Q_1 \cap \ldots \cap Q_n \qquad \text{with } \sqrt{Q_i} = P_i \text{ for } i = 1, \ldots, n$$

be a minimal primary decomposition of $I$. Suppose that $P_i$ is an embedded prime ideal of $I$. Prove that there are infinitely many different choices of an alternative $P_i$-primary ideal of $R$ which can be substituted for $Q_i$ in the above decomposition so that the result is still a minimal primary decomposition of $I$.

Thus Exercise 8.36 shows that the primary components corresponding to embedded prime ideals in a minimal primary decomposition of a proper ideal $I$ of a commutative Noetherian ring are most definitely *not* uniquely determined by $I$.

**8.37 EXERCISE.** Let $P$ be a prime ideal of the commutative Noetherian ring $R$. Prove that

$$\bigcap_{n=1}^{\infty} P^{(n)} = \{r \in R : \text{there exists } s \in R \setminus P \text{ such that } sr = 0\}.$$

The fact, proved in 8.21, that every ideal in a commutative Noetherian ring contains a power of its radical enables us to prove that such a ring with the additional property that every prime ideal is maximal must be Artinian. We obtain this result next.

**8.38** PROPOSITION. *Let $R$ be a commutative Noetherian ring in which every prime ideal is maximal. Then*

(i) $R$, *if non-trivial, is semi-local (see 8.26), so that $R$ has only finitely many maximal ideals; and*

(ii) $R$ *is Artinian.*

*Proof.* We can assume that $R$ is not trivial.

(i) Let $M$ be a maximal ideal of $R$. Since every prime ideal of $R$ is maximal, $M$ must be a minimal prime ideal containing 0, and so, by 4.24, we have $M \in \text{ass } 0$. Hence $\text{Spec}(R) \subseteq \text{ass } 0 \subseteq \text{Spec}(R)$, so that $\text{Spec}(R) = \text{ass } 0$ and is finite.

(ii) Let $M_1, \ldots, M_n$ be the maximal ideals of $R$. By 3.49, $\sqrt{0} = \bigcap_{i=1}^{n} M_i$; also, by 8.21, there exists $t \in \mathbb{N}$ such that $(\sqrt{0})^t = 0$. Hence

$$M_1^t \ldots M_n^t \subseteq \left( \bigcap_{i=1}^{n} M_i \right)^t = (\sqrt{0})^t = 0.$$

It follows that $M_1^t \ldots M_n^t R = 0$; since $R$ is Noetherian, we can now deduce from 7.30 that $R$ is an Artinian $R$-module. Thus $R$ is an Artinian ring, and the proof is complete. $\square$

In fact, the converse statement to 8.38(ii), that is, that a commutative Artinian ring is a commutative Noetherian ring in which every prime ideal is maximal, is also true, and we now set off along a path to this result.

**8.39** LEMMA. *Let $R$ be a commutative Artinian ring. Then every prime ideal of $R$ is maximal.*

*Proof.* Let $P \in \text{Spec}(R)$, and set $R' := R/P$. By 3.23, 7.6, 7.14 and 7.26, $R'$ is an Artinian integral domain. We show that $R'$ is a field.

Let $b \in R'$ with $b \neq 0$. Then

$$R'b \supseteq R'b^2 \supseteq \ldots \supseteq R'b^i \supseteq R'b^{i+1} \supseteq \ldots$$

is a descending chain of ideals of $R'$, so that there exists $n \in \mathbb{N}$ such that $R'b^n = R'b^{n+1}$. Hence $b^n = cb^{n+1}$ for some $c \in R'$, so that, since $R'$ is an integral domain, $1 = cb$ and $b$ is a unit of $R'$. Therefore $R' = R/P$ is a field, so that $P$ is maximal by 3.3. $\square$

**8.40** LEMMA. *Let $R$ be a commutative Artinian ring. Then $R$ has only finitely many maximal ideals.*

*Proof.* We can assume that $R$ is non-trivial. Let $\Phi$ be the set of all ideals of $R$ which can be expressed as intersections of a finite number of maximal ideals of $R$. By the minimal condition, $\Phi$ has a minimal member, $J$ say: there exist maximal ideals $M_1, \ldots, M_n$ of $R$ such that $J = M_1 \cap \ldots \cap M_n$. We shall show that $M_1, \ldots, M_n$ are the only maximal ideals of $R$.

To this end, let $M$ be a maximal ideal of $R$. Then

$$J = M_1 \cap \ldots \cap M_n \supseteq M \cap M_1 \cap \ldots \cap M_n \in \Phi,$$

so that, by minimality of $J$ in $\Phi$, we must have

$$J = M_1 \cap \ldots \cap M_n = M \cap M_1 \cap \ldots \cap M_n.$$

Hence $M_1 \cap \ldots \cap M_n \subseteq M$, so that, since $M$ is prime, it follows from 3.55 that $M_j \subseteq M$ for some $j \in \mathbb{N}$ with $1 \leq j \leq n$. Since $M_j$ and $M$ are maximal ideals of $R$, we deduce that $M_j = M$.

Hence $M_1, \ldots, M_n$ are the only maximal ideals of $R$. $\square$

**8.41** PROPOSITION. *Let $R$ be a commutative Artinian ring, and let $N = \sqrt{0}$, the nilradical of $R$. There exists $t \in \mathbb{N}$ such that $N^t = 0$, that is, such that $(\sqrt{0})^t = 0$.*

*Proof.* We have a descending chain

$$N \supseteq N^2 \supseteq \ldots \supseteq N^i \supseteq N^{i+1} \supseteq \ldots$$

of ideals of $R$, and so, since $R$ is Artinian, there exists $t \in \mathbb{N}$ such that $N^{t+i} = N^t$ for all $i \in \mathbb{N}$. We aim to show that $N^t = 0$: suppose that this is not the case, and look for a contradiction. Set

$$\Theta = \left\{ I : I \text{ is an ideal of } R \text{ and } IN^t \neq 0 \right\}.$$

Then $N^i \in \Theta$ for all $i \in \mathbb{N}$ since $N^i N^t = N^{t+i} = N^t \neq 0$. Since $R$ is Artinian, it follows from the minimal condition (see 7.6) that $\Theta$ has a minimal member: let $J$ be one such.

Since $JN^t \neq 0$, there exists $a \in J$ such that $aN^t \neq 0$. Therefore $(aR)N^t \neq 0$ and $aR \subseteq J$; hence, by the minimality of $J$ in $\Theta$, we must have $aR = J$.

Next, note that the ideal $aN^t$ $(= (aR)N^t)$ of $R$ satisfies

$$(aN^t)N^t = (aR)N^{2t} = (aR)N^t = JN^t \neq 0.$$

Since $aN^t \subseteq aR = J$, it again follows from minimality that $aN^t = aR = J$. In particular, $a = ab$ for some $b \in N^t \subseteq N$. Thus there exists $v \in \mathbb{N}$ such that $b^v = 0$, and since

$$a = ab = (ab)b = ab^2 = \cdots = ab^v = 0,$$

we obtain $JN^t = (aR)N^t = 0$, a contradiction. Hence $N^t = 0$, and the result is proved. $\square$

**8.42** DEFINITION. An ideal $I$ of a commutative ring $R$ is said to be *nilpotent* precisely when there exists $t \in \mathbb{N}$ such that $I^t = 0$.

Perhaps a few words about what we have proved in 8.41 are appropriate. Of course, each element of the nilradical $N$ of the commutative ring $R$ is nilpotent. We have just seen that, when $R$ is Artinian, then $N$ itself is nilpotent, so that there exists $t \in \mathbb{N}$ such that $N^t = 0$. Note that this means that not only is the $t$-th power of every element of $N$ equal to zero, so that there is a common $t$ which 'works' for every nilpotent element, but also $a_1 \ldots a_t = 0$ for all $a_1, \ldots, a_t \in N$.

By 8.21, the nilradical of a commutative Noetherian ring is nilpotent, and so similar comments apply to this situation.

**8.43** EXERCISE. Let $R_\infty = K[X_1, \ldots, X_n, \ldots]$ be the ring of polynomials with coefficients in a field $K$ in the countably infinite family of indeterminates $(X_i)_{i \in \mathbb{N}}$: see 2.21 and 8.16.

Let $I$ be the ideal of $R_\infty$ generated by $\{X_i^i : i \in \mathbb{N}\}$. Show that the nilradical of the ring $R_\infty/I$ is not nilpotent.

**8.44** THEOREM. *A commutative Artinian ring $R$ is Noetherian.*

*Proof.* Of course, we can assume that $R$ is not trivial. By 8.39, every prime ideal of $R$ is maximal. By 8.40, $R$ has only finitely many maximal ideals. Let $M_1, \ldots, M_n$ be the maximal ideals of $R$. By 3.49, $\sqrt{0} = \bigcap_{i=1}^n M_i$; also, by 8.41, there exists $t \in \mathbb{N}$ such that $(\sqrt{0})^t = 0$. Hence

$$M_1^t \ldots M_n^t \subseteq \left( \bigcap_{i=1}^n M_i \right)^t = (\sqrt{0})^t = 0.$$

It follows that $M_1^t \ldots M_n^t R = 0$; since $R$ is Artinian, we can now deduce from 7.30 that $R$ is Noetherian. $\square$

We can now combine together 8.38, 8.39 and 8.44 to obtain the following.

**8.45** COROLLARY. *Let $R$ be a commutative ring. Then $R$ is Artinian if and only if $R$ is Noetherian and every prime ideal of $R$ is maximal.* $\square$

**8.46** EXERCISE. Let $I$ be a proper ideal of the commutative Noetherian ring $R$. Prove that the $R$-module $R/I$ has finite length if and only if ass $I$ consists of maximal ideals of $R$.

**8.47** EXERCISE. Let $Q$ be a $P$-primary ideal of the commutative Noetherian ring $R$. A *primary chain for $Q$ of length $t$* is an ascending chain

$$Q_0 \subset Q_1 \subset \ldots \subset Q_{t-1} \subset Q_t$$

of (distinct) $P$-primary ideals of $R$ in which $Q_0 = Q$ and $Q_t = P$.

Prove that there is a bound on the lengths of primary chains for $Q$, find the least upper bound $u$ for the set of all these lengths, and show that every maximal primary chain for $Q$ (that is, a primary chain for $Q$ which cannot be extended to a longer one by the insertion of an extra term) has length exactly $u$.

**8.48** EXERCISE. Let $A$ be an Artinian module over the commutative ring $R$, and let $a \in A$ with $a \neq 0$. Show that there exist a finite set $\{M_1, \ldots, M_n\}$ of maximal ideals of $R$ and a $t \in \mathbf{N}$ such that

$$(M_1 \ldots M_n)^t a = 0.$$

**8.49** EXERCISE. Let $A$ be a non-zero Artinian module over the commutative ring $R$.

(i) Show that, for each maximal ideal $M$ of $R$,

$$\Gamma_M(A) := \bigcup_{n \in \mathbf{N}} (0 :_A M^n)$$

is a submodule of $A$.

(ii) Let $\mathrm{Max}(R)$ denote the set of all maximal ideals of $R$. Show that the sum $\sum_{M \in \mathrm{Max}(R)} \Gamma_M(A)$ of submodules of $A$ is direct.

(iii) Show that there are only finitely many maximal ideals $M$ of $R$ for which $\Gamma_M(A) \neq 0$. Denote the distinct such maximal ideals by $M_1, \ldots, M_n$, and show that

$$A = \Gamma_{M_1}(A) \oplus \cdots \oplus \Gamma_{M_n}(A).$$

**8.50** EXERCISE. Let $R$ be a non-trivial commutative Artinian ring. Prove that there exist Artinian local rings $R_1, \ldots, R_n$ such that $R$ is isomorphic to the direct product ring $\prod_{i=1}^n R_i$ (see 2.6).

Show further that if $S_1, \ldots, S_m$ are Artinian local rings such that $R \cong \prod_{i=1}^m S_i$, then $m = n$ and there is a permutation $\sigma$ of $\{1, \ldots, n\}$ such that $R_i \cong S_{\sigma(i)}$ for all $i = 1, \ldots, n$.

# Chapter 9

# More module theory

This chapter is concerned more with useful techniques rather than with results which are interesting in their own right. Enthusiasm to reach exciting theorems has meant that certain important technical matters which are desirable for the efficient further development of the subject have been postponed. For example, we met Nakayama's Lemma in 8.24, but we did not develop in Chapter 8 the useful applications of Nakayama's Lemma to finitely generated modules over quasi-local rings. Also, although we undertook a thorough study of rings of fractions and their ideals in Chapter 5, we have still not developed the natural extension of that theory to modules of fractions. Again, the theory of primary decomposition of ideals discussed in Chapter 4 has an extension to modules which is related to the important concept of Associated prime ideal of a module over a commutative Noetherian ring.

These topics will be dealt with in this chapter. In addition, the theory of modules of fractions leads on to the important idea of the support of a module, and this is another topic which we shall explore in this chapter.

We begin with some consequences of Nakayama's Lemma.

**9.1 REMARKS.** Let $M$ be a module over the commutative ring $R$. By a *minimal generating set for $M$* we shall mean a subset, say $\Delta$, of $M$ such that $\Delta$ generates $M$ but no proper subset of $\Delta$ generates $M$.

Observe that if $M$ is finitely generated, by $g_1, \ldots, g_n$ say, then a minimal generating set $\Delta$ for $M$ must be finite: this is because each $g_i$ $(1 \leq i \leq n)$ can be expressed as

$$g_i = \sum_{\delta \in \Delta} r_{i\delta} \delta$$

with $r_{i\delta} \in R$ for all $\delta \in \Delta$ and almost all the $r_{i\delta}$ zero, so that the finite subset $\Delta'$ of $\Delta$ given by

$$\Delta' = \bigcup_{i=1}^{n} \{\delta \in \Delta : r_{i\delta} \neq 0\}$$

also generates $M$. (Recall that 'almost all' is an abbreviation for 'all except possibly finitely many'.)

However, even a finitely generated $R$-module $M$ may have two minimal generating sets which have different cardinalities; that is, the number of elements in one minimal generating set for $M$ need not be the same as the number in another. To give an example of this phenomenon, consider the $\mathbb{Z}$-module $\mathbb{Z}/6\mathbb{Z}$: it is easy to check that $\{1 + 6\mathbb{Z}\}$ and $\{2 + 6\mathbb{Z}, 3 + 6\mathbb{Z}\}$ are both minimal generating sets for $\mathbb{Z}/6\mathbb{Z}$.

Of course, this unpleasant situation does not occur in vector-space theory: a minimal generating set for a finitely generated vector space over a field forms a basis. We show now that it cannot occur for finitely generated modules over quasi-local rings. An interesting aspect of the discussion is that we use Nakayama's Lemma to reduce to a situation where we can use vector-space theory.

**9.2** Corollary of Nakayama's Lemma. *Let $G$ be a finitely generated module over the commutative ring $R$, and let $I$ be an ideal of $R$ such that $I \subseteq \mathrm{Jac}(R)$. Let $H$ be a submodule of $G$ such that $H + IG = G$. Then $H = G$.*

*Proof.* Suppose that $G$ is generated by $g_1, \ldots, g_n$. Then $G/H$ is generated by $g_1 + H, \ldots, g_n + H$, and so is a finitely generated $R$-module. Now, since $H + IG = G$, we have $I(G/H) = (H + IG)/H = G/H$, and so it follows from Nakayama's Lemma 8.24 that $G/H = 0$. Hence $H = G$. $\square$

Many applications of Nakayama's Lemma are to situations where the underlying ring $R$ is quasi-local: then $\mathrm{Jac}(R)$ is just the unique maximal ideal of $R$. This is the case in the application below.

**9.3** Proposition. *Let $R$ be a quasi-local ring having maximal ideal $M$ and residue field $K = R/M$. Let $G$ be a finitely generated $R$-module. Since the $R$-module $G/MG$ is annihilated by $M$, it has, by 6.19, a natural structure as a module over $R/M$, that is, as a $K$-space. Let $g_1, \ldots, g_n \in G$. Then the following statements are equivalent:*

    (i) *$G$ is generated by $g_1, \ldots, g_n$;*

    (ii) *the $R$-module $G/MG$ is generated by $g_1 + MG, \ldots, g_n + MG$;*

    (iii) *the $K$-space $G/MG$ is generated by $g_1 + MG, \ldots, g_n + MG$.*

*Furthermore, the $K$-space $G/MG$ has finite dimension, and the number of elements in each minimal generating set for the $R$-module $G$ is equal to* $\text{vdim}_K G/MG$.

*Proof.* The equivalence of statements (ii) and (iii) is clear from the fact, explained in 6.19, that the $R$-module and $K$-space structures of $G/MG$ are related by the rule that $r(g + MG) = (r + M)(g + MG)$ for all $r \in R$ and all $g \in G$. (See also 6.23.) It is also clear that (i) implies (ii).

(ii) $\Rightarrow$ (i) Assume that the $R$-module $G/MG$ is generated by the elements $g_1 + MG, \ldots, g_n + MG$. Let $H = Rg_1 + \cdots + Rg_n$; we show that $G = H + MG$. Let $g \in G$. Then there exist $r_1, \ldots, r_n \in R$ such that

$$g + MG = r_1(g_1 + MG) + \cdots + r_n(g_n + MG),$$

so that $g - \sum_{i=1}^n r_i g_i \in MG$. It follows that $G \subseteq H + MG$, so that $G = H + MG$ since the reverse inclusion is trivial.

Since $M = \text{Jac}(R)$, it now follows from 9.2 that $G = H$, so that $G$ is generated by $g_1, \ldots, g_n$.

We have now proved the equivalence of statements (i), (ii) and (iii). Since $G$ is a finitely generated $R$-module, it follows from this equivalence that $G/MG$ is a finitely generated $K$-space, that is, a finite-dimensional $K$-space. Note also that, by 9.1, each minimal generating set for $G$ is finite.

Let $\{g'_1, \ldots, g'_w\}$ be a minimal generating set for $G$ having $w$ elements. It follows from the equivalence of statements (i) and (iii) that

$$\{g'_1 + MG, \ldots, g'_w + MG\}$$

(has $w$ elements and) is a generating set for the $K$-space $G/MG$ with the property that no proper subset of it generates $G/MG$. Hence $(g'_i + MG)_{i=1}^w$ is a basis for the $K$-space $G/MG$ and so $w = \text{vdim}_K G/MG$. $\square$

Now we move on to consider modules of fractions: this theory is a natural extension of our work on rings of fractions in Chapter 5. Our first result on this topic is sufficiently reminiscent of material in 5.1, 5.2, 5.3 and 5.4 that its proof can be safely left to the reader.

**9.4** PROPOSITION, TERMINOLOGY and NOTATION. *Let $S$ be a multiplicatively closed subset of the commutative ring $R$, and let $M$ be an $R$-module. The relation $\sim$ on $M \times S$ defined by, for $(m, s), (n, t) \in M \times S$,*

$$(m, s) \sim (n, t) \iff \exists\, u \in S \text{ with } u(tm - sn) = 0$$

*is an equivalence relation on $M \times S$; for $(m, s) \in M \times S$, the equivalence class of $\sim$ which contains $(m, s)$ is denoted by $m/s$.*

*The set $S^{-1}M$ of all equivalence classes of $\sim$ has the structure of a module over the ring $S^{-1}R$ of fractions of $R$ with respect to $S$ under operations for which*

$$\frac{m}{s} + \frac{n}{t} = \frac{tm + sn}{st}, \qquad \frac{r}{s}\frac{n}{t} = \frac{rn}{st}$$

*for all $m, n \in M$, $s, t \in S$ and $r \in R$. The $S^{-1}R$-module $S^{-1}M$ is called the* module of fractions *of $M$ with respect to $S$; its zero element is $0_M/1$, and this is equal to $0_M/s$ for all $s \in S$.*

**9.5** ♯EXERCISE. Prove 9.4.

**9.6** REMARKS. Let the situation be as in 9.4 and 9.5.

(i) Observe that, for $m \in M$ and $s \in S$, we have $m/s = 0_{S^{-1}M}$ if and only if there exists $t \in S$ such that $tm = 0$.

(ii) The map $g : M \to S^{-1}M$ defined by $g(m) = m/1$ for all $m \in M$ is a homomorphism of $R$-modules when $S^{-1}M$ is regarded as an $R$-module by restriction of scalars using the natural ring homomorphism $R \to S^{-1}R$. By (i), $\operatorname{Ker} g = \{m \in M : \text{ there exists } s \in S \text{ such that } sm = 0\}$.

(iii) Expertise gained in Chapter 5 in the manipulation of formal fractions will stand the reader in good stead for work with modules of fractions. For instance, to add two elements of $S^{-1}M$ which are on a common denominator, one only has to add together the two numerators and use the same denominator.

(iv) We shall use obvious extensions of notation introduced in 5.17 and 5.20: thus, in the special case in which $S = \{t^n : n \in \mathsf{N}_0\}$ for a fixed $t \in R$, the module $S^{-1}M$ will sometimes be denoted by $M_t$; and in the special case in which $S = R \setminus P$ for some prime ideal $P$ of $R$, the module $S^{-1}M$ will usually be denoted by $M_P$. In the latter case, $M_P$ is referred to as the *localization of $M$ at $P$*.

Of course, as well as modules over a commutative ring $R$, we have studied $R$-homomorphisms between such modules: we are now going to investigate how such an $R$-homomorphism induces an $S^{-1}R$-homomorphism between the corresponding $S^{-1}R$-modules (for a multiplicatively closed subset $S$ of $R$).

**9.7** LEMMA and NOTATION. *Suppose that $f : L \to M$ is a homomorphism of modules over the commutative ring $R$ and let $S$ be a multiplicatively closed subset of $R$. Then $f$ induces an $S^{-1}R$-homomorphism*

$$S^{-1}f : S^{-1}L \to S^{-1}M$$

*for which $S^{-1}f(a/s) = f(a)/s$ for all $a \in L$, $s \in S$.*

*Proof.* Suppose that $a, b \in L$ and $s, t \in S$ are such that $a/s = b/t$ in $S^{-1}L$. Then there exists $u \in S$ such that $u(ta - sb) = 0$; apply $f$ to both sides of this equation to deduce that $u(tf(a) - sf(b)) = 0$ in $M$; hence $f(a)/s = f(b)/t$ in $S^{-1}M$. It thus follows that there is indeed a mapping $S^{-1}f : S^{-1}L \to S^{-1}M$ given by the formula in the statement of the lemma. It is now routine to check that $S^{-1}f$ is an $S^{-1}R$-homomorphism: recall that two members of $S^{-1}L$ can always be put on a common denominator. $\square$

**9.8** ♮EXERCISE. Let $L, M, N$ be modules over the commutative ring $R$, and let $S$ be a multiplicatively closed subset of $R$. Let $f, f' : L \to M$ and $g : M \to N$ be $R$-homomorphisms. Show that

(i) $S^{-1}(f + f') = S^{-1}f + S^{-1}f'$;

(ii) $S^{-1}z$, where $z$ denotes the zero homomorphism (see 6.27) from $L$ to $M$, is the zero homomorphism from $S^{-1}L$ to $S^{-1}M$;

(iii) $S^{-1}(g \circ f) = S^{-1}g \circ S^{-1}f$;

(iv) $S^{-1}(\mathrm{Id}_M) = \mathrm{Id}_{S^{-1}M}$; and

(v) if $f$ is an $R$-isomorphism, then $S^{-1}f$ is an $S^{-1}R$-isomorphism.

Readers familiar with some basic ideas from homological algebra may realise that, now we have 9.8(i), (iii) and (iv), we have shown that '$S^{-1}$' can be thought of as an additive, covariant functor from the category of all $R$-modules and $R$-homomorphisms to the category of all $S^{-1}R$-modules and $S^{-1}R$-homomorphisms. In the same spirit, the next result shows that this functor is exact. However, it is not expected that the reader should know anything about homological algebra, and he or she can think of the (important) next lemma as showing that '$S^{-1}$' preserves exactness of sequences of modules.

**9.9** LEMMA. *Let*

$$L \xrightarrow{f} M \xrightarrow{g} N$$

*be an exact sequence of modules and homomorphisms over the commutative ring $R$ and let $S$ be a multiplicatively closed subset of $R$. Then*

$$S^{-1}L \xrightarrow{S^{-1}f} S^{-1}M \xrightarrow{S^{-1}g} S^{-1}N$$

*is exact too.*

*Proof.* Since $\mathrm{Ker}\, g = \mathrm{Im}\, f$, we have $g \circ f = 0$; hence, by 9.8(ii) and (iii),

$$S^{-1}g \circ S^{-1}f = S^{-1}(g \circ f) = S^{-1}(0) = 0,$$

so that $\mathrm{Im}\, S^{-1}f \subseteq \mathrm{Ker}\, S^{-1}g$.

To prove the reverse inclusion, let $\lambda \in \operatorname{Ker} S^{-1}g$, so that there exist $m \in M$ and $s \in S$ such that $\lambda = m/s$ and $g(m)/s = (S^{-1}g)(m/s) = 0$. Therefore, by 9.6(i), there exists $t \in S$ such that $tg(m) = 0$; hence $tm \in \operatorname{Ker} g = \operatorname{Im} f$, and so there exists $a \in L$ such that $tm = f(a)$. Thus

$$\lambda = \frac{m}{s} = \frac{tm}{ts} = \frac{f(a)}{ts} = S^{-1}f\left(\frac{a}{ts}\right) \in \operatorname{Im} S^{-1}f$$

and the proof is complete. $\square$

**9.10** CONVENTION. Let $L$ be a submodule of the module $M$ over the commutative ring $R$ and let $S$ be a multiplicatively closed subset of $R$. Let $u : L \to M$ denote the inclusion $R$-homomorphism. From 9.9, we see that the induced $S^{-1}R$-homomorphism $S^{-1}u : S^{-1}L \to S^{-1}M$ is injective: we frequently use $S^{-1}u$ to identify $S^{-1}L$ with the $S^{-1}R$-submodule

$$\left\{\lambda \in S^{-1}M : \lambda = \frac{a}{s} \text{ for some } a \in L \text{ and } s \in S\right\}$$

of $S^{-1}M$.

Of course, an ideal $I$ of $R$ is a submodule of $R$ when $R$ is regarded as an $R$-module in the natural way, and we can apply the above convention to $I$ and $R$: it identifies the $S^{-1}R$-module $S^{-1}I$ with the ideal

$$\left\{\lambda \in S^{-1}R : \lambda = \frac{r}{s} \text{ for some } r \in I \text{ and } s \in S\right\}$$

of $S^{-1}R$. Now, by 5.25, the latter ideal is just the extension $I^e$ of $I$ to $S^{-1}R$ under the natural ring homomorphism: thus, with our convention, $S^{-1}I = I^e$. We shall sometimes use $S^{-1}I$ as an alternative notation for $IS^{-1}R$.

The warning in 5.27 shows that, in general, for a submodule $L$ of $M$, it is *not* the case that *every* formal fraction representation of an element $\lambda$ of the submodule $S^{-1}L$ of $S^{-1}M$ as $\lambda = m/s$ with $m \in M$ and $s \in S$ will have its numerator $m$ in $L$: all we know in general is that such a $\lambda$ has at least one formal fraction representation with numerator in $L$. The reader is warned to be cautious over this point!

The above convention is employed in the next exercise and in the lemma which follows it.

**9.11** ‡EXERCISE. Let $L_1, L_2$ be submodules of the module $M$ over the commutative ring $R$ and let $S$ be a multiplicatively closed subset of $R$. Let $I$ be an ideal of $R$, and let $r \in R$. Use the extension notation of 2.41 in relation to the natural ring homomorphism $R \to S^{-1}R$. Show that

   (i) $S^{-1}(IM) = I^e S^{-1}M$;

  (ii) $S^{-1}(rM) = (r/1)S^{-1}M$;

 (iii) $S^{-1}(L_1 + L_2) = S^{-1}L_1 + S^{-1}L_2$;

 (iv) $S^{-1}(L_1 \cap L_2) = S^{-1}L_1 \cap S^{-1}L_2$;

  (v) if $\Omega$ is an $S^{-1}R$-submodule of $S^{-1}M$, then $G := \{g \in M : g/1 \in \Omega\}$ is a submodule of $M$ and $S^{-1}G = \Omega$;

 (vi) if $M$ is a Noetherian $R$-module, then $S^{-1}M$ is a Noetherian $S^{-1}R$-module; and

(vii) if $M$ is an Artinian $R$-module, then $S^{-1}M$ is an Artinian $S^{-1}R$-module.

**9.12 LEMMA.** *Let $L, N$ be submodules of the module $M$ over the commutative ring $R$ and let $S$ be a multiplicatively closed subset of $R$. Use the extension notation of 2.41 in relation to the natural ring homomorphism $R \to S^{-1}R$.*

  (i) *There is an isomorphism of $S^{-1}R$-modules*

$$S^{-1}M/S^{-1}L \longrightarrow S^{-1}(M/L)$$
$$(m/s) + S^{-1}L \longmapsto (m+L)/s.$$

 (ii) *If $N$ is finitely generated, then $(L :_R N)^e = (S^{-1}L :_{S^{-1}R} S^{-1}N)$.*

(iii) *If $M$ is finitely generated, then $(\mathrm{Ann}_R(M))^e = \mathrm{Ann}_{S^{-1}R}(S^{-1}M)$.*

*Proof.* (i) The exact sequence

$$0 \longrightarrow L \overset{u}{\longrightarrow} M \overset{\pi}{\longrightarrow} M/L \longrightarrow 0,$$

in which $u$ is the inclusion homomorphism and $\pi$ is the canonical epimorphism, induces, by 9.9, an exact sequence of $S^{-1}R$-modules and $S^{-1}R$-homomorphisms

$$0 \longrightarrow S^{-1}L \overset{S^{-1}u}{\longrightarrow} S^{-1}M \overset{S^{-1}\pi}{\longrightarrow} S^{-1}(M/L) \longrightarrow 0.$$

Now, in view of our convention 9.10, the map $S^{-1}u$ is just the inclusion homomorphism; also, $S^{-1}\pi(m/s) = \pi(m)/s = (m+L)/s$ for all $m \in M$ and $s \in S$. The claim therefore follows from the First Isomorphism Theorem for modules 6.33.

  (ii) Set $I := (L :_R N)$. Because $IN \subseteq L$, it follows from 9.11(i) that $I^e(S^{-1}N) = S^{-1}(IN) \subseteq S^{-1}L$, and so

$$(L :_R N)^e \subseteq (S^{-1}L :_{S^{-1}R} S^{-1}N).$$

We have not yet used the fact that $N$ is finitely generated; however, we use it to prove the reverse inclusion. Suppose that $N$ is finitely generated by

$g_1, \ldots, g_t$, and let $\lambda \in (S^{-1}L :_{S^{-1}R} S^{-1}N)$. We can write $\lambda = r/s$ for some $r \in R$ and $s \in S$. Now, for each $i = 1, \ldots, t$, we have $(r/s)(g_i/1) \in S^{-1}L$, so that there exist $a_i \in L$ and $s_i \in S$ such that

$$\frac{r}{s}\frac{g_i}{1} = \frac{rg_i}{s} = \frac{a_i}{s_i}$$

and there exists $s_i' \in S$ such that $s_i'(s_i rg_i - sa_i) = 0$. Let

$$s'' = s_1 \ldots s_t s_1' \ldots s_t', \in S.$$

Then $s''rg_i \in L$ for all $i = 1, \ldots, t$, so that, since $g_1, \ldots, g_t$ generate $N$, we have $s''r \in (L :_R N)$. Hence

$$\lambda = \frac{r}{s} = \frac{s''r}{s''s} \in (L :_R N)^e.$$

Hence $(L :_R N)^e \supseteq (S^{-1}L :_{S^{-1}R} S^{-1}N)$, as required.

(iii) This now follows from part (ii). $\square$

**9.13** EXERCISE. Let $L$ be a submodule of the module $M$ over the commutative ring $R$, let $I$ be a finitely generated ideal of $R$, and let $S$ be a multiplicatively closed subset of $R$. Use the extension notation of 2.41 in relation to the natural ring homomorphism $R \to S^{-1}R$. Show that

$$S^{-1}(L :_M I) = (S^{-1}L :_{S^{-1}M} I^e)$$

and deduce that $S^{-1}(0 :_M I) = (0 :_{S^{-1}M} I^e)$.

So far, our work on modules of fractions has involved little more than natural extensions to module theory of some of the ideas which we studied in Chapter 5. But now we are going to use the concept of localization to introduce the important idea of the support of a module.

**9.14** DEFINITION. Let $G$ be a module over the commutative ring $R$. The *support of $G$*, denoted by $\operatorname{Supp}(G)$ (or $\operatorname{Supp}_R(G)$ when it is desired to emphasize the underlying ring) is defined to be the set $\{P \in \operatorname{Spec}(R) : G_P \neq 0\}$. (Here, $G_P$ denotes the localization of $G$ at $P$ described in 9.6(iv).)

A hint about the importance of the concept of support is given by the next lemma.

**9.15** LEMMA. *Let $G$ be a module over the commutative ring $R$. Then the following statements are equivalent:*

   (i) *$G = 0$;*
   (ii) *$G_P = 0$ for all $P \in \operatorname{Spec}(R)$, that is, $\operatorname{Supp}(G) = \emptyset$; and*
   (iii) *$G_M = 0$ for all maximal ideals $M$ of $R$.*

*Proof.* It is clear that (i) $\Rightarrow$ (ii) and (ii) $\Rightarrow$ (iii). So we assume that (iii) is true and deduce the truth of (i). Suppose that $G \neq 0$ and look for a contradiction. Then there exists $g \in G$ with $g \neq 0$: thus $(0 : g)$ is a proper ideal of $R$, and so, by 3.10, there exists a maximal ideal $M$ of $R$ such that $(0 : g) \subseteq M$. But then it follows from 9.6(i) that $g/1 \neq 0$ in $G_M$. With this contradiction the proof is complete. $\square$

Let $f : L \to M$ be a homomorphism of modules over the commutative ring $R$ and let $S$ be a multiplicatively closed subset of $R$. In the special case in which $S = R \setminus P$ for some prime ideal $P$ of $R$, we shall write $f_P$ for the induced $R_P$-homomorphism $S^{-1}f : S^{-1}L = L_P \to S^{-1}M = M_P$ of 9.7. This notation is employed in the next corollary and the exercise which follows it.

**9.16** COROLLARY. *Let $f : L \to G$ be a homomorphism of modules over the commutative ring $R$. Then the following statements are equivalent:*
(i) *$f$ is injective;*
(ii) *$f_P : L_P \to G_P$ is injective for every $P \in \operatorname{Spec}(R)$; and*
(iii) *$f_M : L_M \to G_M$ is injective for every maximal ideal $M$ of $R$.*

*Proof.* (i) $\Rightarrow$ (ii) This is immediate from 9.9 because the sequence

$$0 \longrightarrow L \xrightarrow{f} G$$

is exact.
(ii) $\Rightarrow$ (iii) This is obvious because every maximal ideal of $R$ is prime.
(iii) $\Rightarrow$ (i) Let $K := \operatorname{Ker} f$. There is an exact sequence

$$0 \longrightarrow K \xrightarrow{u} L \xrightarrow{f} G$$

of $R$-modules and $R$-homomorphisms in which $u$ is the inclusion homomorphism. Let $M$ be a maximal ideal of $R$. By 9.9, the induced sequence

$$0 \longrightarrow K_M \xrightarrow{u_M} L_M \xrightarrow{f_M} G_M$$

is exact. But, by assumption, $f_M$ is injective, and so $\operatorname{Im} u_M = \operatorname{Ker} f_M = 0$; hence $u_M$ is the zero homomorphism and $K_M = \operatorname{Ker} u_M = 0$. This is true for each maximal ideal of $R$, and so it follows from 9.15 that $K = \operatorname{Ker} f = 0$. Hence $f$ is injective. $\square$

**9.17** ‡EXERCISE. *Let $f : L \to G$ be a homomorphism of modules over the commutative ring $R$. Show that the following statements are equivalent:*
(i) *$f$ is surjective;*
(ii) *$f_P : L_P \to G_P$ is surjective for every $P \in \operatorname{Spec}(R)$; and*
(iii) *$f_M : L_M \to G_M$ is surjective for every maximal ideal $M$ of $R$.*

**9.18** EXERCISE. Let $f : L \to G$ be a homomorphism of modules over the commutative ring $R$, and let $S$ be a multiplicatively closed subset of $R$. Show that, when the convention of 9.10 is employed, $S^{-1}(\operatorname{Ker} f) = \operatorname{Ker}(S^{-1}f)$.

**9.19** EXERCISE. Let $R$ be a commutative ring and let

$$0 \longrightarrow L \xrightarrow{f} M \xrightarrow{g} N \longrightarrow 0$$

be an exact sequence of $R$-modules and $R$-homomorphisms. Prove that

$$\operatorname{Supp}(M) = \operatorname{Supp}(L) \cup \operatorname{Supp}(N).$$

In the case of a finitely generated module $M$ over a commutative ring $R$, we can describe $\operatorname{Supp}(M)$ in terms of concepts we have already discussed in this book.

**9.20** LEMMA. *Let $G$ be a finitely generated module over the commutative ring $R$. Then*

$$\operatorname{Supp}(G) = \{P \in \operatorname{Spec}(R) : P \supseteq (0 : G)\} = \operatorname{Var}(\operatorname{Ann}(G)).$$

*(The notation $\operatorname{Var}(I)$ for an ideal $I$ of $R$ was introduced in 3.48.)*

*Proof.* Suppose $P \in \operatorname{Supp}(G)$. Then we must have $P \supseteq (0 : G)$, since otherwise there exists $r \in (0 : G) \setminus P$ and $rg = 0$ for all $g \in G$, so that $g/s = 0$ in $G_P$ for all $g \in G$ and $s \in R \setminus P$, by 9.6(i).

Although we have not made use of the fact that $G$ is finitely generated so far in this proof, we do in order to prove the reverse inclusion. Let $P \in \operatorname{Var}(\operatorname{Ann}(G))$. Suppose that $P \notin \operatorname{Supp}(G)$ and look for a contradiction. Suppose that $g_1, \ldots, g_n$ generate $G$. For each $i = 1, \ldots, n$, the element $g_i/1 = 0$ in $G_P$, and so there exists $s_i \in R \setminus P$ such that $s_i g_i = 0$. Set

$$s = s_1 \ldots s_n, \ \in R \setminus P.$$

Then $sg_i = 0$ for all $i = 1, \ldots, n$, so that, since $g_1, \ldots, g_n$ generate $G$, we must have $sg = 0$ for all $g \in G$. Therefore $s \in \operatorname{Ann}(G) \setminus P$ and $P \not\supseteq \operatorname{Ann}(G)$. This contradiction shows that $P \in \operatorname{Supp}(G)$ and completes the proof. $\square$

**9.21** EXERCISE. Give an example to show that the conclusion of Lemma 9.20 need not hold if the hypothesis that the $R$-module $G$ is finitely generated is omitted.

**9.22** EXERCISE. Let $G$ be a finitely generated module over the commutative ring $R$ and let $S$ be a multiplicatively closed subset of $R$. Show that

$$\text{Supp}_{S^{-1}R}(S^{-1}G) = \{PS^{-1}R : P \in \text{Supp}_R(G) \text{ and } P \cap S = \emptyset\}.$$

**9.23** EXERCISE. Let $G$ be a finitely generated module over the commutative ring $R$, and let $I$ be an ideal of $R$. Show that

$$\sqrt{\text{Ann}_R(G/IG)} = \sqrt{(I + \text{Ann}_R(G))}.$$

(You might find 3.48 and Nakayama's Lemma 8.24 helpful.)

**9.24** ♯EXERCISE. Let $G$ be a module over the commutative ring $R$, let $S$ be a multiplicatively closed subset of $R$ and let $P$ be a prime ideal of $R$ such that $P \cap S = \emptyset$. Note that $PS^{-1}R \in \text{Spec}(S^{-1}R)$, by 5.32(ii). Show that there is an isomorphism

$$G_P \xrightarrow{\cong} S^{-1}G_{PS^{-1}R}$$

of $R_P$-modules when $S^{-1}G_{PS^{-1}R}$ is regarded as an $R_P$-module by means of an appropriate isomorphism $R_P \to S^{-1}R_{PS^{-1}R}$ (see 5.45).

Prove that

$$\text{Supp}_{S^{-1}R}(S^{-1}G) = \{PS^{-1}R : P \in \text{Supp}_R(G) \text{ and } P \cap S = \emptyset\}.$$

The reader has now been exposed to a fairly comprehensive grounding in the basic theory of modules of fractions (although it should be pointed out that a complete coverage of this topic should certainly involve reference to tensor products), and much of the theory of rings of fractions developed in Chapter 5 has been extended to modules. The reader may have been wondering whether the theory of primary decomposition of ideals developed in Chapter 4 has any analogue for modules: in fact, it does, and we are going to turn our attention to this now. There is a very straightforward generalization of the work in Chapter 4 which applies to proper submodules of a Noetherian module over a commutative ring $R$; however, we shall merely sketch a path through this by means of a series of exercises.

There are two reasons for this policy. One is that working through the details for himself or herself should act as good revision for the reader and strengthen his or her grasp of the work in Chapter 4; the other is that the concept of primary decomposition for proper submodules of Noetherian modules (as opposed to the theory for *ideals*) does not nowadays appear very much in research papers on commutative algebra, whereas the idea of 'Associated prime' of a module over a commutative Noetherian ring (see 9.32) seems (to this author at least) to be used far more frequently.

**9.25** EXERCISE. Let $Q$ be an ideal of the commutative ring $R$. Show that $Q$ is primary if and only if $R/Q \neq 0$ and, for each $a \in \mathrm{Zdv}_R(R/Q)$, there exists $n \in \mathbf{N}$ such that $a^n(R/Q) = 0$. Show also that $\sqrt{Q} = \sqrt{\mathrm{Ann}_R(R/Q)}$.

Exercise 9.25 provides some motivation for the definition in 9.26 below.

**9.26** DEFINITIONS and ♯EXERCISE. Let $M$ be a module over the commutative ring $R$. A submodule $Q$ of $M$ is said to be a *primary submodule of $M$* precisely when $M/Q \neq 0$ and, for each $a \in \mathrm{Zdv}_R(M/Q)$, there exists $n \in \mathbf{N}$ such that $a^n(M/Q) = 0$.

Let $Q$ be a primary submodule of $M$. Show that $P := \sqrt{\mathrm{Ann}_R(M/Q)}$ is a prime ideal of $R$: in these circumstances, we say that $Q$ is a *P-primary submodule of $M$*, or that $Q$ is *P-primary in $M$*.

Show that, if $Q_1, \ldots, Q_n$ (where $n \in \mathbf{N}$) are $P$-primary submodules of $M$, then so too is $\bigcap_{i=1}^n Q_i$.

**9.27** DEFINITIONS and ♯EXERCISE. Let $M$ be a module over the commutative ring $R$, and let $G$ be a proper submodule of $M$. A *primary decomposition of $G$ in $M$* is an expression for $G$ as an intersection of finitely many primary submodules of $M$. Such a primary decomposition

$$G = Q_1 \cap \ldots \cap Q_n \qquad \text{with } Q_i \ P_i\text{-primary in } M \ (1 \leq i \leq n)$$

of $G$ in $M$ is said to be *minimal* precisely when

(i) $P_1, \ldots, P_n$ are $n$ different prime ideals of $R$; and

(ii) for all $j = 1, \ldots, n$, we have

$$Q_j \not\supseteq \bigcap_{\substack{i=1 \\ i \neq j}}^{n} Q_i.$$

We say that $G$ is a *decomposable submodule of $M$* precisely when it has a primary decomposition in $M$.

Show that a decomposable submodule of $M$ has a minimal primary decomposition in $M$.

**9.28** ♯EXERCISE: THE FIRST UNIQUENESS THEOREM. Let $M$ be a module over the commutative ring $R$, and let $G$ be a decomposable submodule of $M$. Let

$$G = Q_1 \cap \ldots \cap Q_n \qquad \text{with } Q_i \ P_i\text{-primary in } M \ (1 \leq i \leq n)$$

and

$$G = Q_1' \cap \ldots \cap Q_{n'}' \qquad \text{with } Q_i' \ P_i'\text{-primary in } M \ (1 \leq i \leq n')$$

be two minimal primary decompositions of $G$ in $M$. Prove that $n = n'$ and

$$\{P_1, \ldots, P_n\} = \{P'_1, \ldots, P'_n\}.$$

**9.29** EXERCISE: THE SECOND UNIQUENESS THEOREM. Let $M$ be a module over the commutative ring $R$, and let $G$ be a decomposable submodule of $M$. Let

$$G = Q_1 \cap \ldots \cap Q_n \qquad \text{with } Q_i \ P_i\text{-primary in } M \ (1 \leq i \leq n)$$

and

$$G = Q'_1 \cap \ldots \cap Q'_n \qquad \text{with } Q'_i \ P_i\text{-primary in } M \ (1 \leq i \leq n)$$

be two minimal primary decompositions of $G$ in $M$. (We have here made use of the result of the First Uniqueness Theorem 9.28.)

Suppose that $P_j$ is a minimal member of $\{P_1, \ldots, P_n\}$ with respect to inclusion. Prove that $Q_j = Q'_j$.

**9.30** EXERCISE. Let $M$ be a module over the commutative ring $R$, and let $G, H$ be proper submodules of $M$ with $G \supseteq H$. For a submodule $F$ of $M$ which contains $H$, denote $F/H$ by $\overline{F}$. Let $Q_1, \ldots, Q_n$ be submodules of $M$ which contain $H$. Show that

$$G = Q_1 \cap \ldots \cap Q_n \qquad \text{with } Q_i \ P_i\text{-primary in } M \ (1 \leq i \leq n)$$

is a primary decomposition of $G$ in $M$ if and only if

$$\overline{G} = \overline{Q_1} \cap \ldots \cap \overline{Q_n} \qquad \text{with } \overline{Q_i} \ P_i\text{-primary in } \overline{M} \ (1 \leq i \leq n)$$

is a primary decomposition of $G/H$ in $M/H$, and that one of these primary decompositions is minimal if and only if the other is.

**9.31** ♯EXERCISE. Let $M$ be a Noetherian module over the commutative ring $R$, and let $G$ be a proper submodule of $M$. Show that $G$ is a decomposable submodule of $M$, so that, by 9.27, $G$ has a minimal primary decomposition in $M$.

Let

$$G = Q_1 \cap \ldots \cap Q_n \qquad \text{with } Q_i \ P_i\text{-primary in } M \ (1 \leq i \leq n)$$

be a minimal primary decomposition of $G$ in $M$, and let $P \in \mathrm{Spec}(R)$. Prove that $P$ is one of $P_1, \ldots, P_n$ if and only if there exists $\lambda \in M/G$ such that $(0 :_R \lambda) = \mathrm{Ann}_R(\lambda) = P$. Prove also that

$$P_1 \cup \ldots \cup P_n = \mathrm{Zdv}_R(M/G).$$

The reader should note that, by 7.22(i), the conclusions of 9.31 apply to a proper submodule $G$ of a finitely generated module $M$ over a commutative Noetherian ring $R$. He or she might also like to compare 9.31 with 8.22. In fact, these two results provide motivation for the definition of Associated prime ideal of a module over the commutative Noetherian ring $R$. We turn to this next.

**9.32** DEFINITION. Let $M$ be a module over the commutative Noetherian ring $R$, and let $P \in \operatorname{Spec}(R)$. We say that $P$ is an *Associated prime (ideal)* of $M$ precisely when there exists $m \in M$ with $(0 : m) = \operatorname{Ann}(m) = P$.

Observe that, if $m \in M$ has $(0 : m) = P$ as above, then $m \neq 0$. The set of Associated prime ideals of $M$ is denoted by $\operatorname{Ass}(M)$ (or $\operatorname{Ass}_R(M)$ if it is desired to emphasize the underlying ring concerned).

Note that, if $M$ and $M'$ are isomorphic $R$-modules, then $\operatorname{Ass}(M) = \operatorname{Ass}(M')$.

**9.33** REMARKS. Let $M$ be a module over the commutative Noetherian ring $R$.

(i) Suppose that $I$ is a proper ideal of $R$, so that $I$ is decomposable, by 4.35, and we can form the finite set ass $I$ of associated prime ideals of $I$ for primary decomposition. Observe that, by 8.22, for a prime ideal $P$ of $R$, we have

$$P \in \operatorname{ass} I \quad \Longleftrightarrow \quad P \in \operatorname{Ass}_R(R/I).$$

The reader should notice the use of the upper case 'A' in 9.32 in the terminology 'Associated prime' and the notation '$\operatorname{Ass}(M)$': this is to try to avoid confusion with the concept of associated prime ideal of a decomposable ideal introduced in 4.19. The relationship between the two concepts should be clear from the above comment.

(ii) Suppose that $M$ is finitely generated and that $G$ is a proper submodule of $M$. By 9.31, $G$ is a decomposable submodule of $M$: let

$$G = Q_1 \cap \ldots \cap Q_n \qquad \text{with } Q_i \ P_i\text{-primary in } M \ (1 \leq i \leq n)$$

be a minimal primary decomposition of $G$ in $M$. It also follows from 9.31 that, for a prime ideal $P$ of $R$, we have

$$P \in \{P_1, \ldots, P_n\} \quad \Longleftrightarrow \quad P \in \operatorname{Ass}_R(M/G).$$

In particular, when $M \neq 0$, we have that $P \in \operatorname{Ass}(M)$ if and only if $P$ is one of the prime ideals which 'occurs' in each minimal primary decomposition of the zero submodule in $M$.

(iii) It should be noted that there is no requirement in Definition 9.32 that the module $M$ be finitely generated; indeed, the theory of Associated prime ideals of arbitrary modules over a commutative Noetherian ring is quite extensive, as we shall see in the next few results.

(iv) For $m \in M$, note that $Rm \cong R/(0:m)$ (as $R$-modules), by 7.24. Also, for an ideal $I$ of $R$, $\text{Ann}_R(1+I) = I$, and, of course, the $R$-module $R/I$ is cyclic and generated by $1+I$. It follows that, for $P \in \text{Spec}(R)$,

$$P \in \text{Ass}(M) \quad \Longleftrightarrow \quad M \text{ has a submodule } G \cong R/P.$$

**9.34** LEMMA. *Let $M$ be a non-zero module over the commutative Noetherian ring $R$. Then each maximal member of the non-empty set*

$$\Theta := \{\text{Ann}(m) : m \in M \text{ and } m \neq 0\}$$

*of ideals of $R$ (and $\Theta$ will have at least one maximal member, since $R$ is Noetherian) is prime, and so belongs to $\text{Ass}(M)$.*

*Proof.* Suppose $P = (0:m)$, where $m \in M$ and $m \neq 0$, is a maximal member of $\Theta$. Since $m \neq 0$, we have $P \subset R$. Suppose that $a, b \in R$ are such that $b \in R \setminus P$ but $ab \in P$. Thus $abm = 0$, but $bm \neq 0$. Now $(0:m) \subseteq (0:bm)$, and $(0:bm) \in \Theta$. Therefore, by the maximality of $P = (0:m)$ in $\Theta$, we must have

$$P = (0:m) = (0:bm).$$

Since $abm = 0$, we have $a \in P$. Hence $P \in \text{Spec}(R)$, as claimed. $\square$

**9.35** COROLLARY. *Let $M$ be a module over the commutative Noetherian ring $R$. Then $\text{Ass}(M) \neq \emptyset$ if and only if $M \neq 0$.*

*Proof.* This is now immediate from 9.32 and 9.34: a zero $R$-module cannot have an Associated prime simply because it does not have any non-zero element, while 9.34 shows that a non-zero $R$-module does have an Associated prime. $\square$

**9.36** COROLLARY. *Let $M$ be a module over the commutative Noetherian ring $R$. Then*

$$\text{Zdv}(M) = \bigcup_{P \in \text{Ass}(M)} P.$$

*Note.* The reader should compare this result with 8.19 and 9.31.

*Proof.* Let $P \in \text{Ass}(M)$, so that there exists $m \in M$ with $(0:m) = P$. Since $m \neq 0$, it is clear that $P$ consists of zerodivisors on $M$.

On the other hand, consider $r \in \mathrm{Zdv}(M)$, so that there exists $m' \in M$ with $m' \neq 0$ but $rm' = 0$. Hence

$$\Theta' := \{\mathrm{Ann}(m) : m \in M, \ m \neq 0 \ \text{and} \ r \in \mathrm{Ann}(m)\}$$

is a non-empty subset of the set $\Theta$ of 9.34. Since $R$ is Noetherian, $\Theta'$ will have at least one maximal member, $P$ say, and $P$ will also be a maximal member of $\Theta$; by 9.34, we must have $P \in \mathrm{Ass}(M)$, and since $r \in P$ it follows that we have proved that $\mathrm{Zdv}(M) \subseteq \bigcup_{P' \in \mathrm{Ass}(M)} P'$. $\square$

**9.37 REMARK.** Let $M$ be a module over the commutative Noetherian ring $R$, and let $S$ be a multiplicatively closed subset of $R$. Let $m \in M$ and let $s \in S$. Of course, $(0 :_R m) = (0 :_R Rm)$ and, for $m/1, \ m/s \in S^{-1}M$,

$$(0 :_{S^{-1}R} m/1) = (0 :_{S^{-1}R} S^{-1}R(m/1)) = (0 :_{S^{-1}R} m/s)$$

because $s/1$ is a unit of $S^{-1}R$. Now, with the convention of 9.10, the $S^{-1}R$-submodule $S^{-1}(Rm)$ of $S^{-1}M$ is just $S^{-1}R(m/1)$; it therefore follows from 9.12(iii) that

$$(0 :_R m)S^{-1}R = (0 :_R Rm)S^{-1}R = (0 :_{S^{-1}R} S^{-1}R(m/1))$$
$$= (0 :_{S^{-1}R} m/1) = (0 :_{S^{-1}R} m/s).$$

**9.38 LEMMA.** *Let $M$ be a module over the commutative Noetherian ring $R$, and let $S$ be a multiplicatively closed subset of $R$. Then*

$$\mathrm{Ass}_{S^{-1}R}(S^{-1}M) = \left\{ PS^{-1}R : P \in \mathrm{Ass}_R(M) \ \text{and} \ P \cap S = \emptyset \right\}.$$

*Proof.* It should be observed that $S^{-1}R$ is a commutative Noetherian ring, by 8.3, and so $\mathrm{Ass}_{S^{-1}R}(S^{-1}M)$ is defined.

Let $P \in \mathrm{Ass}_R(M)$ be such that $P \cap S = \emptyset$. Then $PS^{-1}R \in \mathrm{Spec}(S^{-1}R)$ by 5.32(ii), and there exists $m \in M$ such that $P = (0 :_R m)$. It follows from 9.37 that $PS^{-1}R = (0 :_{S^{-1}R} m/1) \in \mathrm{Spec}(S^{-1}R)$, and so $PS^{-1}R \in \mathrm{Ass}_{S^{-1}R}(S^{-1}M)$.

Conversely, suppose that $\mathcal{P} \in \mathrm{Ass}_{S^{-1}R}(S^{-1}M)$. Since $\mathcal{P}$ is a prime ideal of $S^{-1}R$, it follows from 5.32(iv) that there is a (unique) $P \in \mathrm{Spec}(R)$ such that $P \cap S = \emptyset$ and $\mathcal{P} = PS^{-1}R$. Also, there exist $m \in M$ and $s \in S$ such that $\mathcal{P} = (0 :_{S^{-1}R} m/s)$. Since $s/1$ is a unit of $S^{-1}R$, we therefore have

$$(0 :_{S^{-1}R} m/1) = \mathcal{P} = PS^{-1}R.$$

Since $R$ is Noetherian, the ideal $P$ is finitely generated, by $p_1, \ldots, p_n$ say. Thus $p_i m/1 = 0_{S^{-1}M}$ for all $i = 1, \ldots, n$. Hence, for each $i = 1, \ldots, n$, there exists $s_i \in S$ such that $s_i p_i m = 0$. Set $t := s_1 \ldots s_n \ (\in S)$, and observe that $t p_i m = 0$ for all $i = 1, \ldots, n$, so that

$$P \subseteq (0 :_R tm).$$

Our aim now is to establish the reverse inclusion.

Accordingly, let $r \in (0 :_R tm)$. Thus $rtm = 0$, so that $(rt/1)(m/1) = 0_{S^{-1}M}$. Hence $(rt/1) \in (0 :_{S^{-1}R} m/1) = PS^{-1}R$. Therefore, by 5.30, $rt \in P$; since $P$ is prime and $t \in S \subseteq R \setminus P$, we have $r \in P$. We have therefore proved that $P \supseteq (0 :_R tm)$, so that

$$(0 :_R tm) = P \in \text{Spec}(R).$$

Hence $P \in \text{Ass}(M)$. $\square$

The next theorem is often useful in applications of Associated primes.

**9.39** THEOREM. *Let $M$ be a module over the commutative Noetherian ring $R$. Then $\text{Ass}(M) \subseteq \text{Supp}(M)$ and every minimal member of $\text{Supp}(M)$ (with respect to inclusion) belongs to $\text{Ass}(M)$.*

*Proof.* Let $P \in \text{Ass}(M)$. Now, by 9.33(iv), there is a submodule $G$ of $M$ such that $G \cong R/P$. Hence there is an exact sequence

$$0 \longrightarrow R/P \longrightarrow M$$

of $R$-modules and $R$-homomorphisms; by 9.9, the induced sequence

$$0 \longrightarrow (R/P)_P \longrightarrow M_P$$

is also exact, and since 9.20 shows that $(R/P)_P \neq 0$, we deduce that $M_P \neq 0$ and $P \in \text{Supp}(M)$.

Next, let $P'$ be a minimal member of $\text{Supp}(M)$. By 9.24,

$$\text{Supp}_{R_{P'}}(M_{P'}) = \{QR_{P'} : Q \in \text{Supp}_R(M) \text{ and } Q \subseteq P'\},$$

and this is just $\{P'R_{P'}\}$ because $P'$ is a minimal member of $\text{Supp}(M)$.

Now $M_{P'}$ is a non-zero module over the Noetherian ring $R_{P'}$, and so it follows from 9.35 and our proof above of the first part of this theorem that

$$\emptyset \neq \text{Ass}_{R_{P'}}(M_{P'}) \subseteq \text{Supp}_{R_{P'}}(M_{P'}) = \{P'R_{P'}\}.$$

Hence $\text{Ass}_{R_{P'}}(M_{P'}) = \{P'R_{P'}\}$, so that, by 9.38 and 5.33, we must have $P' \in \text{Ass}(M)$. $\square$

**9.40** EXERCISE. Let $M$ be a non-zero finitely generated module over the commutative Noetherian ring $R$. Prove that there is an ascending chain

$$M_0 \subset M_1 \subset \ldots \subset M_{n-1} \subset M_n$$

of submodules of $M$ such that $M_0 = 0$, $M_n = M$ and, for each $i = 1, \ldots, n$, there exists $P_i \in \operatorname{Spec}(R)$ with $M_i/M_{i-1} \cong R/P_i$.

**9.41** EXERCISE. Let $P$ be a prime ideal of the commutative Noetherian ring $R$. Show that $\operatorname{Ass}_R(R/P) = \{P\}$.

**9.42** EXERCISE. Let

$$0 \longrightarrow L \dashrightarrow M \longrightarrow N \longrightarrow 0$$

be a short exact sequence of modules and homomorphisms over the commutative Noetherian ring $R$. Prove that

$$\operatorname{Ass}(L) \subseteq \operatorname{Ass}(M) \subseteq \operatorname{Ass}(L) \cup \operatorname{Ass}(N).$$

**9.43** EXERCISE. Let $M$ be an Artinian module over the commutative Noetherian ring $R$. Show that every member of $\operatorname{Ass}(M)$ is a maximal ideal of $R$.

**9.44** EXERCISE. Let $M$ be a module over the commutative Noetherian ring $R$. Prove that

$$\operatorname{Ass}(M) = \bigcup_{\substack{m \in M \\ m \neq 0}} \operatorname{ass}(0 : m).$$

**9.45** FURTHER STEPS. Part way through this chapter it was mentioned that a complete coverage of modules of fractions should involve reference to tensor products. One of the reasons for this is that, given a multiplicatively closed subset $S$ of a commutative ring $R$ and an $R$-module $M$, the $S^{-1}R$-module $S^{-1}M$ can 'essentially' be produced from $M$ and the ring of fractions $S^{-1}R$ by means of a tensor product construction: to be precise, the tensor product $M \otimes_R S^{-1}R$, automatically an $R$-module, actually inherits a natural structure as $S^{-1}R$-module, and as such it is naturally isomorphisc to $S^{-1}M$. Also, $S^{-1}R$, viewed as an $R$-module by means of the natural homomorphism, is another example of a 'flat' $R$-module: mention of flat modules was made in 8.32. These points serve to reinforce the comments about tensor products made in 6.60 and 8.32.

The comment, made after Exercise 9.8, to the effect that '$S^{-1}$' can be thought of as a 'functor of modules' also hints at links between homological algebra and our subject; this will be amplified a little in Chapter 10.

# Chapter 10

# Modules over principal ideal domains

In this chapter, we shall show how some of the techniques we have developed for handling modules over commutative Noetherian rings so far in this book can be put to good use to provide proofs of some 'classical' theorems. These theorems concern finitely generated modules over principal ideal domains, and include, as special cases, the 'Fundamental Theorem on Abelian Groups' and the 'Jordan Canonical Form Theorem' for square matrices over an algebraically closed field.

Basically, the main results of this chapter show that a finitely generated module over a principal ideal domain can be expressed as a direct sum of cyclic submodules, and that, when certain restrictions are placed on the annihilators of the cyclic summands, such decompositions as direct sums of cyclic modules have certain uniqueness properties.

It is perhaps worth reminding the reader of some basic facts about cyclic modules. Let $R$ be a commutative ring. Recall from 6.12 that an $R$-module $G$ is cyclic precisely when it can be generated by one element; then, $G \cong R/I$ for some ideal $I$ of $R$, by 7.25. Since $\mathrm{Ann}_R(R/I) = I$ and isomorphic $R$-modules have equal annihilators, it follows that a cyclic $R$-module is completely determined up to isomorphism by its annihilator.

Next, we give some indication of the general strategy of the proof in this chapter of the fact that each finitely generated module over a PID $R$ can be expressed as a direct sum of cyclic $R$-modules. The following exercise is a good starting point.

**10.1** ♯EXERCISE. Let $R$ be a commutative ring, let $n \in \mathbb{N}$ and let $F$ be a free $R$-module with a base $(e_i)_{i=1}^n$ of $n$ elements. Let $c_1, \ldots, c_n \in R$. By

6.55, there is an $R$-homomorphism

$$f : F \longrightarrow \bigoplus_{i=1}^{n} R/Rc_i$$

such that $f(\sum_{i=1}^{n} r_i e_i) = (r_1 + Rc_1, \ldots, r_n + Rc_n)$ for all $r_1, \ldots, r_n \in R$, and clearly $f$ is an epimorphism.

Show that $\operatorname{Ker} f$ is generated by $c_1 e_1, \ldots, c_n e_n$. Show further that, if $R$ is an integral domain, then $\operatorname{Ker} f$ is free, and determine $\operatorname{rank}(\operatorname{Ker} f)$ in this case.

Now let us return to our finitely generated module, $G$ say, over our PID $R$. Suppose that $G$ has a generating set with $n$ elements, where $n \in \mathbb{N}$. By 6.57, $G$ can be expressed as a homomorphic image of a free $R$-module $F$ with a base $(e_i)_{i=1}^{n}$ of $n$ elements. Thus there is a submodule $H$ of $F$ such that $F/H \cong G$. If we could find $c_1, \ldots, c_n \in R$ such that $H$ is generated by $c_1 e_1, \ldots, c_n e_n$, then it would follow from 10.1 above that

$$G \cong F/H \cong R/Rc_1 \oplus \cdots \oplus R/Rc_n,$$

a direct sum of cyclic $R$-modules. In general, it is too much to hope that, for a specified base $(e_i)_{i=1}^{n}$ for $F$, it will always be possible to find such $c_1, \ldots, c_n \in R$ with the property that $H$ is generated by $c_1 e_1, \ldots, c_n e_n$: just consider the $\mathbb{Z}$-submodule of $F' := \mathbb{Z} \oplus \mathbb{Z}$ generated by $(1, 3)$ and the base $(\tilde{e}_i)_{i=1}^{2}$ for $F'$ given by $\tilde{e}_1 = (1, 0)$, $\tilde{e}_2 = (0, 1)$. However, in this example, $(1, 3)$ and $(0, 1)$ form another base for $F'$, and this is symptomatic of the general situation: we shall see that, given the submodule $H$ of the above finitely generated free module $F$ over the PID $R$, then it is always possible to find a base $(e_i')_{i=1}^{n}$ for $F$ and $c_1', \ldots, c_n' \in R$ such that $H$ is generated by $c_1' e_1', \ldots, c_n' e_n'$. This result provides the key to some of the main results of the chapter; our proof of it makes significant use of the fact that $R$ is a PID.

**10.2** LEMMA and DEFINITION. *Let $F$ be a non-zero free module with a finite base over the non-trivial commutative ring $R$, and let $(e_i)_{i=1}^{n}$ be a base for $F$. Let $y \in F$, so that $y$ can be uniquely expressed as $y = \sum_{i=1}^{n} r_i e_i$ for suitable $r_1, \ldots, r_n \in R$. Then the ideal $C(y) := \sum_{i=1}^{n} Rr_i$ depends only on $y$ and not on the choice of base for $F$; we call $C(y)$ the content ideal, or just the content, of $y$.*

*Proof.* Let $(e_i')_{i=1}^{n}$ be another base for $F$. (We have here made use of 6.58.) Thus $y$ can be (uniquely) expressed as $y = \sum_{i=1}^{n} r_i' e_i'$ for suitable

$r'_1, \ldots, r'_n \in R$. Now, for each $i = 1, \ldots, n$, we can write

$$e'_i = \sum_{j=1}^{n} a_{ij} e_j \quad \text{and} \quad e_i = \sum_{j=1}^{n} b_{ij} e'_j$$

for suitable $a_{i1}, \ldots, a_{in}, b_{i1}, \ldots, b_{in} \in R$. Hence

$$y = \sum_{i=1}^{n} r_i e_i = \sum_{i=1}^{n} \sum_{j=1}^{n} r_i b_{ij} e'_j = \sum_{j=1}^{n} \left( \sum_{i=1}^{n} r_i b_{ij} \right) e'_j = \sum_{j=1}^{n} r'_j e'_j.$$

Since $(e'_i)_{i=1}^{n}$ is a base for $F$, we have $r'_j = \sum_{i=1}^{n} r_i b_{ij}$ for all $j = 1, \ldots, n$. Hence

$$\sum_{i=1}^{n} R r'_i \subseteq \sum_{i=1}^{n} R r_i.$$

The opposite inclusion is proved by reversing the rôles of the two bases. $\square$

**10.3** LEMMA. *Let $F$ be a non-zero free module with a finite base over the principal ideal domain $R$. Let $y \in F$, and let $c_y$ be a generator of the content ideal $C(y)$ of $y$ (see 10.2), so that $C(y) = Rc_y$. Then there exists a base $(e'_i)_{i=1}^{n}$ for $F$ such that $y = c_y e'_1$.*

*Proof.* Let $(e_i)_{i=1}^{n}$ be a base for $F$. We use induction on $n$. When $n = 1$, the result is easy to prove: $y = re_1$ for some $r \in R$, so that $C(y) = Rr$ and $c_y = ur$ for some unit $u$ of $R$; thus we can take $e'_1 = u^{-1} e_1$, and $e'_1$ forms a base for $F$ with the desired property.

So suppose, inductively, that $n > 1$ and the result has been proved for smaller values of $n$. We can write $y = \sum_{i=1}^{n} r_i e_i$ for suitable $r_1, \ldots, r_n \in R$. Then, in the light of 10.2, we have

$$C(y) = Rr_1 + \cdots + Rr_n = Rc_y.$$

Note that $y = 0$ if and only if $c_y = 0$: clearly, we can assume that $y \neq 0$. Set $z = \sum_{i=2}^{n} r_i e_i$, so that $y = r_1 e_1 + z$. Now $(e_i)_{i=2}^{n}$ is a base for the finitely generated free submodule $F' = \sum_{i=2}^{n} Re_i$ of $F$.

Apply the inductive assumption to the element $z$ of $F'$: there results a base $(e''_i)_{i=2}^{n}$ for $F'$ such that $z = c_z e''_2$, where $c_z$ is a generator of the ideal $\sum_{i=2}^{n} Rr_i$. (Perhaps it is worth pointing out to the reader that the content ideal of $z$ is the same whether we consider $z$ as an element of $F$ or as an element of $F'$.) Now

$$Rc_y = \sum_{i=1}^{n} Rr_i = Rr_1 + \sum_{i=2}^{n} Rr_i = Rr_1 + Rc_z.$$

Thus $r_1 = sc_y$, $c_z = tc_y$ for some $s, t \in R$; also, we can write $c_y = ur_1 + vc_z$ for some $u, v \in R$, so that

$$c_y = usc_y + vtc_y$$

and $1 = us + vt$ in view of our assumption that $c_y \neq 0$.

Set $e_1' = se_1 + te_2''$, $e_2' = ve_1 - ue_2''$ and $e_i' = e_i''$ for all $i = 3, \ldots, n$. Now

$$ue_1' + te_2' = e_1 \qquad \text{and} \qquad ve_1' - se_2' = e_2''$$

because $us + vt = 1$, and so $Re_1 + Re_2'' = Re_1' + Re_2'$. Hence

$$F = Re_1 + \sum_{i=2}^{n} Re_i = Re_1 + \sum_{i=2}^{n} Re_i'' = \sum_{i=1}^{n} Re_i'.$$

Furthermore, $(e_i')_{i=1}^{n}$ is actually a base for $F$, as we now show. Suppose that $r_1', \ldots, r_n' \in R$ are such that $\sum_{i=1}^{n} r_i' e_i' = 0$. Then

$$r_1'(se_1 + te_2'') + r_2'(ve_1 - ue_2'') + \sum_{i=3}^{n} r_i' e_i'' = 0.$$

Since $(e_i'')_{i=2}^{n}$ is a base for $F'$, and $F' \cap Re_1 = 0$, it follows that $r_i' = 0$ for all $i = 3, \ldots, n$,

$$r_1's + r_2'v = 0 \qquad \text{and} \qquad r_1't - r_2'u = 0.$$

It is now a simple matter to use the equation $us + vt = 1$ again to deduce that $r_1' = r_2' = 0$. Hence $(e_i')_{i=1}^{n}$ is a base for $F$, as claimed. Finally, note that

$$c_y e_1' = sc_y e_1 + tc_y e_2'' = r_1 e_1 + c_z e_2'' = r_1 e_1 + z = y.$$

This completes the inductive step, and the proof. $\square$

**10.4** LEMMA. *Let $F$ be a non-zero free module with a finite base over the principal ideal domain $R$. Let $H$ be a submodule of $F$. Recall that $R$ is a Noetherian ring, and let $z \in H$ be such that $C(z)$ (see 10.2) is a maximal member of the set $\{C(y) : y \in H\}$ of all content ideals of elements of $H$. Then*

$$C(z) = \sum_{y \in H} C(y),$$

*so that $C(z) \supseteq C(y)$ for all $y \in H$.*

*Proof.* Let $c_z$ be a generator for $C(z)$. By 10.3, there exists a base $(e_i')_{i=1}^n$ for $F$ such that $z = c_z e_1'$. Let $y \in H$, so that $y = \sum_{i=1}^n r_i e_i'$ for suitable $r_1, \ldots, r_n \in R$.

The ideal $Rc_z + Rr_1 = C(z) + Rr_1$ of $R$ is, of course, principal: let $t$ be a generator for it. Thus $t = uc_z + vr_1$ for some $u, v \in R$. Now

$$uz + vy = uc_z e_1' + v \left( \sum_{i=1}^n r_i e_i' \right) = te_1' + \sum_{i=2}^n vr_i e_i' \in H,$$

and since (in view of 10.2)

$$C(z) = Rc_z \subseteq Rt \subseteq C(uz + vy),$$

we must have $C(z) = Rt$ by virtue of the choice of $z$. Hence $Rc_z = C(z) = Rt = C(z) + Rr_1$, and $r_1 = wc_z$ for some $w \in R$. Note also that we have proved that $Rr_1 \subseteq C(z)$.

Now consider

$$(1 - w)z + y = (1 - w)c_z e_1' + \sum_{i=1}^n r_i e_i' = c_z e_1' + \sum_{i=2}^n r_i e_i' \in H.$$

We have

$$C(z) = Rc_z \subseteq Rc_z + \sum_{i=2}^n Rr_i = C((1 - w)z + y),$$

and so, by choice of $z$, we obtain $C(z) = Rc_z + \sum_{i=2}^n Rr_i$. Therefore $Rr_i \subseteq C(z)$ for all $i = 2, \ldots, n$. Hence

$$C(y) = \sum_{i=1}^n Rr_i \subseteq C(z),$$

as claimed. Since $y$ was an arbitrary element of $H$, the desired result has been proved. $\square$

Lemmas 10.3 and 10.4 provide the key to the first main theorem of this chapter.

**10.5** THEOREM. *Let $F$ be a non-zero free module with a finite base over the principal ideal domain $R$, and let $H$ be a submodule of $F$. Let $n$ be the rank of $F$. Then there exist a base $(e_i)_{i=1}^n$ for $F$ and elements $a_1, \ldots, a_n \in R$ such that*

$$Ra_1 \supseteq Ra_2 \supseteq \ldots \supseteq Ra_n$$

*(or, equivalently, such that $a_i \mid a_{i+1}$ for all $i = 1, \ldots, n - 1$) and $H$ is generated by $a_1 e_1, \ldots, a_n e_n$.*

*Note.* Of course, some of the $a_i$ in the above statement might be 0; however, the conditions imply that, if $a_j = 0$, then $a_{j+1} = \cdots = a_n = 0$ also.

*Proof.* We argue by induction on $n$. When $n = 1$, the result is essentially immediate from the fact that $R$ is a PID: if the element $e_1$ forms a base for $F$, then $H = IRe_1$ for some ideal $I$ of $R$, and we must have $I = Ra_1$ for some $a_1 \in R$.

So suppose, inductively, that $n > 1$ and the result has been proved for smaller values of $n$. By 10.4, there exists $z \in H$ such that $C(z) \supseteq C(y)$ for all $y \in H$, that is, such that the content ideal (see 10.2) of $z$ contains the content ideal of every element of $H$. Let $c_z$ be a generator of $C(z)$. By 10.3, there exists a base $(e_i')_{i=1}^n$ for $F$ such that $z = c_z e_1'$. Now $(e_i')_{i=2}^n$ is a base for the finitely generated free submodule $F' = \sum_{i=2}^n Re_i'$ of $F$. Let $H' := H \cap F'$.

By the inductive hypothesis applied to the submodule $H'$ of the free $R$-module $F'$ of rank $n-1$, there exist a base $(e_i)_{i=2}^n$ for $F'$ and elements $a_2, \ldots, a_n \in R$ such that

$$Ra_2 \supseteq Ra_3 \supseteq \ldots \supseteq Ra_n$$

and $H'$ is generated by $a_2 e_2, \ldots, a_n e_n$. Set $e_1 = e_1'$ and $a_1 = c_z$.

Since $a_2 e_2 \in H$ and $C(a_2 e_2) = Ra_2$ by 10.2, it follows from the choice of $z$ that

$$Ra_1 = Rc_z = C(z) \supseteq C(a_2 e_2) = Ra_2.$$

Next, note that $(e_i)_{i=1}^n$ is a base for $F$: this is because

$$F = \sum_{i=1}^n Re_i' = Re_1' + \sum_{i=2}^n Re_i' = Re_1 + F' = Re_1 + \sum_{i=2}^n Re_i,$$

while if $r_1, \ldots, r_n \in R$ are such that $\sum_{i=1}^n r_i e_i = 0$, then it follows easily from the facts that $Re_1 \cap F' = 0$ and $(e_i)_{i=2}^n$ is a base for $F'$ that $r_i = 0$ for all $i = 1, \ldots, n$.

Finally, we show that $H$ is generated by $a_1 e_1, a_2 e_2, \ldots, a_n e_n$. Since $a_1 e_1 = c_z e_1' = z \in H$, it is clear that $a_i e_i \in H$ for all $i = 1, \ldots, n$. Suppose that $y \in H$, so that there exist $s_1, \ldots, s_n \in R$ such that $\sum_{i=1}^n s_i e_i = y$. Now

$$Rs_1 \subseteq C(y) \subseteq C(z) = Rc_z$$

by choice of $z$, and so $s_1 = tc_z$ for some $t \in R$. Then

$$y - tz = \left( \sum_{i=1}^n s_i e_i \right) - tc_z e_1 = \sum_{i=2}^n s_i e_i \in F' \cap H = H',$$

and so $y - tz \in \sum_{i=2}^{n} Ra_i e_i$. Since $tz = tc_z e_1' = ta_1 e_1$, it follows that $H$ is generated by $a_1 e_1, a_2 e_2, \ldots, a_n e_n$.

This completes the inductive step, and the proof. $\square$

**10.6 COROLLARY.** *Let $F$ be a free module with a finite base over the principal ideal domain $R$. Then each submodule $H$ of $F$ is free, and has rank $H \leq$ rank $F$.*

*Proof.* The claim is clear when $F = 0$, and so we suppose that $F \neq 0$. Set $n =$ rank $F$. By 10.5, there exists a base $(e_i)_{i=1}^{n}$ for $F$ and elements $a_1, \ldots, a_n \in R$ such that

$$Ra_1 \supseteq Ra_2 \supseteq \ldots \supseteq Ra_n$$

and $H$ is generated by $a_1 e_1, \ldots, a_n e_n$.

Some of $a_1, \ldots, a_n$ could be zero: if any are, let $t$ be the least $j \in \mathbb{N}$ (with $1 \leq j \leq n$) such that $a_j = 0$; otherwise, set $t = n + 1$. Observe that $a_t = a_{t+1} = \ldots = a_n = 0$, and that $a_i \neq 0$ for all $i \in \mathbb{N}$ with $i < t$. Thus, $H$ is generated by $a_1 e_1, \ldots, a_{t-1} e_{t-1}$ (with an obvious interpretation when $t = 1$). Now $H$ is free of rank 0 when $t = 1$, and so we consider the case in which $t > 1$, and show that, then, $(a_i e_i)_{i=1}^{t-1}$ is a base for $H$: this will complete the proof.

Let $r_1, \ldots, r_{t-1} \in R$ be such that $\sum_{i=1}^{t-1} r_i a_i e_i = 0$. Since $(e_i)_{i=1}^{n}$ is a base for $F$, it follows that $r_i a_i = 0$ for all $i = 1, \ldots, t-1$. Since $a_i \neq 0$ for each $i = 1, \ldots, t-1$ and $R$ is an integral domain, it follows that $r_i = 0$ for all $i = 1, \ldots, t-1$. Hence $(a_i e_i)_{i=1}^{t-1}$ is a base for $H$, as claimed. $\square$

The result of the next exercise will be used many times in the sequel without comment.

**10.7 ♯EXERCISE.** Let $M$ be a module over the commutative ring $R$, and let $I_1, \ldots, I_n$ be ideals of $R$. Show that

$$M \cong R/I_1 \oplus \cdots \oplus R/I_n$$

if and only if $M$ is the internal direct sum $C_1 \oplus \cdots \oplus C_n$ of cyclic submodules $C_1, \ldots, C_n$ for which $\text{Ann}(C_i) = I_i$ for all $i = 1, \ldots, n$.

**10.8 THEOREM.** *Each finitely generated module $M$ over the principal ideal domain $R$ is a direct sum of cyclic submodules.*

*More precisely, if the $R$-module $M$ can be generated by $n$ elements, then there exist $a_1, \ldots, a_n \in R$ such that*

$$Ra_1 \supseteq Ra_2 \supseteq \ldots \supseteq Ra_n$$

*and*

$$M \cong R/Ra_1 \oplus R/Ra_2 \oplus \cdots \oplus R/Ra_n.$$

*Notes.* (i) This result shows that, if $M$ can be generated by $n$ elements, then $M$ is isomorphic to the (external) direct sum of $n$ cyclic $R$-modules, so that $M$ is the (internal) direct sum of $n$ cyclic submodules. The reader should note that it is possible that some of the $a_i$ might be units, so that some of the $R/Ra_i$ might be zero, and $M$ could be isomorphic to the direct sum of fewer than $n$ cyclic $R$-modules. Of course, if $a_i$ is a unit for an $i \in \mathbb{N}$ with $1 < i \leq n$, then the conditions in the theorem imply that $Ra_1 = \ldots = Ra_i = R$.

(ii) It is also possible that we might have $a_j = a_{j+1} = \ldots = a_n = 0$ for some $j \in \mathbb{N}$ with $1 \leq j \leq n$: then the cyclic summands $R/Ra_j, \ldots, R/Ra_n$ are all isomorphic to $R$ and $M$ has a free direct summand of rank $n - j + 1$.

*Proof.* By 6.57, there exist a free $R$-module $F$ with a base of $n$ elements and an $R$-module epimorphism $f : F \to M$. Let $H := \operatorname{Ker} f$, so that, by the First Isomorphism Theorem for modules 6.33, we have $M \cong F/H$.

By 10.5, there exist a base $(e_i)_{i=1}^n$ for $F$ and elements $a_1, \ldots, a_n \in R$ such that

$$Ra_1 \supseteq Ra_2 \supseteq \ldots \supseteq Ra_n$$

and $H$ is generated by $a_1 e_1, \ldots, a_n e_n$. It is therefore enough for us to show that

$$F/H \cong R/Ra_1 \oplus \cdots \oplus R/Ra_n.$$

However, this can easily be established by means of the ideas of Exercise 10.1: by 6.55, there is a (unique) $R$-homomorphism $g : F \to \bigoplus_{i=1}^n R/Ra_i$ such that

$$g\left(\sum_{i=1}^n r_i e_i\right) = (r_1 + Ra_1, \ldots, r_n + Ra_n) \quad \text{for all } r_1, \ldots, r_n \in R,$$

and it is clear that $g$ is an epimorphism and that $\operatorname{Ker} g = \sum_{i=1}^n Ra_i e_i = H$, so that the desired result follows from the First Isomorphism Theorem for modules 6.33 again. $\square$

Later in the chapter, we shall explore uniqueness aspects of the direct-sum decomposition given in 10.8, but first we are going to establish another decomposition theorem which shows that a finitely generated module $M$ over the PID $R$ can always be expressed as a direct sum of cyclic submodules with prime power annihilators, that is, with annihilators which are powers of prime ideals of $R$. Recall from 3.34 that, since $R$ is a PID,

$$\operatorname{Spec}(R) = \{0\} \cup \{Rp : p \text{ an irreducible element of } R\}.$$

Thus we are going to show that $M$ is isomorphic to a direct sum of cyclic modules of the form $R$ or $R/Rp^n$ where $p$ is an irreducible element of $R$

and $n \in \mathbb{N}$. In order to achieve this result, it is worthwhile for us to recall some facts about ideals of $R$ which were used to motivate the introduction to primary decomposition given at the beginning of Chapter 4. Some of them are involved in the next result.

**10.9** PROPOSITION. *Let $I$ be a non-zero, proper ideal of the principal ideal domain $R$, so that there exists a non-zero, non-unit $a \in R$ such that $I = Ra$. Since $R$ is, by 3.39, a unique factorization domain, there exist $s \in \mathbb{N}$, irreducible elements $p_1, \ldots, p_s \in R$ such that $p_i$ and $p_j$ are not associates whenever $i \neq j$ $(1 \leq i, j \leq s)$, a unit $u$ of $R$, and $t_1, \ldots, t_s \in \mathbb{N}$ such that $a = up_1^{t_1} \ldots p_s^{t_s}$.*

*Since the family $(Rp_i^{t_i})_{i=1}^s$ of ideals of $R$ is pairwise comaximal (see 3.57), it follows from 3.59(ii) that*

$$I = Ra = Rp_1^{t_1} \cap \ldots \cap Rp_s^{t_s}.$$

*The $R$-homomorphism*

$$f : R \longrightarrow R/Rp_1^{t_1} \oplus \cdots \oplus R/Rp_s^{t_s} =: D$$

*for which $f(r) = (r + Rp_1^{t_1}, \ldots, r + Rp_s^{t_s})$ for all $r \in R$ is epimorphic with kernel $I$, so that*

$$R/Ra \cong R/Rp_1^{t_1} \oplus \cdots \oplus R/Rp_s^{t_s}.$$

*Proof.* The fact that $I = Ra = \bigcap_{i=1}^s Rp_i^{t_i}$ was explained in the introductory remarks at the beginning of Chapter 4, and it is clear from this that $\operatorname{Ker} f = I$.

To prove that $f$ is surjective, it is enough, by 9.17, to show that $f_M : R_M \rightarrow D_M$ is surjective for each maximal ideal $M$ of $R$. Since $R$ is not a field by our assumptions, it follows from 3.34 that $M = Rq$ for some irreducible element $q \in R$. Consider $i \in \mathbb{N}$ with $1 \leq i \leq s$: if $q$ and $p_i$ are not associates, then $Rp_i^{t_i} \not\subseteq Rq = M$, and so $R/Rp_i^{t_i}$ is annihilated by an element of $R \setminus M$.

Two cases arise, according as $M = Rq$ is or is not equal to any of $Rp_1, \ldots, Rp_s$. When $M$ is different from all these ideals, then it is immediate from the preceding paragraph that $D_M = 0$, so that $f_M$ must be surjective. On the other hand, when $M = Rp_i$ for some $1 \leq i \leq s$, then we can deduce, again from the preceding paragraph, that, for all $r_1, \ldots, r_s \in R$ and $u \in R \setminus M$,

$$(r_1 + Rp_1^{t_1}, \ldots, r_s + Rp_s^{t_s}) - f(r_i)$$

is annihilated by an element of $R \setminus M$ (because its $i$-th component is 0), so that, in $D_M$,

$$\frac{(r_1 + Rp_1^{t_1}, \ldots, r_s + Rp_s^{t_s})}{u} = \frac{f(r_i)}{u} \in \operatorname{Im} f_M.$$

Thus $f_M$ is surjective for all maximal ideals $M$ of $R$, and so $f$ is surjective.

Lastly, the final claim follows from the First Isomorphism Theorem for modules 6.33. □

We could have appealed to Exercise 3.60 to show that the homomorphism $f$ in Proposition 10.9 is surjective. However, the above proof was chosen to illustrate the technique of localization.

**10.10** ♯EXERCISE. Let $R$ be a commutative ring, let $h \in \mathbb{N}$ and, for each $i = 1, \ldots, h$, let $(n_i \in \mathbb{N}$ and$)$ $G_{i1}, \ldots, G_{in_i}$ be $R$-modules and set $M_i = \bigoplus_{j=1}^{n_i} G_{ij}$. Show that

$$\bigoplus_{i=1}^{h} M_i \cong G_{11} \oplus \cdots \oplus G_{1n_1} \oplus G_{21} \oplus \cdots \oplus G_{h-1\,n_{h-1}} \oplus G_{h1} \oplus \cdots \oplus G_{hn_h}.$$

Proposition 10.9 now enables us to show that each finitely generated module over a PID is a direct sum of cyclic submodules with prime power annihilators.

**10.11** COROLLARY. *Let $M$ be a finitely generated module over the principal ideal domain $R$. Then $M$ is a direct sum of cyclic submodules with prime power annihilators. More precisely, there exist $h, m \in \mathbb{N}_0$, irreducible elements $p_1, \ldots, p_m \in R$, positive integers $t_1, \ldots, t_m$ and $R$-modules $R_1, \ldots, R_h$ all equal to $R$ such that*

$$M \cong R/Rp_1^{t_1} \oplus \cdots \oplus R/Rp_m^{t_m} \oplus R_1 \oplus \cdots \oplus R_h.$$

*Note.* In the above statement, we do not demand that $p_i$ and $p_j$ are not associates whenever $i \neq j$ $(1 \leq i, j \leq m)$.

*Proof.* By 10.8, there exist $a_1, \ldots, a_n \in R$ such that

$$M \cong R/Ra_1 \oplus R/Ra_2 \oplus \cdots \oplus R/Ra_n.$$

If some $a_j = 0$, then $R/Ra_j \cong R$; if some $a_k$ is a unit of $R$, then $R/Ra_k = 0$; hence the result follows immediately from 10.9 and 10.10. □

Our next major aims in this chapter are the establishment of certain uniqueness properties of certain types of direct-sum decompositions of finitely generated modules over principal ideal domains: we shall concentrate on decompositions which come from 10.8 and 10.11. An illuminating approach to the uniqueness properties can be achieved by use of some functorial ideas in commutative algebra, and this seems a good point at which to introduce such ideas.

**10.12** DEFINITIONS. Let $R$ and $S$ be commutative rings. We shall say that $T$ *is a covariant functor from R-modules to S-modules* precisely when $T$ is a rule which associates to each $R$-module $M$ an $S$-module $T(M)$ and to each homomorphism $f : M \to G$ of $R$-modules an $S$-homomorphism $T(f) : T(M) \to T(G)$ in such a way that the following axioms are satisfied:

(i) whenever $f : M \to G$ and $g : G \to H$ are homomorphisms of $R$-modules, then

$$T(g \circ f) = T(g) \circ T(f) : T(M) \longrightarrow T(H);$$

(ii) for every $R$-module $M$,

$$T(\mathrm{Id}_M) = \mathrm{Id}_{T(M)} : T(M) \longrightarrow T(M).$$

There is also a concept of contravariant functor, which has a similar definition except that it 'reverses the directions of homomorphisms'. To be precise, a *contravariant functor from R-modules to S-modules* is a rule $T'$ which associates to each $R$-module $M$ an $S$-module $T'(M)$ and to each homomorphism $f : M \to G$ of $R$-modules an $S$-homomorphism

$$T'(f) : T'(G) \to T'(M)$$

in such a way that the following axioms are satisfied:

(i) whenever $f : M \to G$ and $g : G \to H$ are homomorphisms of $R$-modules, then

$$T'(g \circ f) = T'(f) \circ T'(g) : T'(H) \longrightarrow T'(M);$$

(ii) for every $R$-module $M$,

$$T'(\mathrm{Id}_M) = \mathrm{Id}_{T'(M)} : T'(M) \longrightarrow T'(M).$$

Furthermore, such a functor $T$ (either covariant or contravariant) is said to be *additive* precisely when the following condition is satisfied: whenever $M$ and $G$ are $R$-modules and $f, f' : M \to G$ are $R$-homomorphisms, then $T(f + f') = T(f) + T(f')$. (The sum of homomorphisms such as $f$ and $f'$ was defined in 6.27.)

The reader should be warned that there is much more to the theory of functors than is suggested by the above Definition 10.12: a proper treatment of the subject would need to discuss the abstract concept of 'category' (the collection of all modules over a commutative ring $R$ and the collection of all $R$-homomorphisms between them provide one example of a category), and also to consider functors of several variables. However, this is an introductory book about commutative algebra and is not intended to be a book about homological algebra, and so we shall content ourselves here with only a few ideas about functors.

**10.13** LEMMA. *Let $R$ and $S$ be commutative rings, and let $T$ be a (co-variant or contravariant) functor from $R$-modules to $S$-modules.*

*(i) If $f : M \to G$ is an isomorphism of $R$-modules, then $T(f)$ is an $S$-isomorphism, and $T(f)^{-1} = T(f^{-1})$.*

*(ii) Assume, in addition, that $T$ is additive. If $z : M \to G$ denotes the zero homomorphism, then $T(z)$ is zero too. Also $T(0) = 0$ (where 0 on the left-hand side of this equation stands for the zero $R$-module).*

*Proof.* We provide a proof for the case in which $T$ is contravariant: the other case is similar, and is left as an exercise for the reader.

(i) We have $f^{-1} \circ f = \mathrm{Id}_M$ and $f \circ f^{-1} = \mathrm{Id}_G$. It therefore follows from the axioms for a functor in 10.12 that

$$T(f) \circ T(f^{-1}) = T(f^{-1} \circ f) = T(\mathrm{Id}_M) = \mathrm{Id}_{T(M)}$$

and $T(f^{-1}) \circ T(f) = \mathrm{Id}_{T(G)}$ similarly. Hence $T(f)$ and $T(f^{-1})$ are inverses of each other, and so both are isomorphisms.

(ii) Since $z + z = z$, it follows from the fact that $T$ is additive that $T(z) + T(z) = T(z + z) = T(z)$, from which we see that $T(z)(g') = 0$ for all $g' \in T(G)$.

Now consider the zero $R$-module 0: for it, we have $\mathrm{Id}_0 = 0$, that is, the identity homomorphism from 0 to itself coincides with the zero homomorphism. Therefore, by the preceding paragraph and the second axiom for a functor, $\mathrm{Id}_{T(0)} = T(\mathrm{Id}_0)$ is the zero homomorphism from $T(0)$ to itself, and so $y = 0$ for all $y \in T(0)$. $\square$

**10.14** ♯EXERCISE. Complete the proof of 10.13 by establishing the results in the case when the functor $T$ is covariant.

It is time that we had some examples of functors.

**10.15** EXAMPLE. Let $S$ be a multiplicatively closed subset of the commutative ring $R$. Then it follows from 9.7 and 9.8 that $S^{-1}$ is an additive covariant functor from $R$-modules to $S^{-1}R$-modules.

**10.16** ♯EXERCISE. Let $R$ be an integral domain. Show that, for each $R$-module $M$,

$$\tau(M) := \{m \in M : \text{ there exists } r \in R \setminus \{0\} \text{ such that } rm = 0\}$$

is a submodule of $M$. Let $f : M \to G$ be a homomorphism of $R$-modules. Show that $f(\tau(M)) \subseteq \tau(G)$, and define $\tau(f) : \tau(M) \to \tau(G)$ by $\tau(f)(m) = f(m)$ for all $m \in \tau(M)$ (so that $\tau(f)$ is, essentially, the restriction of $f$ to $\tau(M)$).

Show that, with these assignments, $\tau$ becomes an additive covariant functor from $R$-modules to $R$-modules. We call $\tau$ the *torsion functor*.

(i) An $R$-module $M$ is said to be *torsion-free* precisely when $\tau(M) = 0$. Show that, for each $R$-module $G$, the module $G/\tau(G)$ is torsion-free.

(ii) Show also that, if $(G_\lambda)_{\lambda \in \Lambda}$ is a non-empty family of $R$-modules, then

$$\tau\left(\bigoplus_{\lambda \in \Lambda} G_\lambda\right) = \bigoplus_{\lambda \in \Lambda} \tau(G_\lambda).$$

**10.17** ♯EXERCISE. Let $R$ be a commutative Noetherian ring and let $I$ be an ideal of $R$. For each $R$-module $M$, let

$$\Gamma_I(M) = \{m \in M : \text{ there exists } n \in \mathbb{N} \text{ such that } I^n \subseteq (0 : m)\}$$
$$= \bigcup_{n \in \mathbb{N}} (0 :_M I^n).$$

Let $f : M \to G$ be a homomorphism of $R$-modules. Show that $f(\Gamma_I(M)) \subseteq \Gamma_I(G)$, and define $\Gamma_I(f) : \Gamma_I(M) \to \Gamma_I(G)$ by $\Gamma_I(f)(m) = f(m)$ for all $m \in \Gamma_I(M)$.

Show that, with these assignments, $\Gamma_I$ becomes an additive covariant functor from $R$-modules to $R$-modules: we call it the *I-torsion functor*.

(i) Show that $\Gamma_I(M/\Gamma_I(M)) = 0$ for every $R$-module $M$.

(ii) Show also that, if $(G_\lambda)_{\lambda \in \Lambda}$ is a non-empty family of $R$-modules, then

$$\Gamma_I\left(\bigoplus_{\lambda \in \Lambda} G_\lambda\right) = \bigoplus_{\lambda \in \Lambda} \Gamma_I(G_\lambda).$$

**10.18** ♯EXERCISE. Let $p, q$ be irreducible elements of the principal ideal domain $R$, and assume that $p$ and $q$ are not associates. Let $n \in \mathbb{N}$. Show that, with the notation of 10.17 above, $\Gamma_{Rp}(R/Rq^n) = 0$.

**10.19** ♯EXERCISE. Let $R$ be commutative ring and let $I$ be an ideal of $R$. For each $R$-module $M$, let $T(M) = (0 :_M I)$. Show that, whenever $f : M \to G$ is a homomorphism of $R$-modules, then $f(T(M)) \subseteq T(G)$, and define $T(f) : T(M) \to T(G)$ by $T(f)(m) = f(m)$ for all $m \in T(M)$.

Show that $T$ is an additive covariant functor from $R$-modules to $R$-modules. Show also that, if $(G_\lambda)_{\lambda \in \Lambda}$ is a non-empty family of $R$-modules, then $T\left(\bigoplus_{\lambda \in \Lambda} G_\lambda\right) = \bigoplus_{\lambda \in \Lambda} T(G_\lambda)$.

**10.20** ♯EXERCISE. (i) Let $a, b$ be non-zero elements in the integral domain $R$. Show that $Rb/Rab \cong R/Ra$. Show also that

$$(0 :_{R/Rab} Ra) = Rb/Rab.$$

(ii) Now let $p$ be an irreducible element of the principal ideal domain $R$, and let $n \in \mathbf{N}$. Show that $\ell_R(R/Rp^n) = n$.

**10.21** ♯EXERCISE. Let $R$ be a principal ideal domain, let $r$ be a non-zero, non-unit element of $R$, and let $p$ be an irreducible element of $R$. Show that

$$(0 :_{R/Rr} Rp) \cong \begin{cases} R/Rp & \text{if } Rp \supseteq Rr, \\ 0 & \text{if } Rp \not\supseteq Rr. \end{cases}$$

(You might find Exercise 10.20 helpful.)

**10.22** EXERCISE. Let $R$ and $S$ be commutative rings, and let $T$ be a covariant (respectively contravariant) additive functor from $R$-modules to $S$-modules. We say that $T$ is *left exact* precisely when the following condition is satisfied: whenever

$$0 \longrightarrow L \xrightarrow{f} M \xrightarrow{g} N \longrightarrow 0$$

is an exact sequence of $R$-modules and $R$-homomorphisms, then the induced sequence

$$0 \longrightarrow T(L) \xrightarrow{T(f)} T(M) \xrightarrow{T(g)} T(N)$$

(respectively

$$0 \longrightarrow T(N) \xrightarrow{T(g)} T(M) \xrightarrow{T(f)} T(L)\,)$$

of $S$-modules and $S$-homomorphisms is also exact.

(i) Let $I$ be an ideal of $R$. Show that the functor $(0 :_{(\bullet)} I)$, that is, the functor of 10.19, is left exact. Show also that, when $R$ is Noetherian, the functor $\Gamma_I$ of 10.17 is left exact.

(ii) Suppose that $R$ is an integral domain. Show that the torsion functor $\tau$ of 10.16 is left exact.

**10.23** PROPOSITION. *Let $R$ and $S$ be commutative rings and let $T$ be an additive (covariant or contravariant) functor from $R$-modules to $S$-modules. Let $n \in \mathbf{N}$ and let $G_1, \ldots, G_n$ be $R$-modules. Then $T(\bigoplus_{i=1}^n G_i) \cong \bigoplus_{i=1}^n T(G_i)$ as $S$-modules.*

*Proof.* This time we deal with the case where $T$ is covariant and leave the contravariant case as an exercise for the reader.

Set $G = \bigoplus_{i=1}^n G_i$, and, for each $i = 1, \ldots, n$, let $p_i : G \to G_i$ be the canonical projection and let $q_i : G_i \to G$ be the canonical injection. Then, by 6.47, we have, for $1 \leq i, j \leq n$,

$$p_i \circ q_j = \begin{cases} \mathrm{Id}_{G_i} & \text{for } i = j, \\ 0 & \text{for } i \neq j, \end{cases}$$

and $\sum_{i=1}^{n} q_i \circ p_i = \mathrm{Id}_G$. Now apply $T$: it follows from 10.13 and the properties of additive functors that, for $1 \leq i, j \leq n$,

$$T(p_i) \circ T(q_j) = \begin{cases} \mathrm{Id}_{T(G_i)} & \text{for } i = j, \\ 0 & \text{for } i \neq j, \end{cases}$$

and

$$\sum_{i=1}^{n} T(q_i) \circ T(p_i) = T\left(\sum_{i=1}^{n} q_i \circ p_i\right) = T(\mathrm{Id}_G) = \mathrm{Id}_{T(G)} .$$

It therefore follows from 6.48(ii) that $T(G) = T(\bigoplus_{i=1}^{n} G_i) \cong \bigoplus_{i=1}^{n} T(G_i)$, as required. $\square$

**10.24** EXERCISE. Complete the proof of 10.23 by establishing the result in the case when the functor $T$ is contravariant.

**10.25** ♯EXERCISE. Let $R$ and $S$ be non-trivial commutative rings and let $T$ be an additive (covariant or contravariant) functor from $R$-modules to $S$-modules. Suppose that, as $S$-modules, $T(R) \cong S$. Let $F$ be a finitely generated free $R$-module of rank $n$. Show that $T(F)$ is a finitely generated free $S$-module of rank $n$.

**10.26** DEFINITION and REMARKS. Let $R$ be an integral domain, and let $M$ be a finitely generated $R$-module. Set $S := R \setminus \{0\}$, a multiplicatively closed subset of $R$, and note that $K := S^{-1}R$ is just the field of fractions of $R$, by 5.5. Now $S^{-1}M$ is a finitely generated module over $S^{-1}R$, that is, a finite-dimensional vector space over $K$. We define the *torsion-free rank*, or simply the *rank, of* $M$ to be $\mathrm{vdim}_K S^{-1}M$, and we denote this by rank $M$ (or $\mathrm{rank}_R M$).

This does not conflict with our earlier use of the word 'rank' in connection with free modules having finite bases: if $F$ is a free $R$-module having a base with $n$ elements, then it is immediate from 10.15 and 10.25 that $S^{-1}F$ is a free $K$-module of rank $n$, so that the torsion-free rank of $F$ in the sense of the preceding paragraph is also $n$.

It should be clear to the reader that if $M$ and $M'$ are isomorphic finitely generated $R$-modules, then rank $M = $ rank $M'$.

**10.27** ♯EXERCISE. Let $M, G_1, \ldots, G_n$ be modules over the commutative ring $R$, and suppose that $M \cong \bigoplus_{i=1}^{n} G_i$. Let $I$ be an ideal of $R$. Show that

$$IM \cong I\left(\bigoplus_{i=1}^{n} G_i\right) = \bigoplus_{i=1}^{n} IG_i.$$

We are now in a position to use our functorial ideas to discuss uniqueness aspects of certain types of decomposition (into direct sums of cyclic submodules) of a finitely generated module over a PID. The following terminology will be helpful.

**10.28** DEFINITION. Let $R$ be a principal ideal domain. We say that $(p_i)_{i=1}^n$ is a *family of pairwise non-associate irreducible elements of $R$* precisely when $p_1, \ldots, p_n$ are irreducible elements of $R$ such that $p_i$ and $p_j$ are not associates for $i \neq j$ $(1 \leq i, j \leq n)$.

**10.29** EXERCISE. Let $R$ be a principal ideal domain, and let $(p_i)_{i=1}^n$ be a family of irreducible elements of $R$. Show that $(p_i)_{i=1}^n$ is a family of pairwise non-associate irreducible elements of $R$ if and only if the family of ideals $(Rp_i)_{i=1}^n$ of $R$ is pairwise comaximal (see 3.57).

Now let $R$ be a PID and let $M$ be a non-zero, finitely generated $R$-module. In 10.8, we showed that there exist $a_1, \ldots, a_n \in R$ such that

$$Ra_1 \supseteq Ra_2 \supseteq \ldots \supseteq Ra_n$$

and

$$M \cong R/Ra_1 \oplus R/Ra_2 \oplus \cdots \oplus R/Ra_n.$$

We also saw in 10.11 that $M$ is isomorphic to a direct sum of finitely many cyclic $R$-modules with prime power annihilators. (Of course, one must remember that the zero ideal 0 of $R$ is prime, and so a non-zero free cyclic summand will be isomorphic to $R$ and will have annihilator which is a prime power!) Our next step is to show that the number of non-zero free cyclic summands in a direct decomposition of either of the above types is uniquely determined by $M$ and is independent of the particular direct decomposition chosen. In fact, this number turns out to be just the torsion-free rank of $M$, and so is certainly an invariant of $M$!

**10.30** THEOREM. *Let $M$ be a non-zero, finitely generated module over the principal ideal domain $R$. Suppose that $h, t \in \mathbf{N}_0$ and there exist $b_1, \ldots, b_h \in R \setminus \{0\}$ such that*

$$M \cong R/Rb_1 \oplus \cdots \oplus R/Rb_h \oplus R_1 \oplus \cdots \oplus R_t,$$

*where $R_1 = \ldots = R_t = R$. (The obvious interpretation is to be made if either $h = 0$ or $t = 0$.) Then*
   (i) $\tau(M) \cong R/Rb_1 \oplus \cdots \oplus R/Rb_h$, *and*
   (ii) $t = \operatorname{rank} M$.

*Proof.* By 10.16 and 10.13,

$$\tau(M) \cong \tau(R/Rb_1) \oplus \cdots \oplus \tau(R/Rb_h) \oplus \tau(R_1) \oplus \cdots \oplus \tau(R_t).$$

But $b_i(R/Rb_i) = 0$ and $b_i \neq 0$, so that $\tau(R/Rb_i) = R/Rb_i$, for each $i = 1, \ldots, h$. Also, $\tau(R) = 0$, and so part (i) is proved.

Let $S := R \setminus \{0\}$, and let $K := S^{-1}R$, the field of fractions of $R$. Note that $S^{-1}(R/Rb_i) = 0$ for each $i = 1, \ldots, h$, by 9.6(i). Hence, by 10.15 and 10.23, there is an isomorphism of $K$-spaces

$$S^{-1}M \cong S^{-1}R_1 \oplus \cdots \oplus S^{-1}R_t,$$

so that $t = \mathrm{vdim}_K S^{-1}M = \mathrm{rank}_R M$. $\square$

Of course, given an ideal $I$ of the PID $R$, and a generator $a$ of $I$, any associate of $a$ is also a generator of $I$. Thus we cannot in general talk about 'the' generator of $I$, as this could be ambiguous. For example, the ideal $27\mathbf{Z}$ of $\mathbf{Z}$ is generated by 27 and by $-27$. Thus, in our discussion of uniqueness of decompositions of a finitely generated $R$-module into direct sums of cyclic submodules, we shall phrase our results in terms of the annihilators of cyclic summands rather than in terms of generators of these ideals.

**10.31** LEMMA. *Let $M$ be a non-zero, finitely generated module over the principal ideal domain $R$. Let*

$$M \cong R_1 \oplus \cdots \oplus R_t \oplus R/Rp_1^{u_{11}} \oplus \cdots \oplus R/Rp_1^{u_{1w(1)}}$$
$$\oplus R/Rp_2^{u_{21}} \oplus \cdots \oplus R/Rp_2^{u_{2w(2)}} \oplus \cdots$$
$$\vdots$$
$$\cdots \oplus R/Rp_n^{u_{n1}} \oplus \cdots \oplus R/Rp_n^{u_{nw(n)}},$$

*where $t, n \in \mathbf{N}_0$, where $R_1 = \ldots = R_t = R$, where $(p_i)_{i=1}^n$ is a family of pairwise non-associate irreducible elements of $R$, and where (for each $i = 1, \ldots, n$) $w(i), u_{i1}, \ldots, u_{iw(i)} \in \mathbf{N}$ are such that*

$$u_{i1} \leq u_{i2} \leq \ldots \leq u_{iw(i)}.$$

*Then $\Gamma_{Rp_i}(M) \cong R/Rp_i^{u_{i1}} \oplus \cdots \oplus R/Rp_i^{u_{iw(i)}}$ (for each $i = 1, \ldots, n$). Furthermore, if $p$ is an irreducible element of $R$ which is not an associate of any of $p_1, \ldots, p_n$, then $\Gamma_{Rp}(M) = 0$.*

*Proof.* Observe that, for any irreducible element $q$ of $R$, we have $\Gamma_{Rq}(R) = 0$. Also $\Gamma_{Rq}(R/Rp_i^u) = 0$ for all $u \in \mathbf{N}$ if $q$ and $p_i$ are not associates, by 10.18. All the claims now follow immediately from Exercise 10.17 and Lemma 10.13. □

Lemma 10.31 will enable us, when presented with a decomposition of $M$ into a direct sum of cyclic submodules with prime power annihilators, to concentrate attention on those cyclic summands of the decomposition whose annihilators are powers of a *fixed* non-zero prime $Rp$: we can just achieve this by application of the functor $\Gamma_{Rp}$. A corollary of the next theorem will show that this is a profitable course of action. The next theorem is, in fact, one of the uniqueness results alluded to earlier.

**10.32** THEOREM. *Let $M$ be a non-zero, finitely generated module over the principal ideal domain $R$. By 10.8, there exist $a_1, \ldots, a_n \in R$ such that*

$$R \supset Ra_1 \supseteq Ra_2 \supseteq \ldots \supseteq Ra_n$$

*and*

$$M \cong R/Ra_1 \oplus R/Ra_2 \oplus \cdots \oplus R/Ra_n.$$

*Then the positive integer $n$ and the family $(Ra_i)_{i=1}^{n}$ of ideals of $R$ are invariants of $M$; in other words, if $b_1, \ldots, b_m \in R$ are such that*

$$R \supset Rb_1 \supseteq Rb_2 \supseteq \ldots \supseteq Rb_m$$

*and*

$$M \cong R/Rb_1 \oplus R/Rb_2 \oplus \cdots \oplus R/Rb_m,$$

*then $n = m$ and $Ra_i = Rb_i$ for all $i = 1, \ldots, n$.*

*Proof.* If $Ra_i = 0$ for all $i = 1, \ldots, n$, then $M$ is free of rank $n$, by 6.58 and 6.53; also $\tau(M) = 0$ by 10.30. It therefore follows from the same result that $Rb_i = 0$ for all $i = 1, \ldots, m$, so that $m = \operatorname{rank} M = n$. Hence the result has been proved in this case. We can therefore assume that $Ra_1 \supset 0$.

Then there exist $v \in \mathbf{N}$ such that $1 \leq v \leq n$ and

$$Ra_1 \supseteq \ldots \supseteq Ra_v \supset 0 = Ra_{v+1} = \ldots = Ra_n.$$

(An obvious interpretation is to be made if $v = n$.) We can make a similar definition for the second direct-sum decomposition: there exists $u \in \mathbf{N}$ such that $1 \leq u \leq m$ and

$$Rb_1 \supseteq \ldots \supseteq Rb_u \supset 0 = Rb_{u+1} = \ldots = Rb_m.$$

By 10.30, $n - v = \operatorname{rank} M = m - u$, and

$$R/Ra_1 \oplus \cdots \oplus R/Ra_v \cong \tau(M) \cong R/Rb_1 \oplus \cdots \oplus R/Rb_u.$$

The general result will therefore follow if we prove the result in the special case in which $v = n$ (when $u = m$ too). Thus we assume for the remainder of this proof that $Ra_n \supset 0$ and $Rb_m \supset 0$.

In this case it is clear from 7.46 that $\ell_R(M)$ is finite, and we argue by induction on this length. When $\ell_R(M) = 1$, it follows from 7.41 and 6.48(i) that $n = m = 1$, so that $R/Ra_1 \cong R/Rb_1$ and

$$Ra_1 = \mathrm{Ann}_R(R/Ra_1) = \mathrm{Ann}_R(R/Rb_1) = Rb_1.$$

So suppose, inductively, that $\ell_R(M) > 1$ and that the result has been proved for non-zero $R$-modules of smaller finite length. We can assume, for the sake of argument, that $n \geq m$. Since $Ra_1$ is a proper, non-zero ideal of $R$, there exists an irreducible element $p$ of $R$ which is a factor of $a_1$. For a non-zero, non-unit element $r \in R$ we have, by 10.21,

$$(0 :_{R/Rr} Rp) \cong \begin{cases} R/Rp & \text{if } Rp \supseteq Rr, \\ 0 & \text{if } Rp \not\supseteq Rr. \end{cases}$$

Hence, by 10.19 and 10.13,

$$(0 :_M Rp) \cong \bigoplus_{i=1}^{n} (0 :_{R/Ra_i} Rp)$$

is isomorphic to the direct sum of $n$ copies of the $R$-module $R/Rp$, and so $\ell_R(0 :_M Rp) = n$. However, it similarly follows from the fact that

$$(0 :_M Rp) \cong \bigoplus_{i=1}^{m} (0 :_{R/Rb_i} Rp)$$

that $\ell_R(0 :_M Rp) = w$, where $w$ is the number of integers $i$ between 1 and $m$ for which $p$ is a factor of $b_i$. Hence $n = w \leq m \leq n$, and so $n = m = w$ and $p$ is a factor of $b_i$ for all $i = 1, \ldots, n = m$.

Thus, for each $i = 1, \ldots, n$, there exist $c_i, d_i \in R$ such that $a_i = pc_i$ and $b_i = pd_i$, and it follows that

$$Rp(R/Ra_i) = Rp/Ra_i \cong R/Rc_i \quad \text{and} \quad Rp(R/Rb_i) \cong R/Rd_i$$

by 10.20. We now use 10.27 to see that

$$pM \cong R/Rc_1 \oplus \cdots \oplus R/Rc_n \cong R/Rd_1 \oplus \cdots \oplus R/Rd_n$$

with

$$Rc_1 \supseteq \ldots \supseteq Rc_n \supset 0 \quad \text{and} \quad Rd_1 \supseteq \ldots \supseteq Rd_n \supset 0.$$

It should also be clear to the reader that $pM$ has smaller length than $M$. Let $h$ (respectively $k$) be the greatest integer $i$ between 0 and $n$ for which $Rc_i = R$ (respectively $Rd_i = R$), with the understanding that $h = 0$ (respectively $k = 0$) if there is no such integer.

Then $h = n$ if and only if $pM = 0$, and this is the case if and only if $k = n$. When this is the case, $c_1, \ldots, c_n, d_1, \ldots, d_n$ are all units of $R$, and $Ra_i = Rb_i = Rp$ for all $i = 1, \ldots, n$, so that the claim is proved in this case. When $h < n$, so that $k < n$ too, we can apply the inductive hypothesis to the two direct-sum decompositions for $pM$ displayed in the preceding paragraph to see that

$$n - h = n - k \quad \text{and} \quad Rc_i = Rd_i \quad \text{for all } i = h + 1, \ldots, n.$$

It follows easily from this that $Ra_i = Rb_i$ for all $i = 1, \ldots, n$.

This completes the inductive step, and so the theorem is proved by induction. $\square$

**10.33** EXERCISE. Let the situation be as in 10.32. Find $\mathrm{Ann}_R(\tau(M))$.

**10.34** COROLLARY. *Let $p$ be an irreducible element of the principal ideal domain $R$. Suppose that $(m, n \in \mathbf{N}$ and$)$ $u_1, \ldots, u_m, v_1, \ldots, v_n \in \mathbf{N}$ are such that*

$$u_1 \le u_2 \le \ldots \le u_m \quad \text{and} \quad v_1 \le v_2 \le \ldots \le v_n$$

*and*

$$R/Rp^{u_1} \oplus \cdots \oplus R/Rp^{u_m} \cong R/Rp^{v_1} \oplus \cdots \oplus R/Rp^{v_n}.$$

*Then $m = n$ and $u_i = v_i$ for all $i = 1, \ldots, n$.*

*Proof.* Note that the hypotheses imply that

$$R \supset Rp^{u_1} \supseteq Rp^{u_2} \supseteq \ldots \supseteq Rp^{u_m} \quad \text{and} \quad R \supset Rp^{v_1} \supseteq Rp^{v_2} \supseteq \ldots \supseteq Rp^{v_n}.$$

The claim is therefore an immediate consequence of Theorem 10.32. $\square$

Theorem 10.32 is one uniqueness theorem for decompositions of finitely generated modules over a principal ideal domain as direct sums of cyclic submodules. We are now in a position to prove a second such theorem, this one being concerned with decompositions using cyclic modules having prime power annihilators.

**10.35** THEOREM. *Let $M$ be a non-zero, finitely generated module over the principal ideal domain $R$. By 10.11, $M$ is isomorphic to a direct sum of cyclic $R$-modules with prime power annihilators. Suppose that*

$$M \cong R_1 \oplus \cdots \oplus R_t \oplus R/Rp_1^{u_{11}} \oplus \cdots \oplus R/Rp_1^{u_{1w(1)}}$$
$$\oplus R/Rp_2^{u_{21}} \oplus \cdots \oplus R/Rp_2^{u_{2w(2)}} \oplus \cdots$$
$$\vdots$$
$$\cdots \oplus R/Rp_n^{u_{n1}} \oplus \cdots \oplus R/Rp_n^{u_{nw(n)}},$$

*where $t, n \in \mathbb{N}_0$, where $R_1 = \ldots = R_t = R$, where $(p_i)_{i=1}^n$ is a family of pairwise non-associate irreducible elements of $R$, and where (for each $i = 1, \ldots, n$) $w(i), u_{i1}, \ldots, u_{iw(i)} \in \mathbb{N}$ are such that*

$$u_{i1} \leq u_{i2} \leq \ldots \leq u_{iw(i)}.$$

*Suppose also that*

$$M \cong R_1 \oplus \cdots \oplus R_s \oplus R/Rq_1^{v_{11}} \oplus \cdots \oplus R/Rq_1^{v_{1y(1)}}$$
$$\oplus R/Rq_2^{v_{21}} \oplus \cdots \oplus R/Rq_2^{v_{2y(2)}} \oplus \cdots$$
$$\vdots$$
$$\cdots \oplus R/Rq_m^{v_{m1}} \oplus \cdots \oplus R/Rq_m^{v_{my(m)}},$$

*where $s, m \in \mathbb{N}_0$, where $R_1 = \ldots = R_s = R$, where $(q_i)_{i=1}^m$ is a family of pairwise non-associate irreducible elements of $R$, and where (for each $i = 1, \ldots, m$) $y(i), v_{i1}, \ldots, v_{iy(i)} \in \mathbb{N}$ are such that*

$$v_{i1} \leq v_{i2} \leq \ldots \leq v_{iy(i)}.$$

*Then $t = s = \operatorname{rank}_R M$ and $n = m$. Moreover, after a suitable reordering of $q_1, \ldots, q_n$ if necessary, the families $(Rp_i)_{i=1}^n$ and $(Rq_i)_{i=1}^n$ of prime ideals of $R$ are equal. Furthermore, when this reordering has been effected, we have $w(i) = y(i)$ for all $i = 1, \ldots, n$, and, for each such $i$, the families $(u_{ij})_{j=1}^{w(i)}$ and $(v_{ij})_{j=1}^{w(i)}$ of positive integers are equal.*

*Proof.* By 10.30, we have $t = \operatorname{rank}_R M = s$; also, we can use 10.31 and 3.34 to see that

$$\{Rp_1, \ldots, Rp_n\} = \{P \in \operatorname{Spec}(R) : P \neq 0 \text{ and } \Gamma_P(M) \neq 0\}.$$

It follows that $n = m$ and we can (if necessary) reorder $q_1, \ldots, q_n$ in such a way that $(Rp_i)_{i=1}^n = (Rq_i)_{i=1}^n$.

Now consider an integer $i$ such that $1 \leq i \leq n$. By 10.17, 10.13 and 10.31, we have

$$R/Rp_i^{u_{i1}} \oplus \cdots \oplus R/Rp_i^{u_{iw(i)}} \cong \Gamma_{Rp_i}(M) \cong R/Rp_i^{v_{i1}} \oplus \cdots \oplus R/Rp_i^{v_{iy(i)}}.$$

(We have here made use of the fact that, since $Rp_i = Rq_i$, we must also have $Rp_i^h = (Rp_i)^h = Rq_i^h$ for all $h \in \mathbf{N}$. Note that we are not claiming that $p_i = q_i$, but only that they generate the same ideal, that is, that they are associates.) In view of the hypotheses on the $u_{ij}$ and $v_{ij}$, it follows from 10.34 that $w(i) = y(i)$ and $u_{ij} = v_{ij}$ for all $j = 1, \ldots, w(i)$. $\square$

**10.36** REMARK. The reader might find it helpful if we make a few comments about what has been achieved in Theorem 10.35. With the notation of that theorem, note that $t = \operatorname{rank}_R M$, that the family $(Rp_i)_{i=1}^n$ of $n$ distinct maximal ideals of $R$, and the families $(u_{ij})_{j=1}^{w(i)}$ of positive integers such that $u_{i1} \leq u_{i2} \leq \ldots \leq u_{iw(i)}$ $(1 \leq i \leq n)$, are all uniquely determined by $M$, and that once we know these invariants we can completely describe $M$ up to $R$-isomorphism.

**10.37** EXERCISE. Let the situation be as in 10.35 and 10.36. Show that, if $\tau(M) \neq 0$, then

$$\operatorname{Ann}_R(\tau(M)) = Rp_1^{u_{1w(1)}} p_2^{u_{2w(2)}} \ldots p_n^{u_{nw(n)}}.$$

One of the most familiar examples of a PID is the ring $\mathbf{Z}$ of integers: by 6.5, the concept of $\mathbf{Z}$-module is exactly the same as the concept of Abelian group. Thus our Theorems 10.32 and 10.35 have consequences for the theory of finitely generated Abelian groups. Before we write down versions of these theorems for the special case in which $R = \mathbf{Z}$, note that the only units of $\mathbf{Z}$ are 1 and $-1$, and that every ideal of $\mathbf{Z}$ has a unique non-negative generator; furthermore, for $a \in \mathbf{N}$, the cyclic $\mathbf{Z}$-module $\mathbf{Z}/a\mathbf{Z}$ is a finite Abelian group of order $a$. These observations mean that the formulations of the next two results are less complicated than those of 10.32 and 10.35 (of which they are, respectively, immediate corollaries).

**10.38** COROLLARY (of 10.8 and 10.32). *Let $G$ be a non-zero, finitely generated Abelian group. Then there is a uniquely determined family $(a_i)_{i=1}^n$ of non-negative integers such that*

$$G \cong \mathbf{Z}/\mathbf{Z}a_1 \oplus \mathbf{Z}/\mathbf{Z}a_2 \oplus \cdots \oplus \mathbf{Z}/\mathbf{Z}a_n,$$

*$a_1 \neq 1$ and $a_i \mid a_{i+1}$ for all $i = 1, \ldots, n-1$.* $\square$

**10.39** THE FUNDAMENTAL THEOREM ON ABELIAN GROUPS. *Let $G$ be a non-zero, finitely generated Abelian group. Then $G$ is isomorphic to a direct sum of free cyclic groups and cyclic groups of prime power orders.*

*In fact, there exist uniquely determined integers $t, n \in \mathbb{N}_0$, a uniquely determined family $(p_i)_{i=1}^{n}$ of prime numbers such that $p_1 < p_2 < \ldots < p_n$ and, for each $i = 1, \ldots, n$, uniquely determined $w(i), u_{i1}, \ldots, u_{iw(i)} \in \mathbb{N}$ such that*

$$u_{i1} \leq u_{i2} \leq \ldots \leq u_{iw(i)},$$

*for which*

$$G \cong \mathbb{Z}_1 \oplus \cdots \oplus \mathbb{Z}_t \oplus \mathbb{Z}/\mathbb{Z}p_1^{u_{11}} \oplus \cdots \oplus \mathbb{Z}/\mathbb{Z}p_1^{u_{1w(1)}}$$
$$\oplus \mathbb{Z}/\mathbb{Z}p_2^{u_{21}} \oplus \cdots \oplus \mathbb{Z}/\mathbb{Z}p_2^{u_{2w(2)}} \oplus \cdots$$
$$\vdots$$
$$\cdots \oplus \mathbb{Z}/\mathbb{Z}p_n^{u_{n1}} \oplus \cdots \oplus \mathbb{Z}/\mathbb{Z}p_n^{u_{nw(n)}},$$

*where $\mathbb{Z}_1 = \ldots = \mathbb{Z}_t = \mathbb{Z}$.* □

**10.40** EXERCISE. Determine the number of distinct isomorphism classes of Abelian groups of order 60; for each such isomorphism class, find a representative which is a direct sum of cyclic groups of prime power orders, and also find a representative which is a direct sum of cyclic groups $\mathbb{Z}/\mathbb{Z}a_i$ satisfying the conditions of Corollary 10.38.

**10.41** FURTHER STEPS. There are other approaches to the fundamental Theorems 10.8, 10.11, 10.32 and 10.35 of this chapter: the interested reader might like to study the matrix-orientated approach of [4, Chapters 7, 8]. The particular approach employed in this book has been selected to try to consolidate and illustrate the use of some of the techniques developed earlier in the book, and because it provided an opportunity to introduce some easy ideas concerning the use of functors in commutative algebra.

It should be mentioned that the few functorial ideas used in this chapter represent (once again!) 'a tip of an iceberg'. In 10.12, we essentially considered functors of one variable from one 'category' of modules to another, but we did this without making the idea of 'category' precise. A proper introduction to functors would need to define 'abstract category', and consider functors of several variables. (Tensor product is an important example of a functor of two variables; another comes from the 'Hom' modules mentioned in 6.60.) The reader should consult texts on homological algebra if he or she wishes to learn more about these topics.

Chapter 11 is mainly concerned with applications of 10.32 and 10.35 to matrix theory.

# Chapter 11

# Canonical forms for square matrices

This is just a short chapter, the aim of which is to indicate to the reader how the direct-sum decomposition theorems of Chapter 10 can be used to derive some basic results about canonical forms for square matrices over fields. It is not the intention to provide here an exhaustive account of the theory of canonical forms, because this is not intended to be a book about linear algebra; but the reader might like to see how the ideas of this book, and in particular those of Chapter 10, can be brought to bear on the theory of canonical forms. None of the material in this chapter is needed in the remainder of the book, so that a reader whose interests are in other areas of commutative algebra can omit this chapter.

Square matrices with entries in a field $K$ are intimately related with endomorphisms of finite-dimensional vector spaces over $K$.

**11.1** DEFINITIONS and REMARKS. Let $M$ be a module over the commutative ring $R$. An *$R$-endomorphism of $M$*, or simply an *endomorphism of $M$*, is just an $R$-homomorphism from $M$ to itself. We denote by $\operatorname{End}_R(M)$ the set of all $R$-endomorphisms of $M$. It is routine to check that $\operatorname{End}_R(M)$ is a ring under the addition defined in 6.27 and 'multiplication' given by composition of mappings: the identity element of this ring is $\operatorname{Id}_M$, the identity mapping of $M$ onto itself, while the zero element of $\operatorname{End}_R(M)$ is the zero homomorphism $0 : M \to M$ defined in 6.27.

The reader should be able to construct easy examples from vector space theory which show that, in general, the ring $\operatorname{End}_R(M)$ need not be commutative.

For each $\psi \in \operatorname{End}_R(M)$ and each $r \in R$, we define $r\psi : M \to M$ by the rule $(r\psi)(m) = r\psi(m)$ for all $m \in M$. It is routine to check that $r\psi$ is again an endomorphism of $M$. Observe that the effect of $r \operatorname{Id}_M$ (for $r \in R$) on an element $m \in M$ is just to multiply $m$ by $r$. Note also that each $\psi \in \operatorname{End}_R(M)$ commutes with $r \operatorname{Id}_M$ for all $r \in R$: in fact, an Abelian group homomorphism $\theta : M \to M$ belongs to $\operatorname{End}_R(M)$ if and only if it commutes with $r \operatorname{Id}_M$ for all $r \in R$.

Denote $\{r \operatorname{Id}_M : r \in R\}$ by $R'$: it is clear that $R'$ is a commutative subring of $\operatorname{End}_R(M)$, and that $\eta : R \to R'$ defined by $\eta(r) = r \operatorname{Id}_M$ for all $r \in R$ is a ring homomorphism. Let $\psi$ be a fixed member of $\operatorname{End}_R(M)$. Then it follows from 1.11 that $R'[\psi]$ is a commutative subring of $\operatorname{End}_R(M)$ and, in fact,

$$R'[\psi] = \left\{ \sum_{i=0}^{t} (r_i \operatorname{Id}_M) \circ \psi^i : t \in \mathbb{N}_0,\ r_0, r_1, \ldots, r_t \in R \right\}$$
$$= \left\{ \sum_{i=0}^{t} r_i \psi^i : t \in \mathbb{N}_0,\ r_0, r_1, \ldots, r_t \in R \right\}.$$

(Of course, $\psi^0 = \operatorname{Id}_M$.)

Next note that, with the notation of the preceding paragraph, it is routine to check that $M$ has the structure of $R'[\psi]$-module with respect to the addition it already possesses (by virtue of its being an $R$-module) and a scalar multiplication of its elements by elements of $R'[\psi]$ given by $\phi.m = \phi(m)$ (the result of application of the mapping $\phi$ to $m$) for all $\phi \in R'[\psi]$ and all $m \in M$.

By 1.13, there is a unique ring homomorphism $\zeta : R[X] \to R'[\psi]$ (where $X$ is an indeterminate) which extends $\eta$ and is such that $\zeta(X) = \psi$. (For $f \in R[X]$, we denote $\zeta(f)$ by $f(\psi)$, for obvious reasons: if $f = \sum_{i=0}^{n} r_i X^i$, then $\zeta(f) = \sum_{i=0}^{n} r_i \psi^i$.) We can now use 6.6 to regard $M$ as an $R[X]$-module by means of $\zeta$.

Let us recapitulate the main consequence of this discussion: given a module $M$ over the commutative ring $R$, and an endomorphism $\psi$ of $M$, we have shown that we can regard $M$ as an $R[X]$-module in such a way that, for $m \in M$, $t \in \mathbb{N}_0$ and $r_0, r_1, \ldots, r_t \in R$,

$$(r_0 + r_1 X + \cdots + r_t X^t)(m) = r_0 m + r_1 \psi(m) + \cdots + r_t \psi^t(m).$$

Let us consider for a moment what 11.1 means for a vector space $V$ over a field $K$, that is, for a $K$-module. Of course, a $K$-endomorphism of $V$ is just a $K$-linear mapping of $V$ into itself: given such a linear mapping $\psi$, our work in 11.1 enables us to regard $V$ as a $K[X]$-module. Now $K[X]$ is a

PID, and in Chapter 10 we proved some important decomposition theorems for finitely generated modules over a PID; when $V$ is finite-dimensional, it will automatically be a finitely-generated $K[X]$-module (because the elements of a $K$-basis for $V$ will generate it as a $K[X]$-module), and so those decomposition theorems can be applied to $V$.

Note that an element of $K[X]$ is a unit of that ring if and only if it is a non-zero element of $K$, and each non-zero ideal of $K[X]$ has a unique *monic* generator. (Recall that a non-zero polynomial of degree $d$ in $K[X]$ is said to be *monic* precisely when its $d$-th coefficient is $1_K$.) If we insist, in our applications of 10.32 and 10.35 to $K[X]$-modules, that generators of relevant ideals be monic or zero, we shall find that simplifications in the statements of the results are possible.

**11.2** REMARKS. Let $V$ be a finite-dimensional vector space over the field $K$, and let $\psi \in \text{End}_K(V)$; regard $V$ as a $K[X]$-module, where $X$ is an indeterminate, in the manner explained in 11.1 using $\psi$, so that $Xv = \psi(v)$ for all $v \in V$.

(i) Note that a subset $U$ of $V$ is a $K[X]$-submodule if and only if it is a $K$-subspace of $V$ and $Xu = \psi(u) \in U$ for all $u \in U$, that is, if and only if $U$ is a $K$-subspace of $V$ which is *invariant under* $\psi$.

(ii) Note that $\text{rank}_{K[X]} V = 0$ and $V$ must have finite length as a $K[X]$-module (by 7.42, for example), and so it follows from 7.46 that $\text{Ann}_{K[X]}(V) \neq 0$, and there is a unique monic polynomial $m_{\psi,V} \in K[X]$ such that $\text{Ann}_{K[X]}(V) = m_{\psi,V} K[X]$. We call $m_{\psi,V}$ the *minimal polynomial of $\psi$ on $V$*.

Observe that $m_{\psi,V}(\psi) = 0$, that if $f \in K[X]$ has the property that $f(\psi) = 0$, then $m_{\psi,V} \mid f$ in $K[X]$, and that $m_{\psi,V}$ is the unique member of least degree in the set $\{f \in K[X] : f \text{ is monic and } f(\psi) = 0\}$. In fact,

$$m_{\psi,V} K[X] = \text{Ann}_{K[X]}(V) = \{f \in K[X] : f(\psi) = 0\}.$$

We shall assume that the reader is familiar with the basic ideas from linear algebra described in Reminder 11.3 below.

**11.3** REMINDERS. Let $K$ be a field and let $h \in \mathbb{N}$. Denote by $M_{h,h}(K)$ the set of all $h \times h$ matrices with entries in $K$. Let $V$ be an $h$-dimensional vector space over $K$, and let $\psi \in \text{End}_K(V)$.

(i) Let $\mathcal{B} = (x_i)_{i=1}^h$ be a basis for $V$. The matrix of (or representing) $\psi$ relative to $\mathcal{B}$ is the matrix $A = A_{\psi,\mathcal{B}} = (a_{ij}) \in M_{h,h}(K)$ for which $\psi(x_j) = \sum_{i=1}^h a_{ij} x_i$ for all $j = 1, \ldots, h$. We shall occasionally write $A_{\psi,\mathcal{B}}$ as $A(\psi, \mathcal{B})$.

The mapping

$$\rho \;:\; \text{End}_K(V) \;\longrightarrow\; M_{h,h}(K)$$
$$\mu \;\longmapsto\; A_{\mu,\mathcal{B}}$$

is an isomorphism both of $K$-spaces and rings.

(ii) We can imitate ideas from 11.1 and, for a fixed $B \in M_{h,h}(K)$, consider the commutative subring $K''[B]$ of $M_{h,h}(K)$, where

$$K'' = \{aI_h : a \in K\} :$$

there is a unique ring homomorphism $\xi : K[X] \to K''[B]$ for which $\xi(a) = aI_h$ for all $a \in K$ and $\xi(X) = B$. For $f \in K[X]$, we denote $\xi(f)$ by $f(B)$.

There is a unique member of least degree in the (non-empty) set

$$\{f \in K[X] : f \text{ is monic and } f(B) = 0\}.$$

We denote this polynomial by $m_B$, and refer to it as the *minimal polynomial of B*. It has the property that, if $f \in K[X]$ is such that $f(B) = 0$, then $m_B \mid f$ in $K[X]$.

(iii) Let $B \in M_{h,h}(K)$. Then $B$ represents $\psi$ relative to some basis for $V$ if and only if there is a non-singular matrix $P \in M_{h,h}(K)$ such that $P^{-1}BP = A_{\psi,\mathcal{B}}$, that is, if and only if $B$ is *similar over* $K$ to the matrix which represents $\psi$ relative to $\mathcal{B}$.

(iv) If $V$ is the direct sum of its non-zero $K$-subspaces $C_1, \ldots, C_n$, and $(b_{ij})_{j=1}^{d_i}$ is a $K$-basis for $C_i$ (for each $i = 1, \ldots, n$), then

$$b_{11}, \ldots, b_{1d_1}, b_{21}, \ldots, b_{n-1,d_{n-1}}, b_{n1}, \ldots, b_{nd_n}$$

form a $K$-basis $\mathcal{B}'$ for $V$. Let $A' = A_{\psi,\mathcal{B}'}$ be the matrix of $\psi$ relative to $\mathcal{B}'$. Then $A'$ has the form

$$A' = \begin{pmatrix} A_1 & 0 & \ldots & 0 \\ 0 & A_2 & \ldots & 0 \\ \vdots & \vdots & & \vdots \\ 0 & 0 & \ldots & A_n \end{pmatrix}$$

for some $A_i \in M_{d_i,d_i}(K)$ $(1 \le i \le n)$ if and only if $C_1, \ldots, C_n$ are all invariant under $\psi$, and, when this is the case, for each $i = 1, \ldots, n$,

$$A_i = A(\psi|_{C_i}, (b_{ij})_{j=1}^{d_i}),$$

the matrix of the restriction of $\psi$ to $C_i$ relative to the basis $(b_{ij})_{j=1}^{d_i}$.

(v) Let $B \in M_{h,h}(K)$. We can consider $XI_h - B$ as an $h \times h$ matrix over the quotient field of $K[X]$. The determinant of this matrix, $\det(XI_h - B)$, is the *characteristic polynomial of B*; it is a monic polynomial in $K[X]$ of degree $h$.

**11.4** EXERCISE. Let $h \in \mathbb{N}$ and let $K$ be a field. Let $B \in M_{h,h}(K)$. Show that $\{f \in K[X] : f(B) = 0\}$ is an ideal of $K[X]$, and that its unique monic generator is $m_B$, the minimal polynomial of $B$.

**11.5** REMARK. Let the situation be as in 11.3(i). Note that it follows from the isomorphism of 11.3(i) that $m_{\psi,V} = m_{A_{\psi,B}}$, that is, the minimal polynomial of $\psi$ on $V$ is equal to the minimal polynomial of the matrix which represents $\psi$ relative to the basis $\mathcal{B}$.

A few words about the strategy which is central to this chapter are perhaps appropriate at this point. Given an endomorphism $\psi$ of the vector space $V$ of finite dimension $h > 0$ over the field $K$, we can regard $V$ as a $K[X]$-module using $\psi$ in the manner of 11.1, so that $Xv = \psi(v)$ for all $v \in V$. By 10.8 or 10.11, we can express $V$ as a direct sum of cyclic $K[X]$-submodules, $C_1, \ldots, C_n$ say, satisfying certain conditions. By 11.2(i), the $C_i$ will be $K$-subspaces of $V$ which are invariant under $\psi$, and $V$ is the internal direct sum of these subspaces $C_1, \ldots, C_n$. It follows from 11.3(iv) that if we make up a basis for $V$ by putting bases for the $C_i$ together, then the matrix representing $\psi$ on $V$ relative to this basis will have a 'block diagonal' form. In the light of this, it is interesting to examine a cyclic $K[X]$-submodule $U$ of $V$, and investigate whether suitable choices of bases for $U$ can lead to satisfactory representing matrices for the restriction of $\psi$ to $U$. We turn to this next.

**11.6** DEFINITION. Let $K$ be a field, and let $X$ be an indeterminate. Let $f \in K[X]$ be a non-constant monic polynomial of degree $d$, say

$$f = a_0 + a_1 X + \cdots + a_{d-1}X^{d-1} + X^d.$$

Then the *companion matrix of $f$* is the matrix $C(f) \in M_{d,d}(K)$ given by

$$C(f) := \begin{pmatrix} 0 & 0 & 0 & \ldots & 0 & -a_0 \\ 1 & 0 & 0 & \ldots & 0 & -a_1 \\ 0 & 1 & 0 & \ldots & 0 & -a_2 \\ \vdots & \vdots & \vdots & & \vdots & \vdots \\ 0 & 0 & 0 & \ldots & 1 & -a_{d-1} \end{pmatrix}.$$

Note that, for $a \in K$, the companion matrix $C(X - a)$ of $X - a \in K[X]$ is just the $1 \times 1$ matrix $(a)$.

**11.7** EXERCISE. Let the situation be as in 11.6. Find the characteristic polynomial and the minimal polynomial of the companion matrix $C(f)$.

**11.8 PROPOSITION.** *Let $V$ be a non-zero, finite-dimensional vector space over the field $K$, and let $\psi \in \mathrm{End}_K(V)$; regard $V$ as a $K[X]$-module, where $X$ is an indeterminate, in the manner explained in 11.1 using $\psi$, so that $Xv = \psi(v)$ for all $v \in V$.*

*Then $V$ is cyclic as $K[X]$-module if and only if there exist a non-constant monic polynomial $f \in K[X]$ and a basis $\mathcal{B}$ for $V$ such that the matrix of $\psi$ relative to $\mathcal{B}$ is the companion matrix $C(f)$ (of 11.6).*

*When these conditions are satisfied, $\mathrm{Ann}_{K[X]}(V) = K[X]f$, and so the polynomial $f$ is uniquely determined by the stated conditions because it is the minimal polynomial $m_{\psi,V}$ of $\psi$ on $V$.*

*Proof.* ($\Rightarrow$) Suppose that $v \in V$ is a generator for the $K[X]$-module $V$. Since $V$ is non-zero and finite-dimensional, by 11.2(ii) we must have

$$0 \subset \mathrm{Ann}_{K[X]}(V) \subset K[X],$$

and so there is a unique (non-constant) monic polynomial

$$f = a_0 + a_1 X + \cdots + a_{d-1} X^{d-1} + X^d \in K[X]$$

such that $\mathrm{Ann}_{K[X]}(V) = K[X]f$. (In fact, $f = m_{\psi,V}$, of course.) We show that $\mathcal{B} := (X^{i-1}v)_{i=1}^d = (\psi^{i-1}(v))_{i=1}^d$ is a basis for $V$.

Since $V$ is generated as a $K[X]$-module by $v$, each element of $V$ has the form $gv$ for some $g \in K[X]$; by the division algorithm for polynomials, there exist $q, r \in K[X]$ with $r$ (either zero or) of degree less than $d$ such that $g = qf + r$, and since $fv = 0$, we have $gv = rv$; it follows easily that $V$ is generated as a $K$-space by the members of $\mathcal{B}$.

Next we show that $\mathcal{B}$ is a linearly independent family. Suppose that $b_0, b_1, \ldots, b_{d-1} \in K$ are such that $\sum_{i=0}^{d-1} b_i X^i v = 0$. Then, since $v$ generates $V$ as $K[X]$-module, it follows that

$$h := b_0 + b_1 X + \cdots + b_{d-1} X^{d-1} \in \mathrm{Ann}_{K[X]}(V) = K[X]f.$$

Since $f$ has degree $d$, it follows that $h = 0$ and $b_i = 0$ for all $i = 0, \ldots, d-1$. Hence $\mathcal{B}$ is a basis for $V$, and, since

$$\psi(X^{d-1}v) = X^d v = -a_0 v - a_1 X v - \cdots - a_{d-1} X^{d-1} v,$$

it is easy to see that the matrix of $\psi$ relative to $\mathcal{B}$ is $C(f)$.

($\Leftarrow$) Suppose that there exist a non-constant monic polynomial

$$f = a_0 + a_1 X + \cdots + a_{d-1} X^{d-1} + X^d \in K[X]$$

and a basis $\mathcal{B}$ for $V$ such that the matrix of $\psi$ relative to $\mathcal{B}$ is the companion matrix $C(f)$. Then $\mathcal{B}$ must have $d$ members: let $\mathcal{B} = (y_i)_{i=1}^d$. The form of

$C(f)$ means that $y_i = \psi^{i-1}(y_1) = X^{i-1}y_1$ for all $i = 2, \ldots, d$, and so $V$ is generated as $K[X]$-module by $y_1$, and is therefore cyclic. Also, since

$$\psi(y_d) = -a_0 y_1 - a_1 y_2 - \cdots - a_{d-1}y_d,$$

it follows that $fy_1 = 0$; moreover, no non-zero polynomial in $K[X]$ of degree smaller than $d$ can annihilate $y_1$ because $(X^{i-1}y_1)_{i=1}^d$ is linearly independent over $K$. Hence $\mathrm{Ann}_{K[X]}(V) = (0 :_{K[X]} y_1) = K[X]f$ and $f = m_{\psi,V}$; therefore the proof is complete. $\square$

**11.9 Definition.** Let $K$ be a field and let $h \in \mathbf{N}$. A matrix $A \in M_{h,h}(K)$ is said to be in *rational canonical form* precisely when there exist (an $n \in \mathbf{N}$ and) non-constant monic polynomials $f_1, \ldots, f_n \in K[X]$ such that $f_i \mid f_{i+1}$ for all $i = 1, \ldots, n-1$, that is, such that

$$K[X]f_1 \supseteq K[X]f_2 \supseteq \ldots \supseteq K[X]f_n,$$

and $A$ has the 'block diagonal' form given by

$$A = \begin{pmatrix} C(f_1) & 0 & \ldots & 0 \\ 0 & C(f_2) & \ldots & 0 \\ \vdots & \vdots & & \vdots \\ 0 & 0 & \ldots & C(f_n) \end{pmatrix}.$$

We are now ready for the Rational Canonical Form Theorem for an endomorphism of a finite-dimensional vector space over a field $K$.

**11.10 The Rational Canonical Form Theorem.** *Let $V$ be a non-zero, finite-dimensional vector space over the field $K$, and let $\psi \in \mathrm{End}_K(V)$. Then there exists a basis $\mathcal{B}$ for $V$ relative to which the matrix of $\psi$ is in rational canonical form. Moreover, there is exactly one rational canonical form matrix which can represent $\psi$ in this way, and this is called the rational canonical matrix of $\psi$.*

*In other words, there exist (an $n \in \mathbf{N}$ and) non-constant monic polynomials $f_1, \ldots, f_n \in K[X]$ such that $f_i \mid f_{i+1}$ for all $i = 1, \ldots, n-1$ and*

$$A_{\psi,\mathcal{B}} = \begin{pmatrix} C(f_1) & 0 & \ldots & 0 \\ 0 & C(f_2) & \ldots & 0 \\ \vdots & \vdots & & \vdots \\ 0 & 0 & \ldots & C(f_n) \end{pmatrix}$$

*for some basis $\mathcal{B}$ for $V$. Moreover, the polynomials $f_1, \ldots, f_n$ which satisfy these conditions are uniquely determined by $\psi$.*

*Proof.* We regard $V$ as a $K[X]$-module, where $X$ is an indeterminate, using $\psi$ in the manner described in 11.1. Then $V$ becomes a non-zero, finitely generated $K[X]$-module of torsion-free rank 0, and we apply Theorem 10.32 to this module. In fact, after the groundwork we have now covered in 11.1, 11.2, 11.3 and 11.8, the Rational Canonical Form Theorem follows from Theorem 10.32 in a very straightforward manner, and the reader is left to convince himself or herself of the details. $\square$

**11.11** ‡EXERCISE. Fill in the details of the proof of 11.10.

**11.12** EXERCISE. Let the situation be as in 11.10. Find the characteristic polynomial and the minimal polynomial of $\psi$ on $V$ in terms of the polynomials $f_1, \ldots, f_n$ which provide the rational canonical matrix of $\psi$.

There is a matrix version of the Rational Canonical Form Theorem, and this follows from 11.10 in conjunction with the basic ideas from linear algebra described in 11.3(i), (iii).

**11.13** COROLLARY: RATIONAL CANONICAL FORMS FOR MATRICES. *Let $K$ be a field and let $h \in \mathbb{N}$; let $A \in M_{h,h}(K)$. Then $A$ is similar (over $K$) to exactly one matrix in rational canonical form, and this matrix is called the rational canonical form of $A$.*

*In other words, there exist an $n \in \mathbb{N}$ and non-constant monic polynomials $f_1, \ldots, f_n \in K[X]$ such that $f_i \mid f_{i+1}$ for all $i = 1, \ldots, n-1$ and*

$$P^{-1}AP = \begin{pmatrix} C(f_1) & 0 & \ldots & 0 \\ 0 & C(f_2) & \ldots & 0 \\ \vdots & \vdots & & \vdots \\ 0 & 0 & \ldots & C(f_n) \end{pmatrix}$$

*for some invertible matrix $P \in M_{h,h}(K)$. Also, the polynomials $f_1, \ldots, f_n$ which satisfy these conditions are uniquely determined by $A$.* $\square$

**11.14** EXERCISE. Let the situation be as in 11.13. Find the characteristic polynomial and the minimal polynomial of $A$ in terms of the polynomials $f_1, \ldots, f_n$ which provide the rational canonical form of $A$.

Deduce the result of the Cayley–Hamilton Theorem, that is, that the minimal polynomial of $A$ is a factor in $K[X]$ of the characteristic polynomial of $A$.

Let $p$ be an irreducible polynomial in $K[X]$. Prove that $p$ is a factor in $K[X]$ of the characteristic polynomial of $A$ if and only if $p$ is a factor in $K[X]$ of the minimal polynomial of $A$.

Perhaps a few words about the general philosophy underlying canonical forms are appropriate at this point. Let the situation be as in 11.13. Of course, similarity is an equivalence relation on $M_{h,h}(K)$; our Corollary 11.13 shows that each similarity class of $h \times h$ matrices over $K$ contains exactly one matrix in rational canonical form, so that two $h \times h$ matrices over $K$ are similar over $K$ if and only if they have the same rational canonical form.

Our Rational Canonical Form Theorem 11.10 is, essentially, a consequence of Theorem 10.32. The reader will perhaps recall that we also proved in 10.35 a similar decomposition theorem concerning expressions for finitely generated modules over a PID $R$ as direct sums of cyclic $R$-modules, but in that result the cyclic $R$-modules considered had prime power annihilators. It is natural to wonder whether 10.35 has any interesting consequences for canonical forms of matrices. In this chapter we are going to explore this, but only in the situation in which the underlying field is algebraically closed: we shall use 10.35 to obtain the Jordan Canonical Form Theorem.

Recall that a field $K$ is said to be *algebraically closed* precisely when every non-constant polynomial in $K[X]$ has a root in $K$. The Fundamental Theorem of Algebra states that the complex field $\mathbf{C}$ is algebraically closed: see, for example, [16, Chapter 18] or [17, Theorem 10.7]. When $K$ is algebraically closed, the monic irreducible polynomials in $K[X]$ are precisely the polynomials of the form $X - a$ with $a \in K$. We therefore investigate cyclic $K[X]$-modules whose annihilators are powers of polynomials of this type.

**11.15** DEFINITION. Let $K$ be a field, let $a \in K$ and let $u \in \mathbf{N}$. Then the $u \times u$ *elementary Jordan $a$-matrix* is the matrix $J(a, u) \in M_{u,u}(K)$ given by

$$J(a,u) := \begin{pmatrix} a & 0 & 0 & \dots & 0 & 0 \\ 1 & a & 0 & \dots & 0 & 0 \\ 0 & 1 & a & \dots & 0 & 0 \\ \vdots & \vdots & \vdots & & \vdots & \vdots \\ 0 & 0 & 0 & \dots & a & 0 \\ 0 & 0 & 0 & \dots & 1 & a \end{pmatrix},$$

so that each of its diagonal entries is equal to $a$, each of its entries on the 'subdiagonal' is equal to 1, and all its other entries are zero.

Note that, in particular, $J(a, 1)$ is the $1 \times 1$ matrix $(a)$.

**11.16** EXERCISE. Let the situation be as in 11.15. Find the characteristic polynomial and the minimal polynomial of the elementary Jordan $a$-matrix $J(a, u)$.

We are now in a position to give a result similar to 11.8.

**11.17** PROPOSITION. *Let $V$ be a non-zero, finite-dimensional vector space over the field $K$, and let $\psi \in \text{End}_K(V)$; regard $V$ as a $K[X]$-module, where $X$ is an indeterminate, in the manner explained in 11.1 using $\psi$, so that $Xv = \psi(v)$ for all $v \in V$. Let $a \in K$ and $u \in \mathbb{N}$.*

*Then $V$ is cyclic as $K[X]$-module with $\text{Ann}_{K[X]}(V) = (X - a)^u K[X]$ if and only if there exists a basis $\mathcal{B}$ for $V$ such that the matrix of $\psi$ relative to $\mathcal{B}$ is the elementary Jordan $a$-matrix $J(a, u)$ (of 11.15).*

*Proof.* ($\Rightarrow$) Suppose that $v \in V$ is a generator for the $K[X]$-module $V$, and that $\text{Ann}_{K[X]}(V) = (X-a)^u K[X]$. We show that $\mathcal{B} := ((X-a)^{i-1}v)_{i=1}^u$ is a basis for $V$.

Since $V$ is generated as a $K[X]$-module by $v$, each element of $V$ has the form $gv$ for some $g \in K[X]$; but the subring $K[X - a]$ of $K[X]$ (see 1.11) contains $K$ and $X = (X - a) + a$, and so it follows from 1.11 that each $g \in K[X]$ can be written as a polynomial expression in $(X - a)$ with coefficients in $K$; since $(X - a)^u v = 0$, it follows that $V$ is generated as a $K$-space by $\{(X - a)^{i-1}v : 1 \le i \le u\}$.

Next we show that $\mathcal{B}$ is a linearly independent family. Suppose that $b_0, b_1, \ldots, b_{u-1} \in K$ are such that $\sum_{i=0}^{u-1} b_i(X - a)^i v = 0$. Suppose that at least one of the $b_i$ is non-zero, and look for a contradiction. Let $j \in \mathbb{N}_0$ be the least integer $i$ such that $b_i \ne 0$; multiply both sides of the last equation by $(X - a)^{(u-1)-j}$, and use the fact that $(X - a)^{u-1}v \ne 0$ to obtain a contradiction.

Hence $\mathcal{B}$ is a basis for $V$; it is easy to see that the matrix of $\psi$ relative to $\mathcal{B}$ is $J(a, u)$.

($\Leftarrow$) This is left as an exercise for the reader. $\square$

**11.18** ‡EXERCISE. Complete the proof of 11.17.

**11.19** DEFINITION. Let $K$ be a field and let $h \in \mathbb{N}$. A matrix $A \in M_{h,h}(K)$ is said to be in *Jordan canonical form* precisely when there exist (an $n \in \mathbb{N}$,) $a_1, \ldots, a_n \in K$ and $u_1, \ldots, u_n \in \mathbb{N}$ such that $A$ has the 'block diagonal' form given by

$$
A = \begin{pmatrix}
J(a_1, u_1) & 0 & \cdots & 0 \\
0 & J(a_2, u_2) & \cdots & 0 \\
\vdots & \vdots & & \vdots \\
0 & 0 & \cdots & J(a_n, u_n)
\end{pmatrix},
$$

in which the matrices on the 'diagonal' are elementary Jordan matrices.

Note that there is no requirement in the above definition that the $a_1, \ldots, a_n$ be distinct; thus, for example,

$$\left( \begin{array}{cc|cc} 2 & 0 & 0 & 0 \\ 1 & 2 & 0 & 0 \\ \hline 0 & 0 & 2 & 0 \\ 0 & 0 & 1 & 2 \end{array} \right)$$

is in Jordan canonical form. Also, a diagonal matrix is in Jordan canonical form.

**11.20** EXERCISE. Find the characteristic polynomial and the minimal polynomial of a matrix (over a field $K$) in Jordan canonical form. You may find it helpful to use rather precise notation, analogous to that used for the statement of Theorem 10.35.

Our Theorem 10.35 now leads to results companion to 11.10 and 11.13. This time we leave the details of the proofs completely to the reader.

**11.21** THE JORDAN CANONICAL FORM THEOREM. *Let $K$ be an algebraically closed field. Let $V$ be a non-zero, finite-dimensional vector space over $K$, and let $\psi \in \mathrm{End}_K(V)$. Then there exists a basis $\mathcal{B}$ for $V$ relative to which the matrix of $\psi$ is in Jordan canonical form. Moreover, the Jordan canonical form matrix which represents $\psi$ in this way is uniquely determined by $\psi$ apart from the order in which the elementary Jordan matrices appear on the 'diagonal'.* □

**11.22** COROLLARY: JORDAN CANONICAL FORMS FOR MATRICES. *Let $K$ be an algebraically closed field and let $h \in \mathbb{N}$; let $A \in M_{h,h}(K)$. Then $A$ is similar (over $K$) to a matrix in Jordan canonical form, and this Jordan canonical form matrix is uniquely determined by $A$ apart from the order in which the elementary Jordan matrices appear on the 'diagonal'.* □

**11.23** ♯EXERCISE. Deduce the results of 11.21 and 11.22 from 10.35 and our work in this chapter.

**11.24** EXERCISE. Let $K$ be an algebraically closed field, and let $a, b, c$ be three distinct elements of $K$. For each of the following choices of the polynomial $\chi \in K[X]$, determine the number of similarity classes of $3 \times 3$ matrices over $K$ which have characteristic polynomial equal to $\chi$; for each such similarity class, find a Jordan canonical form matrix and the unique matrix in rational canonical form which belong to the class.

(i) $\chi = (X - a)(X - b)(X - c)$;

(ii) $\chi = (X - a)^2(X - b)$;

(iii) $\chi = (X - a)^3$.

**11.25** EXERCISE. Let $K$ be an algebraically closed field, and let $A \in M_{3,3}(K)$. Assume that both the characteristic polynomial and minimal polynomial of $A$ are known (and factorized into linear factors). Prove that both the rational canonical form of $A$ and a Jordan canonical form matrix similar to $A$ can be deduced from this information.

**11.26** EXERCISE. Let $K$ be an algebraically closed field, and let $A \in M_{h,h}(K)$, where $h \in \mathbb{N}$. Show that $A$ is similar (over $K$) to its transpose $A^T$.

Is each square matrix $B$ over an arbitrary field $L$ similar (over $L$) to its transpose $B^T$? Justify your response.

**11.27** EXERCISE. Let $K$ be a field, and let $A \in M_{h,h}(K)$, where $h \in \mathbb{N}$. Suppose that the characteristic polynomial of $A$ can be written as a product of linear factors in $K[X]$. Show that $A$ is similar (over $K$) to a matrix in Jordan canonical form, and that this Jordan canonical form matrix is uniquely determined by $A$ apart from the order in which the elementary Jordan matrices appear on the 'diagonal'.

**11.28** FURTHER STEPS. What we have tried to do in this chapter is bring out the relevance to the theory of canonical forms for matrices of the important direct-sum decomposition theorems for finitely generated modules over principal ideal domains established in Chapter 10. As this is not intended to be a book primarily about linear algebra, space has not been devoted to consideration of the problems of finding Jordan and rational canonical forms in specific, practical situations: readers interested in such topics might like to read [4, Chapter 12].

# Chapter 12

# Some applications to field theory

Much of the remainder of this book will be concerned with the dimension theory of commutative Noetherian rings. This theory gives some measure of 'size' to such a ring: the intuitive feeling that the ring $K[X_1, \ldots, X_n]$ of polynomials over a field $K$ in the $n$ indeterminates $X_1, \ldots, X_n$ has, in some sense, 'size' $n$ fits nicely into the dimension theory. However, to make a thorough study of the dimension theory of an integral domain $R$ which is a finitely generated algebra (see 8.9) over a field $K$, it is desirable to understand the idea of the 'transcendence degree' over $K$ of the quotient field $L$ of $R$: roughly, this transcendence degree is the largest integer $i \in \mathbb{N}_0$ such that there exist $i$ elements of $L$ which are algebraically independent over $K$, and it turns out to give an appropriate measure of the dimension of $R$. Accordingly, in this chapter, we are going to develop the necessary background material on transcendence degrees of field extensions.

Thus part of this chapter will be devoted to the development of field-theoretic tools which will be used later in the book. However, there is another aspect to this chapter, as its title possibly indicates: in fact, some of the ideas of Chapter 3 about prime ideals and maximal ideals, together with results like 5.10 and 5.15 concerned with the existence of algebra homomorphisms in situations involving rings of fractions, are very good tools with which to approach some of the basic theory of field extensions, and in the first part of this chapter we shall take such an approach. Thus we begin Chapter 12 with a fairly rapid discussion of topics such as the prime subfield of a field, the characteristic of a field, generation of field extensions, algebraic and transcendental elements and algebraic extensions. We shall see that the above-mentioned ideas from earlier in this book mean that rapid progress can be made through these topics. However, we shall concentrate on topics to which ideas from commutative algebra can be applied or which will be needed later in the book, and so the reader should be warned that this chapter does not represent an exhaustive account of elementary field theory.

**12.1** DEFINITIONS. A subset $F$ of a field $K$ is said to be a *subfield* of $K$ precisely when $F$ is itself a field with respect to the operations in $K$. We shall also describe this situation by saying that '$K$ is an *extension field* of $F$', or '$F \subseteq K$ *is an extension of fields*'.

When this is the case, $1_F = 1_K$, so that $F$ is a subring of $K$ in the sense of 1.4, because $1_F^2 = 1_F = 1_F 1_K$ in $K$.

We say that $K$ is an *intermediate field between F and L* precisely when $F \subseteq K$ and $K \subseteq L$ are extensions of fields.

A mapping $f : K_1 \to K_2$, where $K_1, K_2$ are fields, is a *homomorphism*, or a *field homomorphism*, precisely when it is a ring homomorphism. When this is the case, $\operatorname{Ker} f = \{0_{K_1}\}$ (because it must be a proper ideal of $K_1$), and so $f$ is injective by 2.2.

**12.2** EXAMPLES. Let $K$ be a field and let $X$ be an indeterminate.

(i) Denote by $K(X)$ the field of fractions of the integral domain $K[X]$. The composition $K \to K[X] \to K(X)$ of the natural injective ring homomorphisms enables us to consider $K(X)$ as a field extension of $K$. We refer to $K(X)$ as the *field of rational functions in X with coefficients in K*. A typical element of $K(X)$ can be written in the form $f/g$, where $f, g$ are polynomials in $X$ with coefficients in $K$ and $g \neq 0$.

(ii) Let $m \in K[X]$ be a monic irreducible polynomial in $X$ with coefficients in $K$. By 3.34, the ring $L := K[X]/mK[X]$ is a field, and the composition

$$K \to K[X] \to K[X]/mK[X] = L$$

of the natural ring homomorphisms must be injective (by 12.1) even though the second ring homomorphism is not; this composition enables us to regard $L$ as an extension field of $K$. Observe also that, if we denote by $\alpha$ the natural image $X + mK[X]$ of $X$ in $L$, then $m(\alpha) = 0$: to see this, let $m = \sum_{i=0}^{n} a_i X^i$ (where $a_n = 1$), and note that

$$m(\alpha) = \sum_{i=0}^{n} a_i (X + mK[X])^i = m + mK[X] = 0_L$$

(because an $a \in K$ is identified with its natural image $a + mK[X] \in L$).

Thus, given $K$ and the monic irreducible polynomial $m \in K[X]$, we have constructed a field extension $L$ of $K$ in which $m$ has a root.

**12.3** ♯EXERCISE. Let the situation be as in 12.2(ii), so that $m$ is a monic irreducible polynomial in $K[X]$ of degree $n$ and $\alpha = X + mK[X] \in L := K[X]/mK[X]$, and the latter field is regarded as an extension of $K$ in the manner described in 12.2(ii). Let $\lambda \in L$. Show that there exist uniquely determined $b_0, b_1, \ldots, b_{n-1} \in K$ such that

$$\lambda = \sum_{i=0}^{n-1} b_i \alpha^i;$$

deduce that, when $L$ is regarded as a vector space over $K$ by restriction of scalars, $\operatorname{vdim}_K L = n$.

**12.4** EXERCISE. What can you say about the field $\mathbf{R}[X]/(X^2 + 1)\mathbf{R}[X]$?

**12.5** ♯EXERCISE. Let $K$ be a field and let $X$ be an indeterminate. Let $g \in K[X]$ be a non-constant polynomial. Prove that there exists a field extension $K'$ of $K$ such that $g$ factorizes into linear factors in $K'[X]$.

**12.6** ♯EXERCISE: THE SUBFIELD CRITERION. Let $F$ be a subset of the field $K$. Show that $F$ is a subfield of $K$ if and only if the following conditions hold:

  (i) $1_K \in F$;
  (ii) whenever $a, b \in F$, then $a - b \in F$;
  (iii) whenever $a, b \in F$ with $b \neq 0_K$, then $ab^{-1} \in F$.

Deduce that the intersection of any non-empty family of subfields of $K$ is again a subfield of $K$. Deduce also, that if $F$ is a subfield of $K$, then the intersection of any non-empty family of intermediate fields between $F$ and $K$ is again an intermediate field between $F$ and $K$.

**12.7** REMARKS. Let $R$ be integral domain, having field of fractions $Q(R)$, and let $L$ be a field; suppose that $f : R \to L$ is an injective ring homomorphism.

  (i) Since $f(R \setminus \{0_R\}) \subseteq L \setminus \{0_L\}$, it is immediate from 5.10 and 5.5 that there is induced a homomorphism $f' : Q(R) \to L$ for which $f'(r/r') = f(r)f(r')^{-1}$ for all $r, r' \in R$ with $r' \neq 0$. It is easy to see that $\operatorname{Im} f'$ is the smallest subfield of $L$ which contains $f(R)$.

  (ii) Observe that $Q(R)$ is the only subfield of $Q(R)$ which contains $R$.

**12.8** REMARKS. (i) Let $R$ and $S$ be integral domains with quotient fields $Q(R)$ and $Q(S)$ respectively. Suppose that $f : R \to S$ is a ring isomorphism.

  It follows easily from 12.7(i) applied to the composite ring homomorphism

$$R \xrightarrow{f} S \longrightarrow Q(S)$$

(in which the second homomorphism is the natural one) that $f$ induces an isomorphism $f' : Q(R) \to Q(S)$ for which $f'(r/r') = f(r)/f(r')$ for all $r, r' \in R$ with $r' \neq 0$.

(ii) Let $K, L$ be fields and let $g : K \to L$ be an isomorphism; let $X$ and $Y$ be indeterminates. It is easy to deduce from 1.16 that $g$ induces a ring isomorphism $\tilde{g} : K[X] \to L[Y]$ which extends $g$ and is such that $\tilde{g}(X) = Y$. Application of (i) above to this yields an isomorphism $\bar{g} : K(X) \to L(Y)$ of the fields of rational functions which extends $g$ and is such that $\bar{g}(X) = Y$.

**12.9** THE PRIME SUBFIELD OF A FIELD. We say that a field $F$ is a *prime field* precisely when $F$ has no subfield which is a proper subset of itself (that is, $F$ has no 'proper subfield').

Let $K$ be a field. By the Subfield Criterion 12.6, the intersection of all subfields of $K$ is a subfield of $K$ contained in every other subfield of $K$, and so is the 'smallest' subfield of $K$; it is therefore the unique prime subfield of $K$.

**12.10** THE CHARACTERISTIC OF A FIELD. Let $K$ be a field, and let $\Pi$ be its prime subfield. It is clear from the Subfield Criterion 12.6 that $n1_K \in \Pi$ for all $n \in \mathbf{Z}$. Thus (see 1.10) there is a ring homomorphism $f : \mathbf{Z} \to \Pi$ such that $f(n) = n1_K$ for all $n \in \mathbf{Z}$. By the Isomorphism Theorem 2.13, we have $\mathbf{Z}/\operatorname{Ker} f \cong \operatorname{Im} f$, a subring of the field $\Pi$; hence $\operatorname{Im} f$ is an integral domain and so $\operatorname{Ker} f$ is a prime ideal of $\mathbf{Z}$, by 3.23. Of course, every ideal of $\mathbf{Z}$ is principal, and we know all the prime ideals of $\mathbf{Z}$ (see 3.34): there are two cases to consider, according as $\operatorname{Ker} f = p\mathbf{Z}$ for some prime number $p$ or $\operatorname{Ker} f = 0\mathbf{Z}$.

(i) *When* $\operatorname{Ker} f = \{n \in \mathbf{Z} : n1_K = 0_K\} = p\mathbf{Z}$ *for some prime number* $p$, *we say that* $K$ *has characteristic* $p$. In this case, $p$ is the smallest positive integer $n$ for which $n1_K = 0_K$, and $\operatorname{Im} f \cong \mathbf{Z}/p\mathbf{Z}$ is already a field, so that, as it is a subfield of the prime subfield $\Pi$ of $K$, we must have

$$\Pi = \operatorname{Im} f \cong \mathbf{Z}/p\mathbf{Z}.$$

(ii) *When* $\operatorname{Ker} f = \{n \in \mathbf{Z} : n1_K = 0_K\} = 0\mathbf{Z}$, *we say that* $K$ *has characteristic* 0. In this case, 0 is the only integer $n$ for which $n1_K = 0_K$, and $f : \mathbf{Z} \to \Pi$ is an injective ring homomorphism, so that, by 12.7(i), there is induced an isomorphism $f' : \mathbf{Q} \xrightarrow{\cong} \Pi$.

**12.11** REMARKS. Let $K \subseteq L$ be an extension of fields.

(i) We denote the characteristic of $K$ by char $K$. Observe that it is the unique non-negative generator of the ideal $\{n \in \mathbf{Z} : n1_K = 0_K\}$ of $\mathbf{Z}$.

(ii) Note also that char $K = $ char $L$.

**12.12** EXERCISE. Let $K$ be a field of positive characteristic $p$. Show that the mapping $f : K \to K$ defined by $f(a) = a^p$ for every $a \in K$ is a homomorphism. (This $f$ is called the *Frobenius homomorphism* of $K$.)

Just as the Subring Criterion 1.5 led to the idea of 'ring adjunction' in 1.11 and the Submodule Criterion 6.8 led to the idea of generation of submodules, so the Subfield Criterion 12.6 leads to the concept of 'field adjunction', or generation of intermediate fields.

**12.13** FIELD ADJUNCTION (GENERATION OF INTERMEDIATE FIELDS). Let $F \subseteq K$ be an extension of fields, and let $\Gamma \subseteq K$. Then $F(\Gamma)$ is defined to be the intersection of all subfields of $K$ which contain both $F$ and $\Gamma$, that is, the intersection of all intermediate fields between $F$ and $K$ which contain $\Gamma$. Thus, by the Subfield Criterion 12.6, $F(\Gamma)$ is the smallest intermediate field between $F$ and $K$ which contains $\Gamma$.

We refer to $F(\Gamma)$ as *the field obtained by adjoining* $\Gamma$ *to* $F$, or, alternatively, as *the intermediate field between* $F$ *and* $K$ *generated by* $\Gamma$.

In the special case in which $\Gamma$ is a finite set $\{\alpha_1, \ldots, \alpha_n\}$, we write $F(\Gamma)$ as $F(\alpha_1, \ldots, \alpha_n)$. In fact, we shall say that a field extension $F \subseteq L$ is *finitely generated* precisely when there exist ($h \in \mathbb{N}$ and) $\beta_1, \ldots, \beta_h \in L$ such that $L = F(\beta_1, \ldots, \beta_h)$. (Of course, $F \subseteq F$ is a finitely generated extension, simply because $F = F(a)$ for any $a \in F$.)

Observe that, for an indeterminate $X$, the notation $F(X)$ of 12.2(i) for the field of rational functions in $X$ with coefficients in $F$ is consistent with the notation of the preceding paragraph, because $F(X)$ is the smallest subfield of $F(X)$ which contains both $F$ and $X$.

**12.14** ♯EXERCISE. Let $F \subseteq K$ be an extension of fields and let $\Gamma$, $\Delta$ be subsets of $K$. Show that $F(\Gamma \cup \Delta) = F(\Gamma)(\Delta)$, and

$$F(\Gamma) = \bigcup_{\Omega \subseteq \Gamma,\ |\Omega| < \infty} F(\Omega).$$

**12.15** REMARK. Let $F \subseteq K$ be an extension of fields, and let $\alpha_1, \ldots, \alpha_n \in K$. Let $\sigma \in S_n$, the group of permutations of the set of the first $n$ positive integers. It follows from 12.14 that

$$F(\alpha_1)(\alpha_2)\ldots(\alpha_n) = F(\alpha_1, \ldots, \alpha_n) = F(\alpha_{\sigma(1)}, \ldots, \alpha_{\sigma(n)})$$
$$= F(\alpha_{\sigma(1)})(\alpha_{\sigma(2)})\ldots(\alpha_{\sigma(n)}).$$

Thus adjunction of a finite set $\Gamma$ of elements of $K$ to $F$ can be achieved by adjoining the elements of $\Gamma$ one at a time, and in any order.

This last result 12.15 gives added importance to intermediate fields between $F$ and $K$ of the form $F(\alpha)$, obtained by the adjunction of a single element $\alpha$ of $K$. There is special terminology concerned with such extensions.

**12.16 DEFINITION.** We say that an extension of fields $F \subseteq K$ is *simple* precisely when there exists $\alpha \in K$ such that $K = F(\alpha)$, that is, when $K$ can be obtained by adjoining one of its elements to $F$.

**12.17 REMARK.** Let $K$ be a field and let $X$ be an indeterminate, and let $m$ be a monic irreducible polynomial in $K[X]$. Note that both the field extensions $K \subseteq K(X)$ and $K \subseteq K[X]/mK[X]$ of Exercises 12.2 and 12.3 are simple.

**12.18 REMARKS.** Let $K \subseteq L$ be an extension of fields and let $\theta \in L$. We propose now to investigate the structure of the intermediate field $K(\theta)$ between $K$ and $L$. Relevant to our discussion will be the subring $K[\theta]$ of $L$: see 1.11. Recall that $K[\theta]$ is the smallest subring of $L$ which contains both $K$ and $\theta$. Because a subfield is, in particular, a subring, we must have $K[\theta] \subseteq K(\theta)$, and this observation will help us to explore the structure of $K(\theta)$. Of course, the reader should understand the difference between $K(\theta)$ and $K[\theta]$: the former is the result of 'field adjunction' of $\theta$ to $K$, while the latter is the result of 'ring adjunction' of $\theta$ to $K$.

Note that, by 1.11,

$$K[\theta] = \left\{ \sum_{i=0}^{t} a_i \theta^i : t \in \mathsf{N}_0, \ a_0, \ldots, a_t \in K \right\},$$

and so can be thought of as the set of all 'polynomials' in $\theta$ with coefficients in $K$. Let $X$ be an indeterminate. It follows from 1.13 that there is a unique surjective ring homomorphism $g : K[X] \to K[\theta]$ which is such that $g|_K = \mathrm{Id}_K$ and $g(X) = \theta$; in fact,

$$g \left( \sum_{i=0}^{n} a_i X^i \right) = \sum_{i=0}^{n} a_i \theta^i$$

for all $n \in \mathsf{N}_0$, $a_0, \ldots, a_n \in K$.

**12.19 ♯EXERCISE.** Let $K \subseteq L$ be an extension of fields, and consider elements $\alpha_1, \ldots, \alpha_n \in L$. Show that $K(\alpha_1, \ldots, \alpha_n)$ is isomorphic to the quotient field of $K[\alpha_1, \ldots, \alpha_n]$, and that each element $\beta$ of $K(\alpha_1, \ldots, \alpha_n)$ can be written in the form

$$\beta = f(\alpha_1, \ldots, \alpha_n) g(\alpha_1, \ldots, \alpha_n)^{-1}$$

for some $f, g \in K[X_1, \ldots, X_n]$ with $g(\alpha_1, \ldots, \alpha_n) \neq 0$.

**12.20** ALGEBRAIC AND TRANSCENDENTAL ELEMENTS. Let the situation be as in 12.18, so that $K \subseteq L$ is an extension of fields and $\theta \in L$. Consider the ring homomorphism $g : K[X] \to K[\theta] \subseteq K(\theta)$ for which $g(a) = a$ for all $a \in K$ and $g(X) = \theta$. Now, by the Isomorphism Theorem 2.13, $g$ induces a ring isomorphism

$$\bar{g} : K[X]/\operatorname{Ker} g \longrightarrow \operatorname{Im} g = K[\theta]$$

for which $\bar{g}(h + \operatorname{Ker} g) = g(h)$ for all $h \in K[X]$. Since $K[\theta]$, being a subring of a field, is an integral domain, it follows from 3.23 that $\operatorname{Ker} g$ is a prime ideal of $K[X]$; by 3.34, we therefore have $\operatorname{Ker} g = mK[X]$ for some monic irreducible polynomial $m \in K[X]$ or $\operatorname{Ker} g = 0$. These two possibilities mean that there are two cases to consider.

(i) *When* $\operatorname{Ker} g = \{h \in K[X] : h(\theta) = 0\} = mK[X]$ *for some monic irreducible polynomial* $m \in K[X]$, *we say that* $\theta$ *is algebraic over (or with respect to)* $K$. In this case, there exist non-zero polynomials in $K[X]$ which have $\theta$ as a root, and $m$ is a factor in $K[X]$ of every such polynomial; furthermore, $m$ is the monic polynomial in $K[X]$ of least degree which has $\theta$ as a root (and this condition characterizes $m$ uniquely); and $m$ can also be characterized as the unique monic irreducible polynomial in $K[X]$ which has $\theta$ as a root.

The monic irreducible polynomial $m \in K[X]$ is called the *minimal polynomial of* $\theta$ *with respect to* or *over* $K$.

In this case $\operatorname{Ker} g$ is actually a maximal ideal of $K[X]$ by 3.34, and since

$$\bar{g} : K[X]/mK[X] = K[X]/\operatorname{Ker} g \xrightarrow{\cong} K[\theta],$$

$K[\theta]$ is already a field by 3.3; since $K[\theta] \subseteq K(\theta)$ and $K(\theta)$ is the smallest subfield of $L$ which contains both $K$ and $\theta$, we must have $K[\theta] = K(\theta)$ in this situation. Thus it can happen that field adjunction and ring adjunction lead to the same end result!

We can use the the isomorphism $\bar{g}$ to describe the structure of $K[\theta]$, because we have already investigated the field extension $K \subseteq K[X]/mK[X]$ to some extent in 12.2. Set $\alpha = X + mK[X] \in K[X]/mK[X]$; note that $\bar{g}(a) = a$ for all $a \in K$, and that $\bar{g}(\alpha) = \theta$. Thus, speaking loosely, we perform the arithmetic operations in $K[\theta]$ as in the residue class field $K[X]/mK[X]$, with $\theta$ playing the rôle of $X + mK[X]$.

Denote the degree of $m$ by $n$. By the division algorithm for polynomials, each $h \in K[X]$ can be uniquely written in the form $h = qm + r$ with $q, r \in K[X]$ and $r$ (either zero or) of degree less than $n$. It is easy to see from this (see Exercise 12.3) that each $\lambda \in K[X]/mK[X]$ can be uniquely written in the form

$$\lambda = \sum_{i=0}^{n-1} b_i \alpha^i$$

with $b_0, \ldots, b_{n-1} \in K$; hence each $\mu \in K[\theta]$ can be uniquely written in the form

$$\mu = \sum_{i=0}^{n-1} b_i \theta^i$$

with $b_0, \ldots, b_{n-1} \in K$. Thus, when $K(\theta)$ is regarded as a vector space over $K$ by restriction of scalars, we have

$$\text{vdim}_K K(\theta) = n = \deg m.$$

(ii) *When* $\text{Ker}\, g = \{h \in K[X] : h(\theta) = 0\} = 0$, *we say that* $\theta$ *is transcendental over (or with respect to)* $K$. In this case, the only polynomial in $K[X]$ which has $\theta$ as a root is the zero polynomial, and $g$ is injective. By 12.7, $g$ induces a field isomorphism $g' : K(X) \rightarrow K(\theta)$ for which $g'(a) = a$ for all $a \in K$ and $g'(X) = \theta$. Thus, when $\theta$ is transcendental over $K$, the field $K(\theta)$ behaves like the field of rational functions $K(X)$, with $\theta$ playing the rôle of $X$.

Note also that, in this case, we have $K[\theta] \neq K(\theta)$, since $K[\theta] \cong K[X]$, which is not a field.

Another comment to make in the case when $\theta$ is transcendental over $K$ is that, then, when $K(\theta)$ is regarded as a vector space over $K$ by restriction of scalars,

$$\text{vdim}_K K(\theta) = \text{vdim}_K K(X) = \infty,$$

since, for every $r \in \mathbb{N}$, the family $(X^i)_{i=0}^r$ is linearly independent over $K$.

**12.21** REMARKS. Let $K \subseteq L$ be an extension of fields, and let $\theta \in L$.

(i) It should be clear to the reader from 12.20 that $\theta$ is algebraic with respect to $K$ if and only if there exists a non-zero $h \in K[X]$ for which $h(\theta) = 0$.

Of course, each $a \in K$ is algebraic over $K$, simply because $a$ is a root of the polynomial $X - a \in K[X]$.

(ii) It also follows from 12.20 that $\theta$ is algebraic with respect to $K$ if and only if, when $K(\theta)$ is regarded as a vector space over $K$ by restriction of scalars, $\text{vdim}_K K(\theta)$ is finite.

**12.22** EXERCISE. (This exercise is only for those readers who have studied Chapter 11.) Let $K \subseteq L$ be an extension of fields, and let $\theta \in L$ be algebraic over $K$ with minimal polynomial $m$. Consider the $K$-endomorphism $\psi_\theta$ of the finite-dimensional $K$-space $K(\theta)$ for which $\psi_\theta(b) = \theta b$ for all $b \in K(\theta)$. Show that $m$ is the minimal polynomial of $\psi_\theta$ on $K(\theta)$ in the sense of 11.2(ii).

The comment about vector space dimensions in 12.21(ii) leads naturally to an important aspect of the theory of field extensions, namely the concept of finite field extension. We introduce this now.

**12.23** DEFINITION. Let $F \subseteq K$ be an extension of fields. Now $K$ can be viewed as a vector space over $F$ by restriction of scalars: we denote the dimension of this vector space by $[K : F]$, and call this dimension the *degree of K over F*. Thus $[K : F] = \mathrm{vdim}_F K$.

We say that the field extension $F \subseteq K$ is *finite* (or that $K$ is *finite over F*) precisely when $[K : F]$ is finite.

It follows from 12.21(ii) that, for a field extension $K \subseteq L$ and $\theta \in L$, the element $\theta$ is algebraic over $K$ if and only if $K(\theta)$ is a finite extension of $K$.

**12.24** EXERCISE. Let $F$ be a finite field, that is a, field with a finite number of elements. Show that the number of elements in $F$ is $p^n$ for some prime number $p$ and some positive integer $n$.

**12.25** EXERCISE. (This exercise is for those readers who have encountered the Fundamental Theorem on Abelian Groups, either in Chapter 10 or in other studies.) Prove that the multiplicative group of all non-zero elements of a finite field is cyclic. (This is quite a substantial result, and so some hints are provided. Use the Fundamental Theorem on Abelian Groups 10.39 in conjunction with the fact that, for $n \in \mathbb{N}$, there can be at most $n$ elements in a given field $F$ which are roots of the polynomial $X^n - 1 \in F[X]$; also, 10.9 might be helpful.)

**12.26** DEGREES THEOREM. *Let $F \subseteq K \subseteq L$ be extensions of fields. Then $L$ is finite over $F$ if and only if $L$ is finite over $K$ and $K$ is finite over $F$; furthermore, when this is the case,*

$$[L : F] = [L : K][K : F].$$

*Proof.* ($\Rightarrow$) Assume that $L$ is finite over $F$. Then, since $K$ is an $F$-subspace of $L$, it follows that $K$ is finite over $F$. Also, a basis for $L$ as an $F$-space will automatically be a spanning set for $L$ as a $K$-space, and so $L$ is finite over $K$.

($\Leftarrow$) Assume that $L$ is finite over $K$ and $K$ is finite over $F$. Write $[L : K] = n$ and $[K : F] = m$; let $(\phi_i)_{i=1}^n$ be a basis for the $K$-space $L$ and let $(\psi_j)_{j=1}^m$ be a basis for the $F$-space $K$. We shall show that

$$\mathcal{B} := (\phi_i \psi_j)_{1 \leq i \leq n, 1 \leq j \leq m}$$

is a basis for $L$ as an $F$-space. This will complete the proof, because it will also establish the final formula in the statement.

Suppose that $(a_{ij})_{1 \leq i \leq n, 1 \leq j \leq m}$ is a family of elements of $F$ such that

$$\sum_{i=1}^n \sum_{j=1}^m a_{ij} \phi_i \psi_j = 0.$$

For each $i = 1, \ldots, n$, let $b_i = \sum_{j=1}^m a_{ij} \psi_j \ (\in K)$. Then

$$\sum_{i=1}^n b_i \phi_i = 0,$$

so that, since $(\phi_i)_{i=1}^n$ is linearly independent over $K$, we have

$$0 = b_i = \sum_{j=1}^m a_{ij} \psi_j \quad \text{for all } i = 1, \ldots, n.$$

Hence, since $(\psi_j)_{j=1}^m$ is linearly independent over $F$, we have $a_{ij} = 0$ for all $i = 1, \ldots, n$ and $j = 1, \ldots, m$. Thus the family $\mathcal{B}$ is linearly independent over $F$.

Next, let $c \in L$. Then there exist $d_1, \ldots, d_n \in K$ such that

$$c = \sum_{i=1}^n d_i \phi_i.$$

Also, for each $i = 1, \ldots, n$, there exist $b_{i1}, \ldots, b_{im} \in F$ such that

$$d_i = \sum_{j=1}^m b_{ij} \psi_j.$$

Hence $c = \sum_{i=1}^n \sum_{j=1}^m b_{ij} \phi_i \psi_j$. Thus $\mathcal{B}$ spans $L$ as an $F$-space, and so the proof is complete. $\square$

We have already seen in 12.21(ii) that a simple extension $F \subseteq F(\theta)$, where $\theta$ is algebraic over $F$, is finite. However, there is an even stronger connection between finite extensions and algebraic elements.

**12.27** DEFINITION. Let $F \subseteq K$ be an extension of fields. We say that $K$ is *algebraic over $F$* (or that the extension is *algebraic*) precisely when every element of $K$ is algebraic over $F$.

**12.28** LEMMA. *A finite field extension is algebraic.*

*Proof.* Let $F \subseteq K$ be a finite field extension, and write $[K : F] = n$; let $\theta \in K$. Then $(\theta^i)_{i=0}^n$ is linearly dependent over $F$, and so there is a non-zero polynomial in $F[X]$ which has $\theta$ as a root. $\square$

**12.29** COROLLARY. *Let $F \subseteq K$ be an extension of fields. Then $K$ is a finite extension of $F$ if and only if $K$ can be obtained from $F$ by (field) adjunction of a finite number of elements all of which are algebraic with respect to $F$.*

*Proof.* ($\Rightarrow$) Let $(\phi_i)_{i=1}^n$ be a basis for the $F$-space $K$. Then each $\phi_i$ is algebraic over $F$ by 12.28, and, since

$$F\phi_1 + \cdots + F\phi_n \subseteq F(\phi_1, \ldots, \phi_n)$$

(because $F(\phi_1, \ldots, \phi_n)$ is closed under addition and multiplication), we must have $K = F(\phi_1, \ldots, \phi_n)$.

($\Leftarrow$) Assume that $K = F(\alpha_1, \ldots, \alpha_n)$, where $\alpha_1, \ldots, \alpha_n$ are all algebraic over $F$. We show by induction on $i$ that, for all $i = 1, \ldots, n$, the field $F(\alpha_1, \ldots, \alpha_i)$ is finite over $F$. That this is so when $i = 1$ is immediate from 12.21(ii). So suppose, inductively, that $2 \leq i \leq n$ and we have already proved that $F(\alpha_1, \ldots, \alpha_{i-1})$ is finite over $F$. Since

$$F \subseteq F(\alpha_1, \ldots, \alpha_{i-1}) \subseteq K,$$

it is automatic from 12.21(i) that $\alpha_i$ is algebraic over $F(\alpha_1, \ldots, \alpha_{i-1})$ (because $\alpha_i$ is a root of a non-zero polynomial in $F[X]$). Hence $F(\alpha_1, \ldots, \alpha_i) = F(\alpha_1, \ldots, \alpha_{i-1})(\alpha_i)$ is finite over $F(\alpha_1, \ldots, \alpha_{i-1})$ (by 12.21(ii) again), and so it follows from the Degrees Theorem 12.26 that $F(\alpha_1, \ldots, \alpha_i)$ is finite over $F$. This completes the inductive step. $\square$

**12.30** REMARK. Let $K \subseteq L$ be a simple extension of fields with $L = K(\theta)$ for some $\theta \in L$ which is algebraic over $K$. Suppose that $\mu \in L$ is such that $L = K(\mu)$ also. By 12.21(ii), the extension $K \subseteq L$ is finite, and so $\mu$ is algebraic over $K$ by 12.28. It thus makes sense to describe $L$ as a *simple algebraic extension* of $K$: there is no ambiguity.

**12.31** DEFINITION and EXERCISE. Let $F$ be a field, and let $f \in F[X]$ (where $X$ is an indeterminate) be a non-constant polynomial of degree $n$.

A field extension $F \subseteq K$ is said to be a *splitting field for $f$ over $F$* precisely when

(i) $f$ splits into linear factors in $K[X]$, so that there exist $\alpha, \theta_1, \ldots, \theta_n \in K$ such that
$$f = \alpha(X - \theta_1)\ldots(X - \theta_n) \quad \text{in } K[X],$$
and

(ii) $K$ can be obtained from $F$ by (field) adjunction of all the roots of $f$ in $K$, so that
$$K = F(\theta_1, \ldots, \theta_n).$$

Prove that there exists a splitting field for $f$ over $F$, and that each such splitting field $K_1$ satisfies $[K_1 : F] \leq n!$.

**12.32** EXERCISE. Let $F \subseteq K$ and $K \subseteq L$ be algebraic extensions of fields. Show that the field extension $F \subseteq L$ is algebraic. (You might find 12.29 helpful.)

**12.33** EXERCISE. Find the following degrees (over $\mathbb{Q}$, of various intermediate fields between $\mathbb{Q}$ and $\mathbb{C}$):

(i) $[\mathbb{Q}(\sqrt[3]{3}, \sqrt[5]{5}, \sqrt[7]{7}) : \mathbb{Q}]$;

(ii) $[\mathbb{Q}(\sqrt{2}, i) : \mathbb{Q}]$;

(iii) $[\mathbb{Q}(e^{2\pi i/3}, \sqrt[3]{3}) : \mathbb{Q}]$.

(Remember Eisenstein's Irreducibility Criterion [15, Theorem 2.8.14]!)

**12.34** THEOREM and DEFINITION. *Let $F \subseteq L$ be an extension of fields. Then*
$$K := \{a \in L : a \text{ is algebraic over } F\}$$
*is an intermediate field between $F$ and $L$ called the* algebraic closure of $F$ *in $L$.*

*Proof.* Of course, $F \subseteq K$ since every element of $F$ is algebraic over $F$.

Let $\alpha, \beta \in K$. By 12.29, the extension $F \subseteq F(\alpha, \beta)$ is finite. Now $\alpha - \beta \in F(\alpha, \beta)$, and so it follows from 12.28 that $\alpha - \beta$ is algebraic over $F$ and so belongs to $K$. A similar argument shows that, if $\beta \neq 0$, then $\alpha\beta^{-1} \in K$ too. It therefore follows from the Subfield Criterion 12.6 that $K$ is an intermediate field between $F$ and $L$. $\square$

**12.35** ♯EXERCISE. Let $K \subseteq L$ and $K' \subseteq L'$ be field extensions, and let $\xi : L \to L'$ and $\eta : K \to K'$ be isomorphisms such that $\xi|_K = \eta : K \xrightarrow{\cong} K'$. Let $\theta \in L$ and set $\xi(\theta) = \theta'$.

(i) Let $X$ and $Y$ be indeterminates. Show that $\eta$ induces a ring isomorphism $\tilde{\eta} : K[X] \to K'[Y]$ such that $\tilde{\eta}(a) = \eta(a)$ for all $a \in K$ and $\tilde{\eta}(X) = Y$.

(ii) Show that $\theta$ is algebraic over $K$ if and only if $\theta'$ is algebraic over $K'$, and that, when this is the case and $m$ is the minimal polynomial of $\theta$ with respect to $K$, then $\tilde{\eta}(m)$ is the minimal polynomial of $\theta'$ with respect to $K'$.

**12.36** ‡EXERCISE. Let $K \subseteq L$ and $K' \subseteq L'$ be field extensions, and let $\eta : K \to K'$ be an isomorphism. Let $X$ and $Y$ be indeterminates, and let $\tilde{\eta} : K[X] \to K'[Y]$ be the induced ring isomorphism (see 12.35(i)).

(i) Suppose that $\theta \in L$ is transcendental over $K$ and $\theta' \in L'$ is transcendental over $K'$. Prove that there is a unique isomorphism $\breve{\eta} : K(\theta) \to K'(\theta')$ which extends $\eta$ (that is, which is such that $\breve{\eta}(a) = \eta(a)$ for all $a \in K$) and satisfies $\breve{\eta}(\theta) = \theta'$.

(ii) Suppose that $\mu \in L$ is algebraic over $K$ with minimal polynomial $m$, and $\mu' \in L'$ is algebraic over $K'$ with minimal polynomial $m'$. Suppose also that $\tilde{\eta}(m) = m'$. Prove that there is a unique isomorphism $\hat{\eta} : K(\mu) \to K'(\mu')$ which extends $\eta$ and satisfies $\hat{\eta}(\mu) = \mu'$.

**12.37** EXERCISE. Let $F$ be a field, and let $f \in F[X]$ (where $X$ is an indeterminate) be a non-constant polynomial. Let $F \subseteq K$ and $F \subseteq K'$ be splitting fields (see 12.31) for $f$ over $F$.

Prove that there is an isomorphism $\eta : K \to K'$ such that $\eta\,|_F = \mathrm{Id}_F$, the identity mapping of $F$ onto itself.

This exercise and 12.31 together show that $f$ has, up to isomorphism, essentially one splitting field over $F$.

**12.38** EXERCISE. Let $F$ be a finite field having $p^n$ elements, where $p, n \in \mathbb{N}$ with $p$ prime (see 12.24). Prove that

$$a^{p^n} = a \quad \text{for all } a \in F.$$

(Do not forget Lagrange's Theorem from elementary group theory!)

**12.39** EXERCISE. Let $p, n \in \mathbb{N}$ with $p$ prime. Prove that two finite fields both having $p^n$ elements must be isomorphic. (Here is a hint: Exercises 12.31, 12.37 and 12.38 might be helpful.)

**12.40** EXERCISE. Let $K$ be a field and let $X$ be an indeterminate. Given $f = \sum_{i=0}^{n} a_i X^i \in K[X]$, we define the *formal derivative* $f'$ of $f$ by

$$f' = \sum_{i=1}^{n} i a_i X^{i-1} \in K[X].$$

(Of course, for $m \in \mathbb{N}$ and $h \in K[X]$, we interpret $mh$ as in 6.5.)

(i) Show that $(f+g)' = f'+g'$ and $(fg)' = fg'+f'g$ for all $f, g \in K[X]$.

(ii) Let $a \in K$ and let $f$ be a non-constant polynomial in $K[X]$. We say that $a$ is a *multiple* or *repeated root of* $f$ precisely when $(X - a)^2$ is a factor of $f$ in $K[X]$. Show that $f$ has a multiple root in some extension field of $K$ if and only if $GCD(f, f')$ has degree greater than 0.

**12.41** EXERCISE. Let $p, n \in \mathbb{N}$ with $p$ prime. Prove that there exists a finite field with exactly $p^n$ elements. (Here are some hints: consider a splitting field for $X^{p^n} - X$ over $\mathbb{Z}/p\mathbb{Z}$, and use Exercise 12.40.)

This exercise and 12.39 together show that there is, up to isomorphism, exactly one finite field having $p^n$ elements.

**12.42** FURTHER STEPS. We shall not, in this book, proceed further down the roads towards Galois theory, ruler and compass constructions, solution of equations by radicals, and the theory of finite fields: we shall have to leave the interested reader who has not learnt about these topics in other studies to explore them with the aid of other texts, such as [16]. It is perhaps worth pointing out, however, that the construction of isomorphisms as in 12.36(ii) is very relevant to Galois theory, and the result on formal derivatives in 12.40(ii) can play a significant rôle in the theory of separable algebraic field extensions: see, for example, [16, Chapter 8].

Something else which is not proved in this book is the fact that every field has an extension field which is algebraically closed: two references for this are [1, Chapter 1, Exercise 13] and [18, Chapter II, Theorem 32].

We now proceed with our programme aimed at transcendence degrees. We are going to associate with each finitely generated field extension $F \subseteq K$ a non-negative integer tr.deg$_F K$, called the 'transcendence degree of $K$ over $F$', in such a way that tr.deg$_F K = 0$ when $K$ is algebraic over $F$ and tr.deg$_F F(X) = 1$ for an indeterminate $X$. As was mentioned at the beginning of this chapter, this transcendence degree will give, roughly speaking, a measure of the largest integer $i \in \mathbb{N}_0$ such that there exist $i$ elements of $K$ which are algebraically independent over $F$. However, we shall also see that the theory of transcendence degrees of finitely generated field extensions has some similarities with the theory of dimensions of finite-dimensional vector spaces.

We shall need the idea of algebraically independent elements of $K$ (over $F$): recall from 1.14 that a family $(\alpha_i)_{i=1}^n$ of elements of $K$ is said to be algebraically independent over $F$ precisely when the only polynomial $\Theta \in F[X_1, \ldots, X_n]$ which is such that $\Theta(\alpha_1, \ldots, \alpha_n) = 0$ is the zero polynomial of $F[X_1, \ldots, X_n]$.

Note in particular that, when $n = 1$, $(\alpha_i)_{i=1}^1$ is algebraically independent over $F$ if and only if $\alpha_1$ is transcendental over $F$. We adopt the convention that the empty family of elements of $K$ is considered to be algebraically independent over $F$.

**12.43** REMARK. Let $F \subseteq K$ be an extension of fields, and suppose that the family $(\alpha_i)_{i=1}^n$ of elements of $K$ is algebraically independent over $F$. Let $X_1, \ldots, X_n$ be indeterminates, and denote by $F(X_1, \ldots, X_n)$ the quotient field of the integral domain $F[X_1, \ldots, X_n]$. (It follows from 12.7(ii) that this notation is consistent with that introduced in 12.13.) By 1.16, there is a unique ring isomorphism

$$h : F[\alpha_1, \ldots, \alpha_n] \longrightarrow F[X_1, \ldots, X_n]$$

such that $h(\alpha_i) = X_i$ for all $i = 1, \ldots, n$ and $h|_F : F \to F$ is the identity map. By 12.7(i) and 12.8(i), the ring isomorphism $h$ can be extended to an isomorphism (of fields)

$$h' : F(\alpha_1, \ldots, \alpha_n) \longrightarrow F(X_1, \ldots, X_n),$$

such that $h'(\alpha_i) = X_i$ for all $i = 1, \ldots, n$ and $h'|_F : F \to F$ is the identity map.

**12.44** DEFINITION. Let $F \subseteq K$ be an extension of fields, and let

$$\lambda, \alpha_1, \ldots, \alpha_n \in K.$$

We say that $\lambda$ is *algebraically dependent on* $\alpha_1, \ldots, \alpha_n$ *relative to* (or *over*) $F$ precisely when $\lambda$ is algebraic over $F(\alpha_1, \ldots, \alpha_n)$.

Occasionally, we shall need to interpret the above terminology in the case where $n$ has the value $0$: then it is to be taken to mean simply that $\lambda$ is algebraic over $F$.

**12.45** LEMMA. *Let $F \subseteq K$ be an extension of fields, and let $\alpha_1, \ldots, \alpha_n \in K$. Then the family $(\alpha_i)_{i=1}^n$ of elements of $K$ is algebraically independent over $F$ if and only if none of the $\alpha_i$ is algebraically dependent on the other $n - 1$ relative to $F$, that is, if and only if, for each $i = 1, \ldots, n$, the element $\alpha_i$ is transcendental over $F(\alpha_1, \ldots, \alpha_{i-1}, \alpha_{i+1}, \ldots, \alpha_n)$.*

*Note.* When $n = 1$, we interpret $F(\alpha_1, \ldots, \alpha_{i-1}, \alpha_{i+1}, \ldots, \alpha_n)$ as $F$, of course.

*Proof.* The result is clear when $n = 1$, and so we suppose that $n > 1$.

($\Rightarrow$) Suppose that $(\alpha_i)_{i=1}^n$ is algebraically independent over $F$ but that, for some $i \in \mathbb{N}$ with $1 \leq i \leq n$, the element $\alpha_i$ is algebraic over $E := F(\alpha_1, \ldots, \alpha_{i-1}, \alpha_{i+1}, \ldots, \alpha_n)$. Thus there exist $h \in \mathbb{N}$ and $e_0, \ldots, e_{h-1} \in E$ such that

$$\alpha_i^h + e_{h-1}\alpha_i^{h-1} + \cdots + e_1\alpha_i + e_0 = 0.$$

Now use 12.19 and 'clear denominators' (so to speak) to see that there exist polynomials $g_0, g_1, \ldots, g_h \in F[X_1, \ldots, X_{i-1}, X_{i+1}, \ldots, X_n]$ such that

$$\sum_{j=0}^{h} g_j(\alpha_1, \ldots, \alpha_{i-1}, \alpha_{i+1}, \ldots, \alpha_n)\alpha_i^j = 0$$

and $g_h(\alpha_1, \ldots, \alpha_{i-1}, \alpha_{i+1}, \ldots, \alpha_n) \neq 0$. This contradicts the algebraic independence of $(\alpha_i)_{i=1}^{n}$ over $F$.

($\Leftarrow$) Suppose that $(\alpha_i)_{i=1}^{n}$ is not algebraically independent over $F$. Thus there exists $\Theta \in F[X_1, \ldots, X_n]$ such that $\Theta(\alpha_1, \ldots, \alpha_n) = 0$ but $\Theta \neq 0$. Clearly, no such $\Theta$ can be a constant polynomial. Let $r$ be the least integer such that $1 \leq r \leq n$ and there exists $\Phi \in F[X_1, \ldots, X_r]$ such that $\Phi(\alpha_1, \ldots, \alpha_r) = 0$ but $\Phi \neq 0$. Since $\alpha_1$ has to be transcendental over $F$ by hypothesis, we must have $r \geq 2$.

By choice of $r$, at least one of the non-zero monomials of $\Phi$ must involve $X_r$. Thus there exist $h \in \mathbb{N}$ and $\Psi_0, \ldots, \Psi_h \in F[X_1, \ldots, X_{r-1}]$ such that

$$\Phi = \sum_{j=0}^{h} \Psi_j X_r^j$$

and $\Psi_h \neq 0$. Now

$$\sum_{j=0}^{h} \Psi_j(\alpha_1, \ldots, \alpha_{r-1})\alpha_r^j = \Phi(\alpha_1, \ldots, \alpha_r) = 0.$$

By choice of $r$, we must have $\beta := \Psi_h(\alpha_1, \ldots, \alpha_{r-1}) \neq 0$, and if we divide the last displayed equation through by $\beta$, we obtain a contradiction to the hypothesis that $\alpha_r$ is transcendental over $F(\alpha_1, \ldots, \alpha_{r-1}, \alpha_{r+1}, \ldots, \alpha_n)$.

This contradiction completes the proof. $\square$

**12.46** PROPOSITION. *Let $F \subseteq K$ be a field extension, and let*

$$\alpha_1, \ldots, \alpha_n, \beta_1, \ldots, \beta_m, \gamma \in K.$$

*Suppose that, for each $j = 1, \ldots, m$, the element $\beta_j$ is algebraically dependent on $\alpha_1, \ldots, \alpha_n$ relative to $F$, and suppose also that $\gamma$ is algebraically dependent on $\beta_1, \ldots, \beta_m$ relative to $F$. Then $\gamma$ is algebraically dependent on $\alpha_1, \ldots, \alpha_n$ relative to $F$.*

*Proof.* Since $\beta_j$ is algebraic over $F(\alpha_1, \ldots, \alpha_n)$ (for each $j = 1, \ldots, m$), it follows from 12.29 that $F(\alpha_1, \ldots, \alpha_n)(\beta_1, \ldots, \beta_m)$ is a finite extension of $F(\alpha_1, \ldots, \alpha_n)$. Also, since $\gamma$ is algebraic over $F(\beta_1, \ldots, \beta_m)$, so that it must automatically be algebraic over $F(\alpha_1, \ldots, \alpha_n)(\beta_1, \ldots, \beta_m)$, we deduce that

$$F(\alpha_1, \ldots, \alpha_n)(\beta_1, \ldots, \beta_m)(\gamma)$$

is a finite extension of $F(\alpha_1, \ldots, \alpha_n)(\beta_1, \ldots, \beta_m)$. We now use the Degrees Theorem 12.26 to see that $F(\alpha_1, \ldots, \alpha_n)(\beta_1, \ldots, \beta_m)(\gamma)$ is a finite extension of $F(\alpha_1, \ldots, \alpha_n)$, so that $F(\alpha_1, \ldots, \alpha_n)(\gamma)$ is a finite extension of $F(\alpha_1, \ldots, \alpha_n)$ by the same result. Hence, by 12.21(ii), $\gamma$ is algebraic over $F(\alpha_1, \ldots, \alpha_n)$, that is, $\gamma$ is algebraically dependent on $\alpha_1, \ldots, \alpha_n$ relative to $F$. $\square$

**12.47** LEMMA. *Let $F \subseteq K$ be a field extension, and let $\alpha_1, \ldots, \alpha_n, \beta \in K$. Suppose that $\beta$ is algebraically dependent on $\alpha_1, \ldots, \alpha_n$ relative to $F$, but is not algebraically dependent on $\alpha_1, \ldots, \alpha_{n-1}$ relative to $F$. Then $\alpha_n$ is algebraically dependent on $\alpha_1, \ldots, \alpha_{n-1}, \beta$ relative to $F$.*

*Note.* An obvious interpretation has to be made in the case in which $n = 1$.

*Proof.* Let $E := F(\alpha_1, \ldots, \alpha_{n-1})$, with the understanding that $E = F$ if $n = 1$. By hypothesis, $\beta$ is algebraic over $E(\alpha_n)$, and so there exist $h \in \mathbb{N}$ and $c_0, \ldots, c_{h-1} \in E(\alpha_n)$ such that

$$\beta^h + c_{h-1}\beta^{h-1} + \cdots + c_1\beta + c_0 = 0.$$

We can now use 12.19 to see that there exist $g_0, g_1, \ldots, g_h \in E[X_1]$ such that $g_h(\alpha_n) \neq 0$ and

$$g_h(\alpha_n)\beta^h + g_{h-1}(\alpha_n)\beta^{h-1} + \cdots + g_1(\alpha_n)\beta + g_0(\alpha_n) = 0.$$

It follows that the polynomial

$$f := g_h X_2^h + g_{h-1} X_2^{h-1} + \cdots + g_1 X_2 + g_0 \in E[X_1, X_2]$$

is non-zero, although $f(\alpha_n, \beta) = 0$. Now we can rewrite $f$ as

$$f := g'_{h'} X_1^{h'} + g'_{h'-1} X_1^{h'-1} + \cdots + g'_1 X_1 + g'_0,$$

where $h' \in \mathbb{N}$ and $g'_0, g'_1, \ldots, g'_{h'} \in E[X_2]$ and not all the $g'_i$ are 0. Since $\beta$ is not algebraically dependent on $\alpha_1, \ldots, \alpha_{n-1}$ relative to $F$, it follows that not all the $g'_i(\beta)$ are 0. But

$$0 = f(\alpha_n, \beta) = g'_{h'}(\beta)\alpha_n^{h'} + g'_{h'-1}(\beta)\alpha_n^{h'-1} + \cdots + g'_1(\beta)\alpha_n + g'_0(\beta),$$

and so we see that $\alpha_n$ is algebraic over $E(\beta)$, that is, $\alpha_n$ is algebraically dependent on $\alpha_1, \ldots, \alpha_{n-1}, \beta$ relative to $F$. $\square$

**12.48** COROLLARY. *Let $F \subseteq K$ be a field extension, and let $\alpha_1, \ldots, \alpha_n \in K$. Suppose that the family $(\alpha_i)_{i=1}^{n-1}$ is algebraically independent over $F$, but that $(\alpha_i)_{i=1}^{n}$ is not. Then $\alpha_n$ is algebraically dependent on $\alpha_1, \ldots, \alpha_{n-1}$ relative to $F$.*

*Proof.* By 12.45, one of the $\alpha_i$ is algebraically dependent on the remaining $n - 1$ relative to $F$. Thus either $\alpha_n$ is algebraically dependent on $\alpha_1, \ldots, \alpha_{n-1}$ relative to $F$, which is what we want, or there is a $j \in \mathbf{N}$ with $1 \leq j \leq n - 1$ such that $\alpha_j$ is algebraically dependent on

$$\alpha_1, \ldots, \alpha_{j-1}, \alpha_{j+1}, \ldots, \alpha_n$$

relative to $F$; in the second case, since $\alpha_j$ is not algebraically dependent on $\alpha_1, \ldots, \alpha_{j-1}, \alpha_{j+1}, \ldots, \alpha_{n-1}$ relative to $F$ by 12.45, it follows from Lemma 12.47 that $\alpha_n$ is algebraically dependent on $\alpha_1, \ldots, \alpha_{n-1}$ relative to $F$. $\square$

**12.49** DEFINITION. Let $F \subseteq K$ be a field extension, and let

$$\alpha_1, \ldots, \alpha_n, \beta_1, \ldots, \beta_m \in K.$$

We say that the families $(\alpha_i)_{i=1}^{n}$ and $(\beta_i)_{i=1}^{m}$ are *algebraically equivalent relative to $F$* precisely when $\beta_j$ is algebraically dependent on $\alpha_1, \ldots, \alpha_n$ relative to $F$ for all $j = 1, \ldots, m$ and $\alpha_i$ is algebraically dependent on $\beta_1, \ldots, \beta_m$ relative to $F$ for all $i = 1, \ldots, n$.

**12.50** ♯EXERCISE. Let $F \subseteq K$ be an extension of fields. Show that the relation 'is algebraically equivalent to relative to $F$' is an equivalence relation on the set of all finite families of elements of $K$.

**12.51** COROLLARY. *Let $F \subseteq K$ be an extension of fields, and let $(\alpha_i)_{i=1}^{n}$ be a family of elements of $K$. Then there exist $s \in \mathbf{N}_0$ with $s \leq n$, and $s$ different integers $i_1, \ldots, i_s$ between 1 and $n$ such that the (possibly empty) family $(\alpha_{i_j})_{j=1}^{s}$ is algebraically independent over $F$ and algebraically equivalent to $(\alpha_i)_{i=1}^{n}$ relative to $F$.*

*Proof.* Choose $s \in \mathbf{N}_0$, with $s \leq n$, for which there exist $s$ different integers $i_1, \ldots, i_s$ between 1 and $n$ such that the family $(\alpha_{i_j})_{j=1}^{s}$ is algebraically independent over $F$, and such that $s$ is as large as possible subject to these conditions. It then follows from 12.48 that, for every $i = 1, \ldots, n$, whether or not $i$ is one of the $i_j$, the element $\alpha_i$ is algebraically dependent on $\alpha_{i_1}, \ldots, \alpha_{i_s}$ relative to $F$. $\square$

**12.52** EXCHANGE THEOREM. *Let $F \subseteq K$ be an extension of fields, and let $(\alpha_i)_{i=1}^n, (\beta_j)_{j=1}^m$ be families of elements of $K$. Suppose that $(\beta_j)_{j=1}^m$ is algebraically independent over $F$, and that, for each $j = 1, \ldots, m$, the element $\beta_j$ is algebraically dependent on $\alpha_1, \ldots, \alpha_n$ relative to $F$.*

*Then $m \leq n$ and there exist $m$ different integers $i_1, \ldots, i_m$ between 1 and $n$ such that the family $(\gamma_i)_{i=1}^n$, where (for $1 \leq i \leq n$)*

$$\gamma_i = \begin{cases} \beta_j & \text{if } i = i_j \text{ for some } j \text{ with } 1 \leq j \leq m, \\ \alpha_i & \text{otherwise,} \end{cases}$$

*is algebraically equivalent to $(\alpha_i)_{i=1}^n$ relative to $F$.*

*Proof.* We argue by induction on $m$, the result being clear in the case in which $m = 0$. Suppose, inductively, that $m > 0$ and the result has been proved for smaller values of $m$.

Application of the inductive hypothesis to the family $(\beta_j)_{j=1}^{m-1}$ shows that $m - 1 \leq n$ and there exist $m - 1$ different integers $i_1, \ldots, i_{m-1}$ between 1 and $n$ such that the family $(\delta_i)_{i=1}^n$, where (for $1 \leq i \leq n$)

$$\delta_i = \begin{cases} \beta_j & \text{if } i = i_j \text{ for some } j \text{ with } 1 \leq j \leq m-1, \\ \alpha_i & \text{otherwise,} \end{cases}$$

is algebraically equivalent to $(\alpha_i)_{i=1}^n$ relative to $F$.

Note that, by 12.46, $\beta_m$ is algebraically dependent on $\delta_1, \ldots, \delta_n$ relative to $F$. Choose $s \in \mathbb{N}$ as small as possible such that $\beta_m$ is algebraically dependent, relative to $F$, on $\delta_{k_1}, \ldots, \delta_{k_s}$ (where $k_1, \ldots, k_s$ are $s$ different integers between 1 and $n$). By 12.45, we can assume that

$$k_s \in \{1, \ldots, n\} \setminus \{i_1, \ldots, i_{m-1}\}.$$

Hence $m \leq n$. Set $i_m = k_s$; note that $\delta_{k_s} = \alpha_{i_m}$.

Now (for $i = 1, \ldots, n$) set

$$\gamma_i = \begin{cases} \beta_m & \text{for } i = i_m, \\ \delta_i & \text{otherwise,} \end{cases}$$

so that

$$\gamma_i = \begin{cases} \beta_j & \text{if } i = i_j \text{ for some } j \text{ with } 1 \leq j \leq m, \\ \alpha_i & \text{otherwise.} \end{cases}$$

Now $(\gamma_i)_{i=1}^n$ is algebraically equivalent to $(\delta_i)_{i=1}^n$ relative to $F$, since $\beta_m$ is algebraically dependent on $\delta_1, \ldots, \delta_n$ relative to $F$, while $\alpha_{i_m}$ is algebraically dependent on $\gamma_1, \ldots, \gamma_n$ relative to $F$ by 12.47 and the definition of $s$.

An application of 12.50 now shows that $(\gamma_i)_{i=1}^n$ is algebraically equivalent to $(\alpha_i)_{i=1}^n$ relative to $F$. This completes the inductive step, and the proof. $\square$

**12.53** COROLLARY. *Let $F \subseteq K$ be an extension of fields; let*

$$(\alpha_i)_{i=1}^n, \quad (\beta_j)_{j=1}^m$$

*be families of elements of $K$ which are algebraically equivalent relative to $F$ and which are both algebraically independent over $F$. Then $n = m$.*

*Proof.* This is immediate from the Exchange Theorem 12.52, because that theorem shows that $m \leq n$ and $n \leq m$. $\square$

**12.54** DEFINITIONS. Let $F \subseteq K$ be an extension of fields. A (finite) *transcendence basis for $K$ over $F$* is a (possibly empty) family $(\alpha_i)_{i=1}^n$ of elements of $K$ which is algebraically independent over $F$ and such that every element of $K$ is algebraically dependent on $\alpha_1, \ldots, \alpha_n$ relative to $F$.

It follows from 12.53 that, if there exists a finite transcendence basis for $K$ over $F$, then any two such transcendence bases have the same number of elements, and the number of elements in each transcendence basis for $K$ over $F$ is called the *transcendence degree of $K$ over $F$*, and denoted by $\mathrm{tr.deg}_F K$. In this situation, we say that $K$ *has finite transcendence degree over $F$*; we shall sometimes abbreviate this by writing '$\mathrm{tr.deg}_F K < \infty$'.

**12.55** REMARKS. Let $F \subseteq K$ be an extension of fields.

(i) Note that $\mathrm{tr.deg}_F K = 0$ if and only if $K$ is algebraic over $F$.

(ii) Suppose that $K$ is a finitely generated extension field of $F$, so that there exist $\beta_1, \ldots, \beta_m \in K$ such that $K = F(\beta_1, \ldots, \beta_m)$. Then it follows from 12.51 and 12.46 that there exists a (possibly empty) transcendence basis for $K$ over $F$ made up from elements from the set $\{\beta_1, \ldots, \beta_m\}$. Thus a finitely generated field extension always has finite degree of transcendence.

(iii) If $K$ has finite transcendence degree $n$ over $F$, and $n > 0$, then $K$ can be obtained from $F$ by first adjoining the members of a family $(\alpha_i)_{i=1}^n$ which is algebraically independent over $F$, and then making an algebraic extension: just take a transcendence basis for $K$ over $F$ for $(\alpha_i)_{i=1}^n$. Information about the structure of the intermediate field $F(\alpha_1, \ldots, \alpha_n)$ is provided by 12.43.

(iv) Suppose that $K$ has finite transcendence degree over $F$. We say that $K$ is a *pure transcendental extension of $F$* precisely when $K$ *itself* can be obtained from $F$ by the adjunction of the members of a (possibly empty) transcendence basis for $K$ over $F$. Note, in particular, that $F$ itself is considered to be a pure transcendental extension of $F$.

(v) If $K$ is an algebraic extension of a finitely generated intermediate field $F(\beta_1, \ldots, \beta_m)$ between $F$ and $K$, then it follows from 12.46 and part (ii) above that there is a transcendence basis for $K$ over $F$ composed of elements from the set $\{\beta_1, \ldots, \beta_m\}$.

(vi) Suppose that $K$ has finite transcendence degree $n$ over $F$, and that $\alpha_1, \ldots, \alpha_n \in K$ are such that $K$ is algebraic over $F(\alpha_1, \ldots, \alpha_n)$. Then it follows from part (v) above, and the fact (12.54) that any pair of transcendence bases for $K$ over $F$ have the same number of elements, that the family $(\alpha_i)_{i=1}^n$ must be algebraically independent over $F$, and so already a transcendence basis for $K$ over $F$.

(vii) Suppose that $K$ has finite transcendence degree $n$ over $F$, and that $\mathcal{A} := (\alpha_i)_{i=1}^n$ is a transcendence basis for $K$ over $F$. Suppose also that $\mathcal{B} := (\beta_j)_{j=1}^m$ is a family of elements of $K$ which is algebraically independent over $F$.

Then it follows from 12.46, the Exchange Theorem 12.52 and part (vi) above that $m \leq n$ and $\mathcal{B}$ can be enlarged to a transcendence basis for $K$ over $F$ by the addition of $n - m$ elements from $\mathcal{A}$.

In particular, a family of $n$ $(= \mathrm{tr.deg}_F K)$ elements of $K$ which is algebraically independent over $F$ must already be a transcendence basis for $K$ over $F$.

The reader will probably have noticed similarities between many of the comments in 12.55 and fundamental facts from the theory of linear independence in vector spaces. However, our next theorem is in a different style.

**12.56** THEOREM. *Let $F \subseteq K \subseteq L$ be field extensions, let $(\alpha_i)_{i=1}^n$ be a family of elements of $K$ which is algebraically independent over $F$, and suppose that $(\beta_j)_{j=1}^m$ is a family of elements of $L$ which is algebraically independent over $K$.*

*Then $(\gamma_r)_{r=1}^{n+m}$, where*

$$\gamma_r = \begin{cases} \alpha_r & \text{for } 1 \leq r \leq n, \\ \beta_{r-n} & \text{for } n < r \leq n + m, \end{cases}$$

*is algebraically independent over $F$.*

*Furthermore, if $K$ has finite transcendence degree over $F$ and $L$ has finite transcendence degree over $K$, then $L$ has finite transcendence degree over $F$, and*

$$\mathrm{tr.deg}_F L = \mathrm{tr.deg}_F K + \mathrm{tr.deg}_K L.$$

*Proof.* Suppose that $0 \neq \Theta \in F[X_1, \ldots, X_{n+m}]$ is such that

$$\Theta(\alpha_1, \ldots, \alpha_n, \beta_1, \ldots, \beta_m) = 0,$$

and look for a contradiction. Since $\Theta \neq 0$, there exist a non-empty finite subset $\Lambda$ of $\mathbf{N}_0{}^m$ and polynomials

$$\Phi_{j_1,\ldots,j_m} \in F[X_1, \ldots, X_n] \setminus \{0\} \qquad ((j_1, \ldots, j_m) \in \Lambda)$$

such that

$$\Theta = \sum_{(j_1,\ldots,j_m)\in\Lambda} \Phi_{j_1,\ldots,j_m} X_{n+1}^{j_1} \ldots X_{n+m}^{j_m}.$$

Since $(\alpha_i)_{i=1}^n$ is algebraically independent over $F$, we must have

$$\Phi_{j_1,\ldots,j_m}(\alpha_1, \ldots, \alpha_n) \neq 0 \quad \text{for all } i \in \Lambda.$$

Thus

$$\Psi := \sum_{(j_1,\ldots,j_m)\in\Lambda} \Phi_{j_1,\ldots,j_m}(\alpha_1, \ldots, \alpha_n) X_{n+1}^{j_1} \ldots X_{n+m}^{j_m}$$

is a non-zero polynomial in $K[X_{n+1}, \ldots, X_{n+m}]$. But

$$\Psi(\beta_1, \ldots, \beta_m) = \Theta(\alpha_1, \ldots, \alpha_n, \beta_1, \ldots, \beta_m) = 0,$$

and so we have a contradiction to the fact that $(\beta_j)_{j=1}^m$ is algebraically independent over $K$. This proves the first part of the theorem.

For the second part, take $(\alpha_i)_{i=1}^n$ to be a transcendence basis for $K$ over $F$ and $(\beta_j)_{j=1}^m$ to be a transcendence basis for $L$ over $K$. Define $(\gamma_r)_{r=1}^{n+m}$ as in the statement of the theorem: the first part of the proof shows that this family of elements of $L$ is algebraically independent over $F$, and we aim to complete the proof by showing that it is actually a transcendence basis for $L$ over $F$.

Let $\nu \in L$, so that $\nu$ is algebraic over $K(\beta_1, \ldots, \beta_m)$. Thus there exist $h \in \mathbf{N}$ and $e_0, \ldots, e_{h-1} \in K(\beta_1, \ldots, \beta_m)$ such that

$$\nu^h + e_{h-1}\nu^{h-1} + \cdots + e_1\nu + e_0 = 0.$$

Now use 12.19 and 'clear denominators' to see that there exist polynomials $g_0, g_1, \ldots, g_h \in K[X_{n+1}, \ldots, X_{n+m}]$ such that

$$\sum_{j=0}^h g_j(\beta_1, \ldots, \beta_m)\nu^j = 0$$

and $g_h(\beta_1, \ldots, \beta_m) \neq 0$.

Let $\Delta$ denote the (finite) subset of $K$ consisting of all the non-zero coefficients of the polynomials $g_0, \ldots, g_h$, so that

$$g_0, g_1, \ldots, g_h \in F(\Delta)[X_{n+1}, \ldots, X_{n+m}]$$

and $\nu$ is algebraic over $F(\Delta)(\beta_1, \ldots, \beta_m)$. Since each element of $\Delta$ is algebraically dependent on $\alpha_1, \ldots, \alpha_n$ relative to $F$, it follows from 12.46 that $\nu$ is algebraically dependent on $\alpha_1, \ldots, \alpha_n, \beta_1, \ldots, \beta_m$ relative to $F$. This is enough to complete the proof. $\square$

**12.57** EXERCISE. Let $F \subseteq L$ be an extension of fields such that $L$ has finite degree of transcendence over $F$. Let $K$ be an intermediate field between $F$ and $L$. Show that $\mathrm{tr.deg}_F K < \infty$ and $\mathrm{tr.deg}_K L < \infty$.

**12.58** EXERCISE. Let $S$ be the polynomial ring $\mathbf{R}[X_1, \ldots, X_n]$ over the real field $\mathbf{R}$ in the $n$ indeterminates $X_1, \ldots, X_n$, where $n > 1$. Show that $f := X_1^2 + X_2^2 + \cdots + X_n^2$ is irreducible in $S$, so that, by 3.42, $f$ generates a prime ideal of $S$. Let $F$ denote the quotient field of the integral domain

$$\mathbf{R}[X_1, \ldots, X_n]/(X_1^2 + X_2^2 + \cdots + X_n^2)\mathbf{R}[X_1, \ldots, X_n];$$

note that $F$ can be regarded as an extension field of $\mathbf{R}$ in a natural way. Find $\mathrm{tr.deg}_{\mathbf{R}} F$.

# Chapter 13

# Integral dependence on subrings

Before we are in a position to make the planned applications to dimension theory of our work in Chapter 12 on transcendence degrees of finitely generated field extensions, we really need to study the theory of integral dependence in commutative rings. This can be viewed as a generalization to commutative ring theory of another topic studied in Chapter 12, namely algebraic field extensions.

Let $F \subseteq K$ be an extension of fields (as in 12.1). Recall from 12.21 that an element $\lambda \in K$ is algebraic over $F$ precisely when there exists a monic polynomial $g \in F[X]$ such that $g(\lambda) = 0$. Let $R$ be a subring of the commutative ring $S$. We shall say that an element $s \in S$ is 'integral over $R$' precisely when there exists a monic polynomial $f \in R[X]$ such that $f(s) = 0$. Our main task in this chapter is to develop this and related concepts. Some of our results, such as 13.22 below which shows that $\{s \in S : s \text{ is integral over } R\}$ is a subring of $S$ which contains $R$, will represent generalizations to commutative rings of results in Chapter 12 about field extensions, but these generalizations tend to be harder to prove. (The more straightforward arguments which work for field extensions were included in Chapter 12, as it was thought possible that some readers might only be interested in that situation.)

Indeed, early in the development of the theory of integral dependence, we shall reach a situation where it is desirable to use a variant of what Matsumura calls the 'determinant trick' (see [8, Theorem 2.1]). The proof of this presented in this chapter uses facts from the theory of determinants of square matrices over a commutative ring $R$, including the fact that if $A$

is an $n \times n$ matrix with entries in $R$, then

$$A(\text{Adj } A) = (\text{Adj } A)A = (\det A)I_n.$$

It is expected that this result in the special case in which $R$ is a field will be very familiar to readers, but also that some readers will not previously have thought about matrices with entries from commutative rings which are not even integral domains. (In the special case in which $R$ is an integral domain, we can, of course, use the fact that $R$ can be embedded in its field of fractions to obtain the above result.) The proper response from such a reader would be to at least pause and ask himself or herself just why the above equations remain true in the more general situation. In fact, in keeping with the spirit of this book, we are going to lead the reader, admittedly fairly quickly and along a path consisting mainly of exercises, through the development of the theory of determinants of square matrices with entries from $R$ up to a proof of the above displayed equations. It is in this way that the chapter begins.

**13.1** NOTATION and DEFINITION. Throughout our discussion of matrices with entries in commutative rings, we shall let $n$ denote a positive integer and $R$ denote a commutative ring; $M_{m,n}(R)$, where $m \in \mathbb{N}$ also, will denote the set of all $m \times n$ matrices with entries in (we shall also say 'over') $R$. The symbolism $A = (a_{ij}) \in M_{m,n}(R)$ will mean that, for each $i = 1, \ldots, m$ and $j = 1, \ldots, n$, the $(i, j)$-th entry of $A$, that is, the entry at the intersection of the $i$-th row and $j$-th column, is $a_{ij}$. In complicated situations, we may write $a_{i,j}$ instead of $a_{ij}$.

Multiplication of matrices over $R$ is performed exactly as in the case where $R$ is a field. The transpose of $A \in M_{m,n}(R)$ will be denoted by $A^T$.

We shall use $S_n$ to denote the symmetric group of order $n!$, that is, the group of all permutations of the set of the first $n$ positive integers with respect to composition of permutations.

Let $A = (a_{ij}) \in M_{n,n}(R)$. By the *determinant of $A$*, denoted $\det A$, we mean the element of $R$ given by

$$\det A = \sum_{\sigma \in S_n} (\text{sgn } \sigma)a_{1,\sigma(1)}a_{2,\sigma(2)} \cdots a_{n,\sigma(n)},$$

where $\text{sgn } \sigma$, for $\sigma \in S_n$, denotes the sign of $\sigma$.

**13.2** ‡EXERCISE. Let $A = (a_{ij}) \in M_{n,n}(R)$.

(i) Show that $\det A^T = \det A$.

(ii) Show that, for $r \in R$ and an integer $i$ between 1 and $n$, if $B \in M_{n,n}(R)$ is obtained from $A$ by multiplication of all the entries in the $i$-th row of $A$ by $r$, then $\det B = r \det A$. Deduce that $\det(rA) = r^n \det A$.

Obtain similar results for columns.

(iii) Show that, for a row matrix $v \in M_{1,n}(R)$ and an integer $i$ between 1 and $n$, if $C$ is obtained from $A$ by substitution of $v$ for the $i$-th row of $A$, and $D$ is obtained from $A$ by addition of $v$ to the $i$-th row of $A$, then

$$\det D = \det A + \det C.$$

Obtain a similar result for columns.

(iv) Show that, if $A$ is either upper triangular or lower triangular, then $\det A = a_{11}a_{22} \ldots a_{nn}$, the product of the diagonal entries of $A$.

**13.3** LEMMA. *Suppose that $B = (b_{ij}) \in M_{n,n}(R)$ has its $k$-th and $l$-th rows equal, where $1 \leq k < l \leq n$. Then $\det B = 0$.*

*A similar result holds for columns.*

*Proof.* (I am grateful to D. A. Jordan for pointing out the argument in the following proof to me.) Let $A_n$ denote the alternating group of order $n!/2$. Since the mapping $\phi : A_n \to S_n \setminus A_n$ defined by $\phi(\sigma) = \sigma \circ (kl)$ for all $\sigma \in A_n$ is a bijection, we have

$$\det B = \sum_{\sigma \in S_n} (\text{sgn }\sigma) b_{1,\sigma(1)} \ldots b_{n,\sigma(n)}$$

$$= \sum_{\sigma \in A_n} b_{1,\sigma(1)} \ldots b_{n,\sigma(n)} - \sum_{\sigma \in A_n} b_{1,(\sigma \circ (kl))(1)} \ldots b_{n,(\sigma \circ (kl))(n)}$$

$$= \sum_{\sigma \in A_n} b_{1,\sigma(1)} \ldots b_{k,\sigma(k)} \ldots b_{l,\sigma(l)} \ldots b_{n,\sigma(n)}$$

$$\quad - \sum_{\sigma \in A_n} b_{1,\sigma(1)} \ldots b_{k,\sigma(l)} \ldots b_{l,\sigma(k)} \ldots b_{n,\sigma(n)}$$

$$= 0$$

since the $k$-th and $l$-th rows of $B$ are equal.

The proof of the result for columns is left as an exercise. $\square$

**13.4** ♯EXERCISE. Prove the result for columns in Lemma 13.3.

**13.5** ♯EXERCISE. Suppose that $n > 1$, and $A \in M_{n,n}(R)$. Let $k, l \in \mathbb{N}$ with $1 \leq k, l \leq n$ and $k \neq l$. Let $r \in R$. Show that, if $B \in M_{n,n}(R)$ can be obtained from $A$ by the addition of $r$ times the $k$-th row of $A$ to the $l$-th, then $\det B = \det A$.

Establish also the analogous result for columns.

**13.6** COROLLARY. *Suppose that $n > 1$, let $A \in M_{n,n}(R)$ and let $B \in M_{n,n}(R)$ be obtained from $A$ by the interchange of the $i$-th and $j$-th rows of $A$, where $1 \leq i < j \leq n$. Then $\det B = - \det A$.*

*A similar result holds for columns.*

*Proof.* We can interchange the $i$-th and $j$-th rows of $A$ by a succession of four operations of the types considered in 13.2(ii) and 13.5: first add the $j$-th row to the $i$-th; then subtract the (new) $i$-th row from the $j$-th in the result; then add the (new) $j$-th row to the $i$-th in the result; and finally multiply the $j$-th row by $-1$. The claim therefore follows from Exercises 13.2(ii) and 13.5.

The result for columns is proved similarly.  $\square$

**13.7** DEFINITIONS. Suppose that $n > 1$, and $A = (a_{ij}) \in M_{n,n}(R)$. Let $k, l \in \mathbb{N}$ with $1 \leq k, l \leq n$. Denote by $\tilde{A}_{kl}$ the $(n-1) \times (n-1)$ submatrix of $A$ obtained by deleting the $k$-th row and $l$-th column of $A$; we call $\det \tilde{A}_{kl}$ the $(k, l)$-*th minor of* $A$, and

$$\mathrm{cof}_{kl}(A) := (-1)^{k+l} \det \tilde{A}_{kl}$$

the $(k, l)$-*th cofactor of* $A$.

The matrix $\mathrm{Adj}\, A := (d_{ij}) \in M_{n,n}(R)$ for which $d_{ij} = \mathrm{cof}_{ji}(A)$ for all $i, j = 1, \ldots, n$ is called the *adjoint* or *adjugate matrix* of $A$. Thus $\mathrm{Adj}\, A$ is, roughly speaking, the transpose of the matrix of cofactors of $A$.

The adjoint of a $1 \times 1$ matrix over $R$ is to be interpreted as $I_1$, the $1 \times 1$ identity matrix over $R$.

**13.8** LEMMA. *Let the situation be as in* 13.7. *Then*

$$B := \begin{pmatrix} a_{1,1} & \cdots & a_{1,l-1} & a_{1,l} & a_{1,l+1} & \cdots & a_{1,n} \\ \vdots & & \vdots & \vdots & \vdots & & \vdots \\ a_{k-1,1} & \cdots & a_{k-1,l-1} & a_{k-1,l} & a_{k-1,l+1} & \cdots & a_{k-1,n} \\ 0 & \cdots & 0 & 1 & 0 & \cdots & 0 \\ a_{k+1,1} & \cdots & a_{k+1,l-1} & a_{k+1,l} & a_{k+1,l+1} & \cdots & a_{k+1,n} \\ \vdots & & \vdots & \vdots & \vdots & & \vdots \\ a_{n,1} & \cdots & a_{n,l-1} & a_{n,l} & a_{n,l+1} & \cdots & a_{n,n} \end{pmatrix}$$

*satisfies* $\det B = \mathrm{cof}_{kl}(A)$.

*Proof.* We use the notation of 13.7. We can use 13.5 to see that $\det B = \det C$, where $C$ is the result of changing the $l$-th column of $B$ to the $k$-th column of the $n \times n$ identity matrix in $M_{n,n}(R)$.

Next, by interchanging (if $k < n$) the $k$-th and $(k+1)$-th rows, and then interchanging the $(k+1)$-th and $(k+2)$-th rows, and so on up to the $(n-1)$-th and $n$-th rows, and then performing a similar $n-l$ interchanges of columns, we can deduce from 13.6 that $\det C = (-1)^{n-k+n-l} \det D$, where

$$D = \left( \begin{array}{c|c} \tilde{A}_{kl} & 0 \\ \hline 0 & 1 \end{array} \right).$$

Set $D = (d_{ij})$. Now

$$\det D = \sum_{\sigma \in S_n} (\operatorname{sgn} \sigma) d_{1,\sigma(1)} \ldots d_{n-1,\sigma(n-1)} d_{n,\sigma(n)}.$$

But $d_{n1} = \ldots = d_{n,n-1} = 0$ and $d_{nn} = 1$, and so only those permutations $\sigma \in S_n$ for which $\sigma(n) = n$ can conceivably contribute to the above sum; for such a $\sigma$, the restriction of $\sigma$ to $\{1, \ldots, n-1\}$ gives rise to a permutation $\sigma' \in S_{n-1}$ having the same sign as $\sigma$. In fact, each $\phi \in S_{n-1}$ arises as $\sigma'$ for exactly one $\sigma \in S_n$. Hence

$$\det D = \sum_{\substack{\sigma \in S_n \\ \sigma(n)=n}} (\operatorname{sgn} \sigma) d_{1,\sigma(1)} \ldots d_{n-1,\sigma(n-1)} d_{n,n}$$

$$= \sum_{\sigma' \in S_{n-1}} (\operatorname{sgn} \sigma') d_{1,\sigma'(1)} \ldots d_{n-1,\sigma'(n-1)}$$

$$= \det \tilde{A}_{kl},$$

so that $\det B = (-1)^{k+l} \det \tilde{A}_{kl} = \operatorname{cof}_{kl}(A)$. $\square$

**13.9 THE EXPANSION THEOREM.** *Suppose that $n > 1$, and $A = (a_{ij}) \in M_{n,n}(R)$. Let $k \in \mathbb{N}$ with $1 \le k \le n$. Then*

$$\det A = \sum_{j=1}^{n} a_{kj} \operatorname{cof}_{kj}(A) \qquad (expansion \ by \ the \ k\text{-}th \ row),$$

$$= \sum_{i=1}^{n} a_{ik} \operatorname{cof}_{ik}(A) \qquad (expansion \ by \ the \ k\text{-}th \ column).$$

*Proof.* We can write

$$A = \begin{pmatrix} a_{11} & a_{12} & \ldots & a_{1n} \\ \vdots & \vdots & & \vdots \\ a_{k1} + 0 & 0 + a_{k2} & \ldots & 0 + a_{kn} \\ \vdots & \vdots & & \vdots \\ a_{n1} & a_{n2} & \ldots & a_{nn} \end{pmatrix}$$

and use 13.2(iii) to express $\det A$ as the sum of two determinants. Now use this argument $n - 2$ more times to see that

$$\det A = \sum_{l=1}^{n} \det \begin{pmatrix} a_{1,1} & \cdots & a_{1,l-1} & a_{1,l} & a_{1,l+1} & \cdots & a_{1,n} \\ \vdots & & \vdots & \vdots & \vdots & & \vdots \\ 0 & \cdots & 0 & a_{k,l} & 0 & \cdots & 0 \\ \vdots & & \vdots & \vdots & \vdots & & \vdots \\ a_{n,1} & \cdots & a_{n,l-1} & a_{n,l} & a_{n,l+1} & \cdots & a_{n,n} \end{pmatrix}.$$

(The matrices on the right-hand side of the above equation only differ from $A$ in (at most) their $k$-th rows.) The formula for the expansion of $\det A$ by the $k$-th row now follows from 13.2(ii) and 13.8.

We leave the proof of the formula for expansion by the $k$-th column as an exercise. $\square$

**13.10** ♯EXERCISE. Complete the proof of 13.9.

**13.11** THE RULE OF FALSE COFACTORS. *Suppose that $n > 1$, and $A = (a_{ij}) \in M_{n,n}(R)$. Let $k, l \in \mathbb{N}$ with $1 \leq k, l \leq n$. Then*

$$\sum_{j=1}^{n} a_{kj} \operatorname{cof}_{lj}(A) = \begin{cases} 0 & \text{if } k \neq l, \\ \det A & \text{if } k = l. \end{cases}$$

*Also*

$$\sum_{i=1}^{n} a_{ik} \operatorname{cof}_{il}(A) = \begin{cases} 0 & \text{if } k \neq l, \\ \det A & \text{if } k = l. \end{cases}$$

*Proof.* The two claims in the case in which $k = l$ are immediate from the Expansion Theorem 13.9, and so we suppose that $k \neq l$. Then $\sum_{j=1}^{n} a_{kj} \operatorname{cof}_{lj}(A)$ is just the expansion along the $l$-th row of the matrix $B \in M_{n,n}(R)$ obtained from $A$ by replacement of the $l$-th row of $A$ by the $k$-th. Since $B$ has two equal rows, it follows from 13.3 that $\det B = 0$. The other formula is proved similarly. $\square$

**13.12** COROLLARY. *Let $A \in M_{n,n}(R)$. Then*

$$A(\operatorname{Adj} A) = (\operatorname{Adj} A)A = (\det A)I_n.$$

*Proof.* This is clear from the definitions in the trivial case in which $n = 1$, and when $n > 1$ the result is immediate from the Rule of False Cofactors 13.11. $\square$

**13.13** EXERCISE. Let $A \in M_{n,n}(R)$. We say that $A$ is *invertible* precisely when there exists $B \in M_{n,n}(R)$ such that $AB = BA = I_n$. Prove that $A$ is invertible if and only if $\det A$ is a unit of $R$. (Do not forget 3.11.)

**13.14** FURTHER STEPS. We have thus achieved the stated aim of this little excursion into the theory of matrices over the commutative ring $R$. Lack of space prevents our pursuing this road further: an interested reader might like to explore the properties of matrices over a PID developed in [4, Chapter 7], and to follow up the connections with our work in Chapter 10.

It should probably also be mentioned that a complete treatment of determinants of matrices over commutative rings should involve exterior algebra, but that that topic, like homological algebra, is beyond the scope of this book.

We now move on to explain the relevance of 13.12 to commutative algebra.

**13.15** PROPOSITION. *Let $R$ be a subring of the commutative ring $S$, and let $M$ be an $S$-module which, when considered as an $R$-module by restriction of scalars (see 6.6), can be finitely generated by $n$ elements (where $n \geq 1$). Let $s \in S$ and let $I$ be an ideal of $R$ such that $sM \subseteq IM$. Then there exist $a_i \in I^i$ for $i = 1, \ldots, n$ such that*

$$s^n + a_1 s^{n-1} + \cdots + a_{n-1} s + a_n \in (0 :_S M).$$

*Proof.* Suppose that $M$ is generated as $R$-module by $g_1, \ldots, g_n$. Then, for each $i = 1, \ldots, n$, there exist $b_{i1}, \ldots, b_{in} \in I$ such that

$$sg_i = \sum_{j=1}^{n} b_{ij} g_j.$$

Write $C = (c_{ij})$ for the $n \times n$ matrix over $S$ given by $c_{ij} = s\delta_{ij} - b_{ij}$ with $\delta_{ij} = 1_S$ or $0_S$ according as $i$ and $j$ are or are not equal (for all $i, j = 1, \ldots, n$). Then

$$\sum_{j=1}^{n} c_{ij} g_j = 0 \qquad \text{for all } i = 1, \ldots, n.$$

Now use the fact (13.12) that $(\operatorname{Adj} C)C = (\det C)I_n$ to deduce that

$$(\det C)g_i = 0 \qquad \text{for all } i = 1, \ldots, n,$$

so that $\det C \in (0 :_S M)$. Finally, it follows from the definition of determinant that

$$\det C = s^n + a_1 s^{n-1} + \cdots + a_{n-1} s + a_n,$$

with $a_i \in I^i$ for $i = 1, \ldots, n$. $\square$

One can use 13.15 to prove Nakayama's Lemma 8.24: if $M$ is a finitely generated module over the commutative ring $R$ and $I$ is an ideal of $R$ such that $M = IM$, then, by 13.15 with $S = R$ and $s = 1$, we see that $M$ is annihilated by an element of the form $1 + a$ with $a \in I$, and if $I \subseteq \text{Jac}(R)$, then $1 + a$ is a unit of $R$ by 3.17.

We are now ready to begin the theory of integral dependence proper.

**13.16** DEFINITIONS. Let $R$ be a subring of the commutative ring $S$, and let $s \in S$. We say that $s$ is *integral over* $R$ precisely when there exist $h \in \mathbf{N}$ and $r_0, \ldots, r_{h-1} \in R$ such that

$$s^h + r_{h-1}s^{h-1} + \cdots + r_1 s + r_0 = 0,$$

that is, if and only if $s$ is a root of a *monic* polynomial in $R[X]$.

Thus, in the case in which $R$ and $S$ are both fields, $s$ is integral over $R$ if and only if it is algebraic over $R$. Thus algebraic field extensions give rise to examples of integral elements. Perhaps it should be pointed out that in the present more general situation of integral dependence on subrings, the insistence on *monic* polynomials assumes greater importance.

Clearly, every element of $R$ is integral over $R$. We say that $S$ is *integral over* $R$ precisely when every element of $S$ is integral over $R$.

We say that a homomorphism $f : R \to R'$ of commutative rings is *integral* if and only if $R'$ is integral over its subring $\text{Im}\, f$.

**13.17** LEMMA. *Let $R$ be a unique factorization domain, and let $K$ be its field of fractions. Let $u \in K$ be integral over $R$. Then $u \in R$.*

*Proof.* Clearly, we can assume that $u \neq 0$. We can write $u = s/t$ with $s, t \in R$ and $t \neq 0$, and, since $R$ is a UFD, we can assume that there is no irreducible element of $R$ which is a factor in $R$ of both $s$ and $t$. Now there exist $h \in \mathbf{N}$ and $r_0, \ldots, r_{h-1} \in R$ such that

$$\frac{s^h}{t^h} + r_{h-1}\frac{s^{h-1}}{t^{h-1}} + \cdots + r_1\frac{s}{t} + r_0 = 0,$$

so that

$$s^h + r_{h-1}s^{h-1}t + \cdots + r_1 st^{h-1} + r_0 t^h = 0.$$

From this we see that every irreducible factor of $t$ must be a factor of $s^h$, and so of $s$. Hence $t$ has no irreducible factor, so that $t$ is a unit of $R$ and $u \in R$. $\square$

**13.18** DEFINITION. We say that a module $M$ over a commutative ring $R$ is *faithful* precisely when $(0 : M) = 0$, that is, if and only if $M$ has zero annihilator.

**13.19** REMARKS. Let $R$ be a subring of the commutative ring $S$.

(i) When, in the sequel, we speak of $S$ as an $R$-module without qualification, it is always to be understood that $S$ is being regarded as an $R$-module by restriction of scalars (see 6.6) using the inclusion ring homomorphism.

(ii) Now suppose, in addition, that $R$ and $S$ are both subrings of the commutative ring $T$; thus $R \subseteq S \subseteq T$. If $T$ is finitely generated by $\{t_1, \ldots, t_n\}$ as an $S$-module, and $S$ is finitely generated by $\{s_1, \ldots, s_m\}$ as an $R$-module, then it is easy to see that $T$ is finitely generated by $\{s_i t_j : 1 \leq i \leq m, \ 1 \leq j \leq n\}$ as an $R$-module. We shall make considerable use of this simple observation in this chapter.

**13.20** PROPOSITION. *Let $R$ be a subring of the commutative ring $S$, and let $u \in S$. Then the following statements are equivalent:*

(i) *$u$ is integral over $R$;*

(ii) *the subring $R[u]$ of $S$ is finitely generated as an $R$-module;*

(iii) *there exists a subring $R'$ of $S$ such that $R[u] \subseteq R'$ and $R'$ is finitely generated as an $R$-module;*

(iv) *there exists a faithful $R[u]$-module which, when regarded as an $R$-module by restriction of scalars, is finitely generated.*

*Proof.* (i) $\Rightarrow$ (ii) Note that $R[u]$ is generated as an $R$-module by $\{u^i : i \in \mathbb{N}_0\}$. Now there exist $h \in \mathbb{N}$ and $r_0, \ldots, r_{h-1} \in R$ such that

$$u^h + r_{h-1}u^{h-1} + \cdots + r_1 u + r_0 = 0.$$

It is therefore enough for us to show that $u^{h+n} \in R1 + Ru + \cdots + Ru^{h-1}$ for all $n \in \mathbb{N}_0$, and this can easily be achieved by induction on $n$ since the above displayed equation shows that

$$u^{h+n} = -r_{h-1}u^{h+n-1} - \cdots - r_1 u^{n+1} - r_0 u^n \qquad \text{for all } n \in \mathbb{N}_0.$$

(ii) $\Rightarrow$ (iii) Just take $R' = R[u]$.

(iii) $\Rightarrow$ (iv) Just take $M = R'$, which is a faithful $R[u]$-module since $a \in (0 :_{R[u]} R')$ implies that $a1_R = 0$.

(iv) $\Rightarrow$ (i) Let $M$ be a faithful $R[u]$-module which is finitely generated as $R$-module. Since $uM \subseteq RM$, we can apply 13.15 with $S = R[u]$ and $I = R$ to see that there exist $n \in \mathbb{N}$ and $a_1, \ldots, a_n \in R$ such that

$$u^n + a_1 u^{n-1} + \cdots + a_{n-1}u + a_n \in (0 :_{R[u]} M) = 0.$$

Hence $u$ is integral over $R$. $\square$

**13.21** COROLLARY. *Let $R$ be a subring of the commutative ring $S$, and let $u_1, \ldots, u_n \in S$ be integral over $R$. Then the subring $R[u_1, \ldots, u_n]$ of $S$ is a finitely generated $R$-module.*

*Proof.* We use induction on $n$; when $n = 1$, the result follows from 13.20. So suppose that $n > 1$ and the result has been proved for smaller values of $n$. By this inductive hypothesis, $R[u_1, \ldots, u_{n-1}]$ is a finitely generated $R$-module, while, since $u_n$ is *a fortiori* integral over $R[u_1, \ldots, u_{n-1}]$, we see from 13.20 that $R[u_1, \ldots, u_n] = R[u_1, \ldots, u_{n-1}][u_n]$ is a finitely generated $R[u_1, \ldots, u_{n-1}]$-module. Hence, by 13.19(ii), $R[u_1, \ldots, u_n]$ is a finitely generated $R$-module, and so the inductive step is complete. $\square$

It follows from 13.21 that, if a commutative ring $S$ is finitely generated as an *algebra* (see 8.9) over its subring $R$, and $S$ is integral over $R$, then $S$ is actually finitely generated as an *R-module*. This interplay between two different notions of 'finitely generated' might help to consolidate the ideas for the reader!

**13.22** COROLLARY and DEFINITIONS. *Let $R$ be a subring of the commutative ring $S$. Then*

$$R' := \{s \in S : s \text{ is integral over } R\}$$

*is a subring of $S$ which contains $R$, and is called the* integral closure *of $R$ in $S$. We say that $R$ is* integrally closed *in $S$ precisely when $R' = R$, that is, if and only if every element of $S$ which is integral over $R$ actually belongs to $R$.*

*Proof.* We have already remarked in 13.16 that every element of $R$ is integral over $R$, and so $R \subseteq R'$. Hence $1_R \in R'$. Now let $a, b \in R'$; then, by 13.21, the ring $R[a, b]$ is a finitely generated $R$-module; hence, by 13.20, $a + b, -a, ab \in R'$. It therefore follows from the Subring Criterion 1.5 that $R'$ is a subring of $S$. $\square$

**13.23** COROLLARY. *Let $R \subseteq S \subseteq T$ with $R$ and $S$ subrings of the commutative ring $T$. Assume that $S$ is integral over $R$ and $T$ is integral over $S$. Then $T$ is integral over $R$.*

*Proof.* Let $t \in T$, so that $t$ is integral over $S$. Now there exist $h \in \mathbb{N}$ and $s_0, \ldots, s_{h-1} \in S$ such that

$$t^h + s_{h-1}t^{h-1} + \cdots + s_1 t + s_0 = 0.$$

Hence $t$ is integral over $C := R[s_0, \ldots, s_{h-1}]$, and so, by 13.20, $C[t]$ is finitely generated as a $C$-module. But $C$ is finitely generated as an $R$-module by

13.21, and so $C[t]$ is finitely generated as an $R$-module by 13.19(ii). Now we can use 13.20 to see that $t$ is integral over $R$. It follows that $T$ is integral over $R$. $\square$

**13.24** COROLLARY. *Let $R$ be a subring of the commutative ring $S$ and let $R'$ be the integral closure of $R$ in $S$. Then $R'$ is integrally closed in $S$.*

*Proof.* Let $R''$ denote the integral closure of $R'$ in $S$. Then, by 13.23, $R''$ is integral over $R$, and so $R' = R''$. $\square$

**13.25** EXERCISE. Let $R, S_1, \ldots, S_n$ ($n \geq 1$) be commutative rings, and suppose that $f_i : R \to S_i$ is an integral ring homomorphism (see 13.16) for each $i = 1, \ldots, n$. Show that the ring homomorphism $f : R \to \prod_{i=1}^{n} S_i$ from $R$ into the direct product ring (see 2.6) for which $f(r) = (f_1(r), \ldots, f_n(r))$ for all $r \in R$ is also integral.

Next we give a technical lemma which investigates the behaviour of the concept of integral dependence under certain familiar ring operations.

**13.26** LEMMA. *Let $R$ be a subring of the commutative ring $S$, and suppose that $S$ is integral over $R$.*

(i) *Let $J$ be an ideal of $S$, and denote by $J^c$ the ideal $R \cap J$ of $R$ (so that we are using the contraction notation of 2.41). Now $J^c$ is the kernel of the composite ring homomorphism*

$$g : R \xrightarrow{\subseteq} S \longrightarrow S/J$$

*(in which the second homomorphism is the natural one), and so, by 2.13, there is induced by $g$ an injective ring homomorphism $\bar{g} : R/J^c \to S/J$ such that $\bar{g}(r + J^c) = r + J$ for all $r \in R$, and this enables us to regard $R/J^c$ as a subring of $S/J$.*
*With this convention, $S/J$ is integral over $R/J^c$.*

(ii) *Let $U$ be a multiplicatively closed subset of $R$. Then there is an injective ring homomorphism $h : U^{-1}R \to U^{-1}S$ for which $h(r/u) = r/u$ for all $r \in R$ and $u \in U$, and, using this, we regard $U^{-1}R$ as a subring of $U^{-1}S$.*
*With this convention, $U^{-1}S$ is integral over $U^{-1}R$.*

*Proof.* Let $s \in S$. Then there exist $n \in \mathbb{N}$ and $r_0, \ldots, r_{n-1} \in R$ such that

$$s^n + r_{n-1}s^{n-1} + \cdots + r_1 s + r_0 = 0.$$

(i) Apply the natural homomorphism from $S$ to $S/J$ to this equation to deduce that $s + J$ is integral over $R/J^c$.

(ii) Let $u \in U$. We deduce from the above displayed equation that

$$\frac{s^n}{u^n} + \frac{r_{n-1}}{u} \frac{s^{n-1}}{u^{n-1}} + \cdots + \frac{r_1}{u^{n-1}} \frac{s}{u} + \frac{r_0}{u^n} = 0,$$

from which we see that $s/u$ is integral over $U^{-1}R$. $\square$

**13.27** COROLLARY. *Let $R$ be a subring of the commutative ring $S$ and let $R'$ be the integral closure of $R$ in $S$. Let $U$ be a multiplicatively closed subset of $R$. Then $U^{-1}R'$ is the integral closure of $U^{-1}R$ in $U^{-1}S$.*

*Proof.* By 13.26(ii), $U^{-1}R'$ is integral over $U^{-1}R$. Suppose that $s \in S$, $u \in U$ are such that the element $s/u$ of $U^{-1}S$ is integral over $U^{-1}R$. Then there exist ($n \in \mathbb{N}$ and) $r_0, \ldots, r_{n-1} \in R$, $u_0, \ldots, u_{n-1} \in U$ such that

$$\frac{s^n}{u^n} + \frac{r_{n-1}}{u_{n-1}} \frac{s^{n-1}}{u^{n-1}} + \cdots + \frac{r_1}{u_1} \frac{s}{u} + \frac{r_0}{u_0} = 0.$$

Let $v := u_{n-1} \ldots u_0 \ (\in U)$. By multiplying the above equation through by $v^n u^n/1$, we can deduce that there exist $r_0', \ldots, r_{n-1}' \in R$ such that

$$\frac{v^n s^n}{1} + \frac{r_{n-1}'}{1} \frac{v^{n-1} s^{n-1}}{1} + \cdots + \frac{r_1'}{1} \frac{vs}{1} + \frac{r_0'}{1} = 0.$$

Hence there exists $v' \in U$ such that

$$v'(v^n s^n + r_{n-1}' v^{n-1} s^{n-1} + \cdots + r_1' vs + r_0') = 0,$$

from which we see that $v'vs$ is integral over $R$; therefore $v'vs \in R'$ and $s/u = v'vs/v'vu \in U^{-1}R'$. This completes the proof. $\square$

**13.28** DEFINITIONS. The integral closure of an integral domain $R$ in its field of fractions is referred to as the *integral closure* of $R$ (without qualification). An integral domain is said to be *integrally closed* (without qualification) precisely when it is integrally closed in its field of fractions (in the sense of 13.22).

More generally, let $R$ be a commutative ring. The set

$$V := \{r \in R : r \text{ is a non-zerodivisor on } R\}$$

is a multiplicatively closed subset of $R$, and the natural ring homomorphism $f : R \to V^{-1}R$ is injective (because, by 5.4(iv), each element of $\mathrm{Ker}\, f$ is annihilated by an element of $V$), and so we can use $f$ to regard $R$ as a subring of $V^{-1}R$. We say that $R$ is *integrally closed* (without qualification) precisely when it is integrally closed in $V^{-1}R$.

**13.29** PROPOSITION. *Let $R$ be an integral domain. Then the following statements are equivalent:*

(i) *$R$ is integrally closed;*

(ii) *$R_P$ is integrally closed for all $P \in \text{Spec}(R)$;*

(iii) *$R_M$ is integrally closed for all maximal ideals $M$ of $R$.*

*Proof.* Let $K$ be the field of fractions of $R$, and let $\bar{R}$ be the integral closure of $R$ (see 13.28). Let $f : R \to \bar{R}$ be the inclusion mapping, so that $R$ is integrally closed if and only if $f$ is surjective. Let $P \in \text{Spec}(R)$. By 13.27, $\bar{R}_P$ is the integral closure of $R_P$ in $K_P$.

However, under the identifications of 5.16, the ring $K_P$ is just $K$ itself, which can also be identified with the quotient field of $R_P$. We thus see that $f_P : R_P \to \bar{R}_P$ is surjective if and only if $R_P$ is integrally closed. The result therefore follows from 9.17. □

When we study in Chapter 14 the dimension theory of a (non-trivial) finitely generated commutative algebra $A$ over a field $K$, we shall be involved not only in the use of transcendence degrees of field extensions, but also with what are called 'chains of prime ideals' of $A$: such a chain is a finite, strictly increasing sequence

$$P_0 \subset P_1 \subset \ldots \subset P_{n-1} \subset P_n$$

of prime ideals of $A$. The length of the chain is the number of 'links', that is, one less than the number of prime ideals; thus the displayed chain above has length $n$. We shall see in Chapter 14 that the dimension of $A$ is actually defined to be the greatest length of such a chain of prime ideals of $A$.

To link this idea to the concept of transcendence degree of field extension, we shall try to 'approximate' $A$ by a subring $B$ which is essentially a polynomial ring over $K$ in a finite number of indeterminates and over which $A$ is integral. For this reason, it becomes of interest and importance to compare chains of prime ideals of $A$ and similar chains for $B$. This explains why the next part of our work in this chapter is concerned with comparisons of the prime ideal structures of $R$ and $S$ when $R$ is a subring of the commutative ring $S$ and $S$ is integral over $R$.

**13.30** PROPOSITION. *Let $R$ be a subring of the integral domain $S$ and suppose that $S$ is integral over $R$. Then $S$ is a field if and only if $R$ is a field.*

*Proof.* ($\Leftarrow$) Suppose that $R$ is a field. Let $s \in S$ with $s \neq 0$. Then there exists a monic polynomial in $R[X]$ which has $s$ as a root: let $n$ be the smallest possible degree of such a monic polynomial. Then $n \in \mathbb{N}$, and there exist $r_0, \ldots, r_{n-1} \in R$ such that

$$s^n + r_{n-1}s^{n-1} + \cdots + r_1 s + r_0 = 0.$$

Thus

$$s(-s^{n-1} - r_{n-1}s^{n-2} - \cdots - r_1) = r_0.$$

Since $S$ is an integral domain and $s \neq 0$, it follows from the choice of $n$ that $r_0 \neq 0$; hence $r_0$ has an inverse in $R$, and it follows from the last displayed equation that

$$-r_0^{-1}(s^{n-1} + r_{n-1}s^{n-2} + \cdots + r_1)$$

is an inverse for $s$ in $S$. Hence $S$ is a field.

($\Rightarrow$) Suppose that $S$ is a field. Let $r \in R$ with $r \neq 0$. Since $S$ is a field, $r$ has an inverse $r^{-1}$ in $S$: we show that $r^{-1} \in R$.

Since $r^{-1}$ is integral over $R$, there exist $h \in \mathbb{N}$ and $u_0, \ldots, u_{h-1} \in R$ such that

$$(r^{-1})^h + u_{h-1}(r^{-1})^{h-1} + \cdots + u_1 r^{-1} + u_0 = 0.$$

Multiply both sides of this equation by $r^{h-1}$ to see that

$$r^{-1} = -(u_{h-1} + \cdots + u_1 r^{h-2} + u_0 r^{h-1}) \in R.$$

This completes the proof. $\square$

**13.31** COROLLARY. *Let $R$ be a subring of the commutative ring $S$, and suppose that $S$ is integral over $R$. Let $Q \in \operatorname{Spec}(S)$, and let $P := Q \cap R = Q^c$ be the contraction of $Q$ to $R$, so that $P \in \operatorname{Spec}(R)$ by 3.27(ii). Then $Q$ is a maximal ideal of $S$ if and only if $P$ is a maximal ideal of $R$.*

*Proof.* By 13.26(i), the integral domain $S/Q$ is integral over $R/P = R/Q^c$ when we regard the latter as a subring of $S/Q$ in the manner of that lemma. It therefore follows from 13.30 that $S/Q$ is a field if and only if $R/P$ is a field, so that, by 3.3, $Q$ is a maximal ideal of $S$ if and only if $P$ is a maximal ideal of $R$. $\square$

Although the following remark is simple, it can be surprisingly helpful.

**13.32** REMARK. Let $f : R \to S$ and $g : S \to T$ be homomorphisms of commutative rings, and let $J$ be an ideal of $T$. Then

$$f^{-1}(g^{-1}(J)) = (g \circ f)^{-1}(J).$$

**13.33** THE INCOMPARABILITY THEOREM. *Let $R$ be a subring of the commutative ring $S$, and suppose that $S$ is integral over $R$. Suppose that $Q, Q' \in \operatorname{Spec}(S)$ are such that $Q \subseteq Q'$ and $Q' \cap R = Q \cap R =: P$, say. Then $Q = Q'$.*

*Note.* The name of this theorem comes from the following rephrasing of its statement: if $Q_1, Q_2$ are two different prime ideals of $S$ which have the same contraction in $R$, then $Q_1$ and $Q_2$ are 'incomparable' in the sense that neither is contained in the other.

*Proof.* Let $U := R \setminus P$. Let $\tau : R \to S$ be the inclusion homomorphism, let $\sigma : U^{-1}R = R_P \to U^{-1}S$ be the induced injective ring homomorphism (see 13.26(ii)), and let $\theta : R \to R_P$ and $\phi : S \to U^{-1}S$ be the canonical homomorphisms. Observe that $\phi \circ \tau = \sigma \circ \theta$.

Consider $QU^{-1}S$ and $Q'U^{-1}S$. By 5.32(ii), these are both prime ideals of $U^{-1}S$, and $QU^{-1}S \subseteq Q'U^{-1}S$. Also, by 5.30,

$$\tau^{-1}(\phi^{-1}(QU^{-1}S)) = \tau^{-1}(Q) = Q \cap R = P = \tau^{-1}(\phi^{-1}(Q'U^{-1}S)).$$

Therefore, by 13.32 and 3.27(ii), the ideals $\sigma^{-1}(QU^{-1}S)$ and $\sigma^{-1}(Q'U^{-1}S)$ of $R_P$ are prime and

$$\theta^{-1}(\sigma^{-1}(QU^{-1}S)) = P = \theta^{-1}(\sigma^{-1}(Q'U^{-1}S)).$$

Hence, by 5.33, $\sigma^{-1}(QU^{-1}S) = PR_P = \sigma^{-1}(Q'U^{-1}S)$. Since $\sigma$ is an integral ring homomorphism (by 13.26), and $PR_P$ is the maximal ideal of the quasi-local ring $R_P$, it follows from 13.31 that $QU^{-1}S$ and $Q'U^{-1}S$ are both maximal ideals of $U^{-1}S$. But $QU^{-1}S \subseteq Q'U^{-1}S$, and so $QU^{-1}S = Q'U^{-1}S$; hence, by 5.33 again, $Q = Q'$. $\square$

**13.34** THE LYING-OVER THEOREM. *Let $R$ be a subring of the commutative ring $S$, and suppose that $S$ is integral over $R$. Let $P \in \operatorname{Spec}(R)$. Then there exists $Q \in \operatorname{Spec}(S)$ such that $Q \cap R = P$, that is, such that $Q$ 'lies over' $P$.*

*Proof.* We use similar notation to that used in the proof of 13.33. Thus, let $U := R \setminus P$, let $\tau : R \to S$ be the inclusion homomorphism, let $\sigma : U^{-1}R = R_P \to U^{-1}S$ be the induced injective ring homomorphism, and let $\theta : R \to R_P$ and $\phi : S \to U^{-1}S$ be the canonical homomorphisms.

Observe that $U^{-1}S$ is not trivial (since, for example, $\sigma$ is injective), and so, by 3.9, there exists a maximal ideal $N$ of $U^{-1}S$. Since $\sigma$ is an integral ring homomorphism by 13.26(ii), it follows from 13.31 that $\sigma^{-1}(N) = PR_P$. Hence, since $\phi \circ \tau = \sigma \circ \theta$, we can deduce from 13.32 and 5.33 that $Q := \phi^{-1}(N) \in \operatorname{Spec}(S)$ and

$$Q \cap R = \tau^{-1}(\phi^{-1}(N)) = \theta^{-1}(\sigma^{-1}(N)) = \theta^{-1}(PR_P) = P. \quad \square$$

**13.35** EXERCISE. Let $R$ be a subring of the commutative ring $S$, and suppose that $S$ is integral over $R$.

(i) Show that, if $r \in R$ is a unit in $S$, then $r$ is a unit in $R$.

(ii) Show that $\operatorname{Jac}(R) = \operatorname{Jac}(S) \cap R$.

**13.36** EXERCISE. Let $R$ be a non-trivial commutative ring. An *automorphism of $R$* is a ring isomorphism of $R$ onto itself. It is easy to see that the set of all automorphisms of $R$ is a group with respect to composition of mappings. Let $G$ be a finite subgroup of this automorphism group of $R$. Show that

$$R^G := \{r \in R : \sigma(r) = r \text{ for all } \sigma \in G\}$$

is a subring of $R$, and that $R$ is integral over $R^G$. (Here is a hint: consider, for $r \in R$, the polynomial $\prod_{\sigma \in G}(X - \sigma(r)) \in R[X]$.)

Now let $P \in \operatorname{Spec}(R^G)$, and let

$$\mathcal{P} := \left\{ Q \in \operatorname{Spec}(R) : Q \cap R^G = P \right\}.$$

Let $Q_1, Q_2 \in \mathcal{P}$. By considering $\prod_{\sigma \in G} \sigma(r)$ for $r \in Q_1$, show that

$$Q_1 \subseteq \bigcup_{\sigma \in G} \sigma(Q_2),$$

and deduce that $Q_1 = \tau(Q_2)$ for some $\tau \in G$.

Deduce that $\mathcal{P}$ is finite.

**13.37** EXERCISE. Let $f : R \to S$ be a homomorphism of commutative rings; use the extension and contraction notation of 2.41. Let $P \in \operatorname{Spec}(R)$. Show that there exists $Q \in \operatorname{Spec}(S)$ such that $Q^c = P$ if and only if $P^{ec} = P$. (Here is a hint: show that, if $P^{ec} = P$, then $P^e \cap f(R \setminus P) = \emptyset$, and recall 3.44.)

**13.38** THE GOING-UP THEOREM. *Let $R$ be a subring of the commutative ring $S$, and suppose that $S$ is integral over $R$. Let $m \in \mathbf{N}_0$ and $n \in \mathbf{N}$ with $m < n$. Let*

$$P_0 \subset P_1 \subset \ldots \subset P_{n-1} \subset P_n$$

*be a chain of prime ideals of $R$ and suppose that*

$$Q_0 \subset Q_1 \subset \ldots \subset Q_{m-1} \subset Q_m$$

*is a chain of prime ideals of $S$ such that $Q_i \cap R = P_i$ for all $i = 0, \ldots, m$. Then it is possible to extend the latter chain by prime ideals $Q_{m+1}, \ldots, Q_n$ of $S$, so that*

$$Q_0 \subset Q_1 \subset \ldots \subset Q_{n-1} \subset Q_n,$$

*in such a way that $Q_i \cap R = P_i$ for all $i = 0, \ldots, n$.*

*Proof.* It should be clear to the reader that we can, by use of induction, reduce to the special case in which $m = 0$ and $n = 1$. As we are then looking for a prime ideal of $S$ which contains $Q_0$, a natural step, in view of the bijective correspondence (see 3.28) between the set of prime ideals of $S$ which contain $Q_0$ and $\mathrm{Spec}(S/Q_0)$, is to consider the residue class ring $S/Q_0$.

Let $\tau : R \to S$ be the inclusion homomorphism, let $\rho : R/P_0 \to S/Q_0$ be the induced injective ring homomorphism (see 13.26(i)), and let

$$\xi : R \to R/P_0 \quad \text{and} \quad \psi : S \to S/Q_0$$

be the canonical homomorphisms. Observe that $\psi \circ \tau = \rho \circ \xi$.

Now $P_1/P_0 \in \mathrm{Spec}(R/P_0)$, by 2.37 and 3.28. Since $\rho$ is an integral ring homomorphism by 13.26(i), it follows from the Lying-over Theorem 13.34 that there exists a prime ideal $Q \in \mathrm{Spec}(S/Q_0)$ such that $\rho^{-1}(Q) = P_1/P_0$. But, by 2.37 and 3.28 again, there exists $Q_1 \in \mathrm{Spec}(S)$ with $Q_1 \supseteq Q_0$ such that $Q = Q_1/Q_0$. Then we have

$$Q_1 \cap R = \tau^{-1}(\psi^{-1}(Q)) = \xi^{-1}(\rho^{-1}(Q)) = \xi^{-1}(P_1/P_0) = P_1.$$

This completes the proof, as it is clear that $Q_0 \subset Q_1$. $\square$

The Going-up Theorem 13.38 will be helpful when we want to compare the dimension of a finitely generated commutative algebra over a field with the dimension of one of its subrings over which it is integral. Also helpful in similar comparisons is the so-called 'Going-down Theorem', which has a rather similar statement to the Going-up Theorem except that the inclusion relations in the chains of prime ideals are reversed, and we shall impose stronger hypotheses on $R$ and $S$: we shall assume not only that the commutative ring $S$ is integral over its subring $R$, but also that $R$ and $S$ are both integral domains and that $R$ is actually integrally closed (see 13.28). These additional hypotheses enable us to use some of the theory of algebraic extensions of fields developed in Chapter 12.

We give now two preparatory results for the Going-down Theorem 13.41.

**13.39** LEMMA. *Let $R$ be a subring of the commutative ring $S$, and suppose that $S$ is integral over $R$. Let $I$ be an ideal of $R$. Then*

$$\sqrt{(IS)} = \{ s \in S : s^n + a_{n-1}s^{n-1} + \cdots + a_1 s + a_0 = 0 \text{ for some}$$
$$n \in \mathbb{N} \text{ and } a_0, \dots, a_{n-1} \in I \}.$$

*Proof.* Let $s \in \sqrt{(IS)}$. Then there exist $(h, n \in \mathbb{N}$ and$)$ $a_1, \dots, a_n \in I$ and $s_1, \dots, s_n \in S$ such that $s^h = \sum_{i=1}^{n} a_i s_i$. By 13.21, the ring $T :=$

$R[s, s_1, \ldots, s_n]$ is a finitely generated $R$-module, and we have $s^h T \subseteq IT$. Now $T$ is a faithful $T$-module, and so it follows from 13.15 that there exist $m \in \mathbb{N}$ and $b_0, \ldots, b_{m-1} \in I$ such that

$$(s^h)^m + b_{m-1}(s^h)^{m-1} + \cdots + b_1 s^h + b_0 = 0,$$

so that

$$s^{hm} + b_{m-1} s^{h(m-1)} + \cdots + b_1 s^h + b_0 = 0.$$

Conversely, suppose that $s \in S$ and

$$s^n + a_{n-1} s^{n-1} + \cdots + a_1 s + a_0 = 0$$

for some $n \in \mathbb{N}$ and $a_0, \ldots, a_{n-1} \in I$. Then $s^n = -\sum_{i=0}^{n-1} a_i s^i \in IS$. $\square$

**13.40** PROPOSITION. *Let $R$ be a subring of the integral domain $S$, and suppose that $S$ is integral over $R$ and that $R$ is integrally closed. Let $K$ be the field of fractions of $R$. Let $I$ be an ideal of $R$ and let $s \in IS$. Then $s$ is algebraic over $K$ and its minimal polynomial over $K$ has the form*

$$X^h + a_{h-1} X^{h-1} + \cdots + a_1 X + a_0,$$

*where $a_0, \ldots, a_{h-1} \in \sqrt{I}$.*

*Proof.* Since $s$ is integral over $R$, it is certainly algebraic over $K$. Let its minimal polynomial over $K$ be

$$f = X^h + a_{h-1} X^{h-1} + \cdots + a_1 X + a_0 \in K[X].$$

We aim to show that $a_0, \ldots, a_{h-1} \in \sqrt{I}$.

By 12.5, there exists a field extension $L$ of the field of fractions of $S$ such that $f$ splits into linear factors in $L[X]$: let $s = s_1, \ldots, s_h \in L$ be such that, in $L[X]$,

$$f = (X - s_1)(X - s_2) \ldots (X - s_h).$$

Equating coefficients shows that each of $a_0, \ldots, a_{h-1}$ can be written as a 'homogeneous polynomial' (in fact, a 'symmetric function') in $s_1, \ldots, s_h$ with coefficients $\pm 1$, and so, in particular, $a_0, \ldots, a_{h-1} \in R[s_1, \ldots, s_h]$.

By 13.39, there exist $m \in \mathbb{N}$ and $b_0, \ldots, b_{m-1} \in I$ such that

$$s^m + b_{m-1} s^{m-1} + \cdots + b_1 s + b_0 = 0.$$

Next, each $s_i$ $(i = 1, \ldots, h)$ is algebraic over $K$ with minimal polynomial $f$, and so it follows from 12.36(ii) that, for each $i = 2, \ldots, h$, there is an isomorphism of fields $\sigma_i : K(s) \to K(s_i)$ such that $\sigma_i(s) = s_i$ and $\sigma_i(a) = a$ for all $a \in K$. Hence

$$s_i^m + b_{m-1}s_i^{m-1} + \cdots + b_1 s_i + b_0 = 0 \qquad \text{for all } i = 1, \ldots, h.$$

(Recall that $s = s_1$.) In particular, all the $s_i$ $(i = 1, \ldots, m)$ are integral over $R$, and so, by 13.21, the ring $R[s_1, \ldots, s_h]$ is a finitely generated $R$-module; 13.20 therefore shows that $a_0, \ldots, a_{h-1}$ are all integral over $R$, since they belong to $R[s_1, \ldots, s_h]$. But $a_0, \ldots, a_{h-1} \in K$ and $R$ is integrally closed; hence $a_0, \ldots, a_{h-1} \in R$.

Set $T := R[s_1, \ldots, s_h]$. By 13.39, $s_1, \ldots, s_h \in \sqrt{(IT)}$. In view of our expressions for $a_0, \ldots, a_{h-1}$ in terms of the $s_i$, it follows from 13.39 again that each $a_i$ is a root of a monic polynomial in $R[X]$ all of whose coefficients (other than the leading one) belong to $I$. Hence, by 13.39 again, and the fact that $a_0, \ldots, a_{h-1} \in R$, we deduce that $a_0, \ldots, a_{h-1} \in \sqrt{I}$. $\square$

We are now in a position to prove the promised Going-down Theorem.

**13.41** THE GOING-DOWN THEOREM. *Let $R$ be a subring of the integral domain $S$, and suppose that $S$ is integral over $R$. Assume that $R$ is integrally closed. Let $m \in \mathbb{N}_0$ and $n \in \mathbb{N}$ with $m < n$. Let*

$$P_0 \supset P_1 \supset \ldots \supset P_{n-1} \supset P_n$$

*be a chain of prime ideals of $R$ and suppose that*

$$Q_0 \supset Q_1 \supset \ldots \supset Q_{m-1} \supset Q_m$$

*is a chain of prime ideals of $S$ such that $Q_i \cap R = P_i$ for all $i = 0, \ldots, m$. Then it is possible to extend the latter chain by prime ideals $Q_{m+1}, \ldots, Q_n$ of $S$, so that*

$$Q_0 \supset Q_1 \supset \ldots \supset Q_{n-1} \supset Q_n,$$

*in such a way that $Q_i \cap R = P_i$ for all $i = 0, \ldots, n$.*

*Proof.* As in the proof of the Going-up Theorem 13.38, we reduce to the special case in which $m = 0$ and $n = 1$. This time, we are looking for a prime ideal of $S$ which is contained in $Q_0$, and 3.44 turns out to be an appropriate tool.

Set $U := S \setminus Q_0$, a multiplicatively closed subset of $S$, and $V := R \setminus P_1$, a multiplicatively closed subset of $R$. Then

$$W := UV = \{uv : u \in U, \ v \in V\}$$

is a multiplicatively closed subset of $S$. Our immediate aim is to prove that $P_1 S \cap W = \emptyset$. We suppose that this is not the case, so that there exists $s \in P_1 S \cap W$, and we look for a contradiction.

Let $K$ be the field of fractions of $R$; of course, we can regard $K$ as a subfield of the field of fractions of $S$. By 13.40, $s$ is algebraic over $K$ and its minimal polynomial over $K$ has the form

$$f = X^h + a_{h-1} X^{h-1} + \cdots + a_1 X + a_0,$$

where ($h \in \mathbb{N}$ and) $a_0, \ldots, a_{h-1} \in \sqrt{(P_1)} = P_1$. Also, since $s \in W$, we can write $s = uv$ for some $u \in U$ and $v \in V$. Now $v \neq 0$, and since

$$u^h v^h + a_{h-1} u^{h-1} v^{h-1} + \cdots + a_1 uv + a_0 = 0,$$

we see that $u = s/v$ is a root of the polynomial

$$g = X^h + \frac{a_{h-1}}{v} X^{h-1} + \cdots + \frac{a_1}{v^{h-1}} X + \frac{a_0}{v^h} \in K[X];$$

moreover, $g$ is irreducible over $K$, because a factorization

$$g = \left( \sum_{i=0}^{k} \alpha_i X^i \right) \left( \sum_{j=0}^{l} \beta_j X^j \right)$$

in $K[X]$ would lead to a factorization

$$f = \left( \sum_{i=0}^{k} \frac{\alpha_i}{v^i} X^i \right) \left( \sum_{j=0}^{l} \beta_j v^{h-j} X^j \right).$$

Therefore $u$ is algebraic over $K$ with minimal polynomial $g$.

It now follows from 13.40 (with $I = R$) that all the coefficients of $g$ actually lie in $R$. Thus, for each $i = 0, \ldots, h-1$, there exists $\rho_i \in R$ with $a_i = v^{h-i} \rho_i$.

But $a_0, \ldots, a_{h-1} \in P_1$ and $v \in R \setminus P_1$. Hence $\rho_i \in P_1$ for all $i = 0, \ldots, h-1$. But

$$g = X^h + \rho_{h-1} X^{h-1} + \cdots + \rho_1 X + \rho_0$$

is the minimal polynomial of $u$ over $K$. We can therefore use 13.39 to see that $u \in \sqrt{(P_1 S)} \subseteq \sqrt{(P_0 S)} \subseteq Q_0$, contrary to the fact that $u \in U$. This contradiction shows that $P_1 S \cap W = \emptyset$.

We can now use 3.44 to see that there exists $Q_1 \in \mathrm{Spec}(S)$ such that $Q_1 \cap W = \emptyset$ and $P_1 S \subseteq Q_1$. Hence $P_1 \subseteq P_1 S \cap R \subseteq Q_1 \cap R$, and since $Q_1 \cap W = \emptyset$ and $V = R \setminus P_1 \subseteq W$, we must have $P_1 = Q_1 \cap R$. Likewise, since $U = S \setminus Q_0 \subseteq W$, we must have $Q_1 \subseteq Q_0$. The proof is complete. $\square$

**13.42** EXERCISE. Let $R$ be a commutative ring, and let $f$ be a non-constant monic polynomial in $R[X]$. Show that there is a commutative ring $R'$ which contains $R$ as a subring and has the property that $f$ can be written as a product of monic linear factors in the polynomial ring $R'[X]$. (Use induction on $\deg f$; consider $R[X]/fR[X]$.)

**13.43** EXERCISE. Let $R$ be a subring of the commutative ring $S$, and let $f, g$ be monic polynomials in $S[X]$. Suppose that all the coefficients of $fg$ are integral over $R$. Show that all the coefficients of $f$ and all the coefficients of $g$ are integral over $R$. (Here is a hint. Use Exercise 13.42 to find a commutative ring $S'$ which contains $S$ as a subring and is such that $f$ and $g$ can both be written as products of monic linear factors in the polynomial ring $S'[X]$. Consider the integral closure of $R$ in $S'$, and use Corollary 13.24.)

**13.44** EXERCISE. Let $R$ be a subring of the commutative ring $S$, and let $f \in S[X_1, \ldots, X_n]$, the ring of polynomials over $S$ in the $n$ indeterminates $X_1, \ldots, X_n$. Show that $f$ is integral over $R[X_1, \ldots, X_n]$ if and only if all the coefficients of $f$ are integral over $R$.

(Again, some suggestions might be helpful. The hard implication is '($\Rightarrow$)', and so we shall concentrate on that. Clearly, one can reduce to the case in which $n = 1$ by an inductive argument: let us write $X = X_1$. We can assume that $f \neq 0$. Let

$$q = Y^m + F_{m-1}Y^{m-1} + \cdots + F_1 Y + F_0 \in R[X][Y]$$

be a monic polynomial in the indeterminate $Y$ with coefficients in $R[X]$ which has $f$ as a root. Choose $h \in \mathbf{N}$ such that

$$h > \max\{\deg f, \deg F_{m-1}, \ldots, \deg F_0\},$$

and consider $g := f - X^h$ ($\in S[X]$), so that $-g$ is a monic polynomial of degree $h$. Observe that $q(Y + X^h) \in R[X][Y]$ has the form

$$Y^m + G_{m-1}Y^{m-1} + \cdots + G_1 Y + G_0$$

for suitable $G_0, \ldots, G_{m-1} \in R[X]$, and satisfies $q(g + X^h) = 0$. Use this to express $G_0$ in terms of $g$ and $G_1, \ldots, G_{m-1}$, and then use Exercise 13.43.)

**13.45** FURTHER STEPS. An important aspect of the theory of integral closure that we have not had space to include here concerns its links with the concept of 'valuation ring': the interested reader will find these connections explored in some of the texts listed in the Bibliography, such as [8, Chapter 4] and [1, Chapter 5].

# Chapter 14

# Affine algebras over fields

An affine algebra over a field $K$ is simply a finitely generated commutative $K$-algebra. We are interested in such algebras not only because they provide a readily available fund of examples of commutative Noetherian rings (see 8.11), but also because such algebras have fundamental importance in algebraic geometry. In this book, we are not going to explore the reasons for this: the interested reader might like to study Miles Reid's book [13] to discover something about the connections.

What we are going to do in this chapter, in addition to developing the dimension theory of affine algebras over fields and linking this with transcendence degrees, is to prove some famous and fundamental theorems about such algebras, such as Hilbert's Nullstellensatz and Noether's Normalization Theorem, which are important tools in algebraic geometry. Although their significance for algebraic geometry will not be fully explored here, they have interest from an algebraic point of view, and we shall see that Noether's Normalization Theorem is a powerful tool in dimension theory.

Our first major landmark in this chapter is the Nullstellensatz. Some preparatory results are given first, and we begin with a convenient piece of terminology.

**14.1** DEFINITION. Let $K$ be a field. An *affine $K$-algebra* is a finitely generated commutative $K$-algebra, that is, a commutative $K$-algebra which is finitely generated as $K$-algebra (see 8.9).

Observe that an affine $K$-algebra as in 14.1 is a homomorphic image of a ring $K[X_1, \ldots, X_n]$ of polynomials over $K$ in $n$ indeterminates $X_1, \ldots, X_n$, for some $n \in \mathbb{N}$, and so is automatically a commutative Noetherian ring: see 8.11.

**14.2** PROPOSITION. *Let $R, S$ be subrings of the commutative ring $T$, with $R \subseteq S \subseteq T$. Suppose that $R$ is a Noetherian ring, that $T$ is finitely generated as an $R$-algebra and that $T$ is finitely generated as an $S$-module. Then $S$ is finitely generated as an $R$-algebra.*

*Proof.* Suppose that $T$ is generated as an $R$-algebra by $\{c_1, \ldots, c_m\}$ and that $T$ is generated as an $S$-module by $\{b_1, \ldots, b_n\}$. Thus there exist $s_{ij}, s_{ijk} \in S$ $(1 \leq i \leq m, \ 1 \leq j, k \leq n)$ such that

$$c_i = \sum_{j=1}^{n} s_{ij} b_j \qquad \text{for all } i = 1, \ldots, m$$

and

$$c_i b_j = \sum_{k=1}^{n} s_{ijk} b_k \qquad \text{for all } i = 1, \ldots, m, \ j = 1, \ldots, n.$$

Let $S_0 = R[\Gamma \cup \Delta]$, where

$$\Gamma = \{s_{ij} : 1 \leq i \leq m, \ 1 \leq j \leq n\},$$

$$\Delta = \{s_{ijk} : 1 \leq i \leq m, \ 1 \leq j, k \leq n\}.$$

Thus $S_0$ is the $R$-subalgebra of $S$ generated by the (finite) set formed by all the $s_{ij}$ and $s_{ijk}$, and so is Noetherian by 8.11.

Let $t \in T$. Then $t$ can be written as a polynomial expression in $c_1, \ldots, c_m$ with coefficients in $R$. Use the above displayed expressions for the $c_i$ and $c_i b_j$ to see that $t$ can be expressed as

$$t = u_1 b_1 + \cdots + u_n b_n + u_{n+1} 1_R \qquad \text{with } u_1, \ldots, u_{n+1} \in S_0.$$

Thus $T$ is a finitely generated $S_0$-module, and so is a Noetherian $S_0$-module by 7.22(i). But $S$ is an $S_0$-submodule of $T$, and so $S$ is a finitely generated $S_0$-module, by 7.13. Hence, if $S$ is generated as an $S_0$-module by the finite set $\Phi$, then $S$ is generated as an $R$-algebra by the finite set $\Gamma \cup \Delta \cup \Phi$. $\square$

**14.3** COROLLARY. *Let $R, S$ be subrings of the commutative ring $T$, with $R \subseteq S \subseteq T$. Suppose that $R$ is a Noetherian ring, that $T$ is finitely generated as an $R$-algebra and that $T$ is integral over $S$. Then $S$ is finitely generated as an $R$-algebra.*

*Proof.* Suppose that $T$ is generated as an $R$-algebra by $\{c_1, \ldots, c_m\}$. Then $T = S[c_1, \ldots, c_m]$, and this is a finitely generated $S$-module by 13.21. The result therefore follows from 14.2. $\square$

**14.4** PROPOSITION. *Let $K$ be a field and let $R$ be an affine $K$-algebra. Suppose that $R$ is a field. Then $R$ is a finite algebraic extension of $K$.*

*Proof.* There exist $r_1, \ldots, r_n \in R$ such that $R = K[r_1, \ldots, r_n]$. Of course, the square brackets here denote 'ring adjunction', but since $R$ is a field, we also have $R = K(r_1, \ldots, r_n)$, where here the round parentheses denote the field adjunction of 12.13. By 12.51, there exist $m \in \mathbb{N}_0$ and $m$ distinct integers $i_1, \ldots, i_m$ between 1 and $n$ such that the family $(r_{i_j})_{j=1}^m$ is algebraically independent over $K$ and algebraically equivalent to $(r_i)_{i=1}^n$ relative to $K$. This means that, after possibly reordering the $r_i$, we can assume that $(r_i)_{i=1}^m$ is algebraically independent over $K$ and that $R$ is an algebraic extension field of $K(r_1, \ldots, r_m)$. It follows from 12.29 that $R$ is actually a finite extension of $K(r_1, \ldots, r_m)$, and so, in order to complete the proof, it is sufficient for us to show that $m = 0$.

We suppose that $m > 0$ and look for a contradiction. It follows from 14.2 that $K(r_1, \ldots, r_m)$ is a finitely generated $K$-algebra, and so we deduce from 12.43 that our supposition has led to the following: there is an $m \in \mathbb{N}$ such that the quotient field $K(X_1, \ldots, X_m)$ of the ring $S := K[X_1, \ldots, X_m]$ of polynomials in the $m$ indeterminates $X_1, \ldots, X_m$ is a finitely generated $K$-algebra. We shall show that this leads to a contradiction.

Suppose that $K(X_1, \ldots, X_m)$ is generated by $\{\alpha_1, \ldots, \alpha_h\}$ as a $K$-algebra. Now $S$ is a UFD (by 1.42). We can assume that, for all $i = 1, \ldots h$, we have $\alpha_i \neq 0$: we can write $\alpha_i = f_i/g_i$ for $f_i, g_i \in S$ with $\mathrm{GCD}(f_i, g_i) = 1$. It is clear that at least one $g_i$ will be non-constant, because $S$ is not a field.

Consider $1/(g_1 \ldots g_h + 1) \in K(X_1, \ldots, X_m)$. Since $K(X_1, \ldots, X_m)$ is generated by the $\alpha_i$, there exists $\Phi \in K[Y_1, \ldots, Y_h]$, the polynomial ring over $K$ in $h$ indeterminates $Y_1, \ldots, Y_h$, such that

$$\frac{1}{(g_1 \ldots g_h + 1)} = \Phi\left(\frac{f_1}{g_1}, \ldots, \frac{f_h}{g_h}\right).$$

Multiply both sides of this equation by $(g_1 \ldots g_h + 1)(g_1 \ldots g_h)^d$, where $d$ is the (total) degree of $\Phi$. There results an equation of the form

$$(g_1 \ldots g_h)^d = (g_1 \ldots g_h + 1)f$$

with $f \in S$. This equation provides a contradiction to the fact that $S$ is a UFD, because an irreducible factor of $(g_1 \ldots g_h + 1)$ cannot be a factor of the left-hand side.

This contradiction shows that $m = 0$, and completes the proof. $\square$

**14.5** EXERCISE. Let $K$ be a field and let $f : R \to S$ be a homomorphism of affine $K$-algebras. Let $M$ be a maximal ideal of $S$. Prove that $f^{-1}(M)$ is a maximal ideal of $R$.

Compare this with 3.27(iii).

14.6 HILBERT'S NULLSTELLENSATZ (ZEROS THEOREM). *Let $K$ be an algebraically closed field and let $R = K[X_1, \ldots, X_n]$, the polynomial ring over $K$ in the $n$ ($> 0$) indeterminates $X_1, \ldots, X_n$. Let $M, J$ be ideals of $R$.*

(i) *The ideal $M$ is a maximal ideal of $R$ if and only if there exist $a_1, \ldots, a_n \in K$ such that*

$$M = (X_1 - a_1, \ldots, X_n - a_n).$$

(ii) *Suppose that $J$ is proper. Then there exist $b_1, \ldots, b_n \in K$ such that $f(b_1, \ldots, b_n) = 0$ for all $f \in J$; that is, there is a 'common zero' of all the polynomials in $J$.*

(iii) *Let*

$$V(J) = \{(c_1, \ldots, c_n) \in K^n : f(c_1, \ldots, c_n) = 0 \text{ for all } f \in J\},$$

*the set of all 'common zeros' of the polynomials in $J$. Then*

$$\{g \in R : g(c_1, \ldots, c_n) = 0 \text{ for all } (c_1, \ldots, c_n) \in V(J)\} = \sqrt{J}.$$

*Proof.* (i) ($\Leftarrow$) It follows from 3.15 that an ideal of $R$ of the form $(X_1 - a_1, \ldots, X_n - a_n)$ is a maximal ideal of $R$.

($\Rightarrow$) Conversely, let $M$ be a maximal ideal of $R$. The composite ring homomorphism $\phi : K \to R \to R/M$, in which the first map is the inclusion homomorphism and the second is the natural surjective ring homomorphism, is a (necessarily injective) homomorphism of fields; since $R$ is an affine $K$-algebra, so to is $R/M$. It therefore follows from 14.4 that, when $R/M$ is regarded as a field extension of $K$ by means of $\phi$, this extension is finite and algebraic. But $K$ is algebraically closed, and so $\phi$ is an isomorphism. Hence, for each $i = 1, \ldots, n$, there exists $a_i \in K$ such that $X_i - a_i \in M$. Thus

$$(X_1 - a_1, \ldots, X_n - a_n) \subseteq M,$$

and since both ideals involved in this display are maximal, it follows that $M = (X_1 - a_1, \ldots, X_n - a_n)$.

(ii) Since $J \subset R$, there exists, by 3.10, a maximal ideal $M$ of $R$ with $J \subseteq M$. By part (i) above, there exist $b_1, \ldots, b_n \in K$ such that

$$M = (X_1 - b_1, \ldots, X_n - b_n).$$

Thus each $f \in J$ can be expressed as

$$f = (X_1 - b_1)f_1 + \cdots + (X_n - b_n)f_n$$

for some $f_1, \ldots, f_n \in R$, so that evaluation at $b_1, \ldots, b_n$ shows that

$$f(b_1, \ldots, b_n) = 0.$$

(iii) Let $f \in \sqrt{J}$, so that there exists $m \in \mathbb{N}$ such that $f^m \in J$. By definition of $V(J)$, we have $(f^m)(c_1, \ldots, c_n) = 0$ for all $(c_1, \ldots, c_n) \in V(J)$. Therefore $(f(c_1, \ldots, c_n))^m = 0$ for all $(c_1, \ldots, c_n) \in V(J)$, and so, since a field has no non-zero nilpotent element, $f(c_1, \ldots, c_n) = 0$ for all $(c_1, \ldots, c_n) \in V(J)$.

We now turn to the converse statement, which is the non-trivial part of the result. Let $f \in R$ be such that $f(c_1, \ldots, c_n) = 0$ for all $(c_1, \ldots, c_n) \in V(J)$. In our attempts to show that $f \in \sqrt{J}$, we can, and do, assume that $f \neq 0$. In the polynomial ring $S := R[X_{n+1}] = K[X_1, \ldots, X_{n+1}]$, consider the ideal $J' := JS + (X_{n+1}f - 1)S$. If this were a proper ideal of $S$, then, by part (ii) above, there would exist $b_1, \ldots, b_{n+1} \in K$ such that $g(b_1, \ldots, b_{n+1}) = 0$ for all $g \in J'$; this would mean, in particular, that $\tilde{g}(b_1, \ldots, b_n) = 0$ for all $\tilde{g} \in J$, so that $(b_1, \ldots, b_n) \in V(J)$; and so we would have

$$0 = b_{n+1}f(b_1, \ldots, b_n) - 1 = 0 - 1,$$

a contradiction! Thus $J' = S$.

Hence there exist $h \in \mathbb{N}$, $q_1, \ldots, q_h, p \in S$ and $f_1, \ldots, f_h \in J$ such that

$$1 = \sum_{i=1}^{h} q_i f_i + (X_{n+1}f - 1)p.$$

By 1.16, there is a ring homomorphism

$$\phi : K[X_1, \ldots, X_{n+1}] \to K(X_1, \ldots, X_n)$$

for which $\phi(a) = a$ for all $a \in K$, $\phi(X_i) = X_i$ for all $i = 1, \ldots, n$ and $\phi(X_{n+1}) = 1/f$. Apply $\phi$ to both sides of the last displayed equation to see that

$$1 = \sum_{i=1}^{h} q_i \left( X_1, \ldots, X_n, \frac{1}{f} \right) f_i.$$

Let $d \in \mathbb{N}$ be at least as big as the maximum of the degrees of $q_1, \ldots, q_h$ when considered as polynomials in $X_{n+1}$ with coefficients in $R$. Multiply both sides of the last equation by $f^d$ to conclude that

$$f^d \in \sum_{i=1}^{h} Rf_i \subseteq J.$$

Hence $f \in \sqrt{J}$, and the proof is complete. $\quad\square$

**14.7** EXERCISE. Let $R$ be a non-trivial commutative ring. We say that $R$ is a *Hilbert ring*, or a *Jacobson ring*, if and only if every prime ideal of $R$ is equal to the intersection of all the maximal ideals of $R$ which contain it.

(i) Show that $\mathbb{Z}$ is a Hilbert ring.

(ii) Is every PID a Hilbert ring? Justify your response.

(iii) Show that, if $R$ is a Hilbert ring and $I$ is an ideal of $R$, then $\sqrt{I}$ is the intersection of all the maximal ideals of $R$ which contain $I$.

(iv) Show that a non-trivial homomorphic image of a Hilbert ring is a Hilbert ring.

(v) Show that a non-trivial affine algebra over an algebraically closed field is a Hilbert ring.

(vi) Suppose that $R$ is a subring of a commutative ring $S$, and that $S$ is integral over $R$. Prove that $S$ is a Hilbert ring if and only if $R$ is.

(vii) Prove that, if $R$ is a Hilbert ring and there exists a non-maximal prime ideal of $R$, then $R$ has infinitely many maximal ideals.

**14.8** FURTHER STEPS. We are not going to explain in this book the geometrical significance of the Nullstellensatz. However, the reader should be aware that that theorem is a corner-stone of affine algebraic geometry over an algebraically closed field: two (of several) texts where he or she can explore this are the books of Kunz [6] and Reid [13].

Also, there is much more to the theory of Hilbert rings than the little covered in Exercise 14.7 above: the interested reader might like to learn more about these rings from [12, Chapter 6] or [5, Section 1-3]. It is perhaps worth pointing out that one can use the theory of Hilbert rings to prove that, if $K$ is an *arbitrary* field and $R = K[X_1, \ldots, X_n]$, the polynomial ring over $K$ in the $n$ indeterminates $X_1, \ldots, X_n$, then each maximal ideal of $R$ can be generated by $n$ elements: in the case of an algebraically closed $K$, this follows from the Nullstellensatz. See [12, Section 6.2, Theorem 3].

Our next topic in this chapter is Noetherian normalization theory. Given a non-trivial affine $K$-algebra $A$, where $K$ is a field, we can, and do, identify $K$ with its image in $A$ under the structural ring homomorphism; with this identification, we shall see that there exists a subring $K[Y_1, \ldots, Y_n]$ of $A$ for which the family $(Y_i)_{i=1}^{n}$ is algebraically independent over $K$ and such that $A$ is integral over $K[Y_1, \ldots, Y_n]$, or, equivalently in these circumstances (see 13.20 and 13.21), $A$ is a finitely generated $K[Y_1, \ldots, Y_n]$-module. This will be a consequence of Noether's Normalization Theorem 14.14 below. In fact, given a proper ideal $I$ of $A$, we shall be able to find such $Y_1, \ldots, Y_n$ with the property that

$$I \cap K[Y_1, \ldots, Y_n] = \sum_{i=d+1}^{n} K[Y_1, \ldots, Y_n]Y_i = (Y_{d+1}, \ldots, Y_n)$$

for some $d \in \mathbb{N}_0$ with $d \leq n$. The proof that this is possible will be achieved gradually after a build-up through increasingly more complex results. The following lemma will help with some of the technical details.

**14.9** LEMMA. *Let $K$ be a field and let $A, B$ be non-trivial affine $K$-algebras. Suppose that $\phi : A \to B$ is a surjective homomorphism of $K$-algebras, and let $I := \operatorname{Ker} \phi$. Let $(Y_i)_{i=1}^n$ be a family of elements of $A$ which is algebraically independent over $K$ and such that $A$ is a finitely generated $K[Y_1, \ldots, Y_n]$-module. Suppose that*

$$I \cap K[Y_1, \ldots, Y_n] = \sum_{i=d+1}^{n} K[Y_1, \ldots, Y_n]Y_i = (Y_{d+1}, \ldots, Y_n)$$

*for some $d \in \mathbb{N}_0$ with $d \leq n$. Then $(\phi(Y_i))_{i=1}^d$ is algebraically independent over $K$ and such that $B$ is a finitely generated $K[\phi(Y_1), \ldots, \phi(Y_d)]$-module.*

*Proof.* The composite $K$-algebra homomorphism

$$\psi : K[Y_1, \ldots, Y_n] \longrightarrow A \xrightarrow{\phi} B$$

(in which the first map is the inclusion homomorphism) has kernel equal to $I \cap K[Y_1, \ldots, Y_n] = (Y_{d+1}, \ldots, Y_n)$, and so, by the Isomorphism Theorem 2.13, there is induced an injective $K$-algebra homomorphism

$$\bar{\psi} : K[Y_1, \ldots, Y_n]/(Y_{d+1}, \ldots, Y_n) \longrightarrow B$$

for which $\bar{\psi}(\tilde{Y}_i) = \phi(Y_i)$ for $i = 1, \ldots, d$ but $\bar{\psi}(\tilde{Y}_i) = 0$ for $i = d + 1, \ldots, n$ (and $\tilde{\ }$ is used to denote natural images in $K[Y_1, \ldots, Y_n]/(Y_{d+1}, \ldots, Y_n)$). It is easy to deduce that $B$ is a finitely generated $K[\phi(Y_1), \ldots, \phi(Y_d)]$-module. On the other hand, consideration of the evaluation homomorphism $\theta : K[Y_1, \ldots, Y_n] \to K[Y_1, \ldots, Y_d]$ at $Y_1, \ldots, Y_d, 0, \ldots, 0$ leads to a $K$-algebra isomorphism

$$\mu : K[Y_1, \ldots, Y_d] \to K[Y_1, \ldots, Y_n]/(Y_{d+1}, \ldots, Y_n)$$

for which $\mu(Y_i) = \tilde{Y}_i$ for $i = 1, \ldots, d$. Since $(Y_i)_{i=1}^d$ is algebraically independent over $K$ and the $K$-algebra homomorphism $\bar{\psi} \circ \mu$ is injective, it follows that $(\phi(Y_i))_{i=1}^d$ is algebraically independent over $K$. $\square$

The next technical lemma is concerned with a change of algebraically independent 'variables' in the polynomial ring $K[X_1, \ldots, X_n]$ over the field $K$ which transforms a previously specified non-constant $f \in K[X_1, \ldots, X_n]$ into a particular form. The proof presented works for any field $K$; an easier argument is available in the case in which $K$ is infinite, and this is the subject of Exercise 14.11.

**14.10** LEMMA. *Let $K$ be a field and let $n \in \mathbb{N}$; let $f \in K[X_1, \ldots, X_n]$ be non-constant. Then there exist $Y_1, \ldots, Y_{n-1} \in K[X_1, \ldots, X_n]$ such that $K[Y_1, \ldots, Y_{n-1}, X_n] = K[X_1, \ldots, X_n]$ and*

$$f = cX_n^m + \sum_{i=0}^{m-1} g_i X_n^i \quad \text{with } m \in \mathbb{N}, \ g_i \in K[Y_1, \ldots, Y_{n-1}], \ c \in K \setminus \{0\}.$$

*Note.* It follows from 12.7(ii) and 12.55(vi) applied to the quotient field $K(X_1, \ldots, X_n)$ of $K[X_1, \ldots, X_n]$ that the family $(Y_i)_{i=1}^n$, where $Y_n = X_n$, is algebraically independent over $K$.

*Proof.* There exists a finite subset $\Lambda$ of $\mathbb{N}_0^n$ and elements

$$a_{i_1, \ldots, i_n} \in K \setminus \{0\} \qquad ((i_1, \ldots, i_n) \in \Lambda)$$

such that

$$f = \sum_{(i_1, \ldots, i_n) \in \Lambda} a_{i_1, \ldots, i_n} X_1^{i_1} \ldots X_n^{i_n}.$$

Let $h - 1$ be the largest integer which occurs as a component of one of the members of $\Lambda$, so that $h \geq 2$. Observe that, for two different members $(i_1, \ldots, i_n)$, $(j_1, \ldots, j_n)$ of $\Lambda$, we must have

$$i_n + i_1 h + \cdots + i_{n-1} h^{n-1} \neq j_n + j_1 h + \cdots + j_{n-1} h^{n-1}.$$

Let $Y_i = X_i - X_n^{h^i}$ for all $i = 1, \ldots, n-1$. Since $X_i = Y_i + X_n^{h^i}$ for all $i = 1, \ldots, n-1$, it is clear that $K[Y_1, \ldots, Y_{n-1}, X_n] = K[X_1, \ldots, X_n]$. Also, in terms of the $Y_i$ and $X_n$, we have

$$f = \sum_{(i_1, \ldots, i_n) \in \Lambda} a_{i_1, \ldots, i_n} (X_n^{h^1} + Y_1)^{i_1} \ldots (X_n^{h^{n-1}} + Y_{n-1})^{i_{n-1}} X_n^{i_n}.$$

If we now set

$$m = \max_{(i_1, \ldots, i_n) \in \Lambda} \left\{ i_n + i_1 h + \cdots + i_{n-1} h^{n-1} \right\},$$

then $f$ has the desired form. $\square$

**14.11** EXERCISE. Let the situation be as in 14.10 and suppose, in addition, that the field $K$ is infinite. Show that there exist $a_1, \ldots, a_{n-1} \in K$ such that the elements $Y_i = X_i - a_i X_n$ $(i = 1, \ldots, n-1)$ of $K[X_1, \ldots, X_n]$ satisfy the conclusions of the lemma. (Write $f$ as a sum of homogeneous polynomials, and use Exercise 1.19. If you still find the exercise difficult, consult [13, pp. 59-60].)

Our next two results are preparatory to Noether's Normalization Theorem 14.14.

**14.12** PROPOSITION. *Let $A = K[X_1, \ldots, X_n]$, the ring of polynomials in the $n$ indeterminates $X_1, \ldots, X_n$ over the field $K$, and let $I = fA$, the principal ideal of $A$ generated by the non-constant $f \in A$. Then there exist $Y_1, \ldots, Y_n \in A$ with $Y_n = f$ such that*

　(i) $(Y_i)_{i=1}^{n}$ *is algebraically independent over $K$;*
　(ii) $A$ *is integral over $K[Y_1, \ldots, Y_n]$; and*
　(iii) $I \cap K[Y_1, \ldots, Y_n] = K[Y_1, \ldots, Y_n]Y_n = (Y_n)$.

*Proof.* By 14.10, there exist $Y_1, \ldots, Y_{n-1} \in K[X_1, \ldots, X_n]$ such that $K[Y_1, \ldots, Y_{n-1}, X_n] = K[X_1, \ldots, X_n]$ and

$$f = cX_n^m + \sum_{i=0}^{m-1} g_i X_n^i \quad \text{with } m \in \mathbf{N},\ g_i \in K[Y_1, \ldots, Y_{n-1}],\ c \in K \setminus \{0\}.$$

Set $Y_n = f$. Then the above displayed equation shows that $X_n$ is integral over $K[Y_1, \ldots, Y_n]$, and so it follows from 13.22 that $K[X_1, \ldots, X_n]$ is integral over $K[Y_1, \ldots, Y_n]$. Hence each $X_i$ $(i = 1, \ldots, n)$ is algebraic over the subfield $K(Y_1, \ldots, Y_n)$ of $K(X_1, \ldots, X_n)$. It therefore follows from 12.55(vi) that $(Y_i)_{i=1}^{n}$ is algebraically independent over $K$.

Lastly, we show that $I \cap K[Y_1, \ldots, Y_n] = K[Y_1, \ldots, Y_n]Y_n$. Of course,

$$I \cap K[Y_1, \ldots, Y_n] \supseteq K[Y_1, \ldots, Y_n]Y_n$$

since $Y_n = f$. Let $p \in I \cap K[Y_1, \ldots, Y_n]$, so that $p = gf$ for some $g \in A$. Since $A$ is integral over $K[Y_1, \ldots, Y_n]$, there exist $h \in \mathbf{N}$ and $q_0, \ldots, q_{h-1} \in K[Y_1, \ldots, Y_n]$ such that

$$g^h + q_{h-1}g^{h-1} + \cdots + q_1 g + q_0 = 0;$$

multiplication by $f^h = Y_n^h$ now yields

$$p^h + q_{h-1}Y_n p^{h-1} + \cdots + q_1 Y_n^{h-1}p + q_0 Y_n^h = 0.$$

But $p, q_0, \ldots, q_{h-1}, Y_n \in K[Y_1, \ldots, Y_n]$, and, by 1.42, this is a UFD since $(Y_i)_{i=1}^{n}$ is algebraically independent over $K$. It follows from this observation that $Y_n \mid p$ in the ring $K[Y_1, \ldots, Y_n]$, and so $p \in K[Y_1, \ldots, Y_n]Y_n$. Hence

$$I \cap K[Y_1, \ldots, Y_n] \subseteq K[Y_1, \ldots, Y_n]Y_n$$

and the proof is complete. $\square$

**14.13** PROPOSITION. *Let $A = K[X_1, \ldots, X_n]$, the ring of polynomials in the $n$ indeterminates $X_1, \ldots, X_n$ over the field $K$, and let $I$ be a proper ideal of $A$. Then there exist $Y_1, \ldots, Y_n \in A$ and $d \in \mathbb{N}_0$ with $0 \le d \le n$ such that*

(i) *$(Y_i)_{i=1}^n$ is algebraically independent over $K$;*
(ii) *$A$ is integral over $K[Y_1, \ldots, Y_n]$; and*
(iii) *$I \cap K[Y_1, \ldots, Y_n] = \sum_{i=d+1}^n K[Y_1, \ldots, Y_n]Y_i = (Y_{d+1}, \ldots, Y_n)$.*

*Proof.* In the case in which $I = 0$, the result is easy: just take $X_i = Y_i$ for all $i = 1, \ldots, n$ and $d = n$. Thus we suppose henceforth in this proof that $I \ne 0$.

We argue by induction on $n$. In the case in which $n = 1$, the non-zero proper ideal $I$ of the PID $A = K[X_1]$ must be generated by a non-constant polynomial, and so the result in this case follows from 14.12. Now suppose, inductively, that $n > 1$ and the result has been proved for smaller values of $n$. Since $I$ is proper and non-zero, there exists a non-constant polynomial $f$ of $A$ with $f \in I$. By 14.12, there exist $Z_1, \ldots, Z_n \in A$ with $Z_n = f$ such that $(Z_i)_{i=1}^n$ is algebraically independent over $K$, the ring $A$ is integral over $K[Z_1, \ldots, Z_n]$, and

$$fA \cap K[Z_1, \ldots, Z_n] = K[Z_1, \ldots, Z_n]Z_n = (Z_n) = (f).$$

Now consider the (necessarily proper) ideal $I \cap K[Z_1, \ldots, Z_{n-1}]$ of the ring $K[Z_1, \ldots, Z_{n-1}]$. Bearing in mind 1.16, we can deduce from either the first paragraph of this proof or the inductive hypothesis that there exist $Y_1, \ldots, Y_{n-1} \in K[Z_1, \ldots, Z_{n-1}]$ and $d \in \mathbb{N}_0$ with $0 \le d < n$ such that $(Y_i)_{i=1}^{n-1}$ is algebraically independent over $K$, the ring $K[Z_1, \ldots, Z_{n-1}]$ is integral over $K[Y_1, \ldots, Y_{n-1}]$, and

$$I \cap K[Y_1, \ldots, Y_{n-1}] = \sum_{i=d+1}^{n-1} K[Y_1, \ldots, Y_{n-1}]Y_i = (Y_{d+1}, \ldots, Y_{n-1}).$$

Put $Y_n = Z_n = f$: we shall show that $Y_1, \ldots, Y_n$ satisfy the conditions stated in the proposition.

First of all, $Z_1, \ldots, Z_{n-1}$ are all integral over

$$K[Y_1, \ldots, Y_n] = K[Y_1, \ldots, Y_{n-1}][Y_n],$$

and clearly so also is $Z_n = Y_n$. We thus deduce from 13.22 that the ring $K[Z_1, \ldots, Z_n]$ is integral over $K[Y_1, \ldots, Y_n]$, and so the ring $A$ is integral over $K[Y_1, \ldots, Y_n]$ in view of 13.23. Thus each of $X_1, \ldots, X_n$ is algebraic over the subfield $K(Y_1, \ldots, Y_n)$ of $K(X_1, \ldots, X_n)$, and so it follows from

12.55(vi) that $(Y_i)_{i=1}^n$ is algebraically independent over $K$. There remains only condition (iii) for us to check: we aim to show that

$$I \cap K[Y_1, \ldots, Y_n] = \sum_{i=d+1}^{n} K[Y_1, \ldots, Y_n]Y_i = (Y_{d+1}, \ldots, Y_n).$$

First, note that $Y_{d+1}, \ldots, Y_{n-1}, Y_n = f \in I$, and so

$$I \cap K[Y_1, \ldots, Y_n] \supseteq \sum_{i=d+1}^{n} K[Y_1, \ldots, Y_n]Y_i.$$

Conversely, let $g \in I \cap K[Y_1, \ldots, Y_n]$. Write $g$ as a polynomial in $Y_1, \ldots, Y_n$ with coefficients in $K$. The sum of those non-zero monomial terms of $g$ which involve $Y_n$ can be written in the form $Y_n g_2$ for some $g_2 \in K[Y_1, \ldots, Y_n]$, and thus $g = g_1 + Y_n g_2$ for some $g_1 \in K[Y_1, \ldots, Y_{n-1}]$. Since $Y_n = f \in I$, we see that

$$g_1 \in I \cap K[Y_1, \ldots, Y_{n-1}] = \sum_{i=d+1}^{n-1} K[Y_1, \ldots, Y_{n-1}]Y_i.$$

Hence

$$g = g_1 + Y_n g_2 \in \sum_{i=d+1}^{n} K[Y_1, \ldots, Y_n]Y_i.$$

Therefore $I \cap K[Y_1, \ldots, Y_n] \subseteq \sum_{i=d+1}^{n} K[Y_1, \ldots, Y_n]Y_i$, and condition (iii) has been verified. This completes the inductive step, and the proof. $\square$

We are now ready to prove a general version of Noether's Normalization Theorem.

**14.14** Noether's Normalization Theorem. *Let $A$ be a non-trivial affine algebra over the field $K$, and let $I$ be a proper ideal of $A$. Then there exist $n, d \in \mathbb{N}_0$ with $0 \leq d \leq n$ and $Y_1, \ldots, Y_n \in A$ such that*
   (i) *$(Y_i)_{i=1}^n$ is algebraically independent over $K$;*
   (ii) *$A$ is integral over $K[Y_1, \ldots, Y_n]$; and*
   (iii) *$I \cap K[Y_1, \ldots, Y_n] = \sum_{i=d+1}^{n} K[Y_1, \ldots, Y_n]Y_i = (Y_{d+1}, \ldots, Y_n).$*

*Proof.* Suppose that $A$ is generated as a $K$-algebra by $h$ elements. We can use 8.10 and 1.16 to see that there is a surjective $K$-algebra homomorphism $\phi : K[X_1, \ldots, X_h] \to A$. Let $J := \operatorname{Ker} \phi$, and let $I' := \phi^{-1}(I)$, a proper ideal of $K[X_1, \ldots, X_h]$ which contains $J$.

By 14.13, there exist $Z_1, \ldots, Z_h \in K[X_1, \ldots, X_h]$ and $n \in \mathbb{N}_0$ with $0 \leq n \leq h$ such that $(Z_i)_{i=1}^h$ is algebraically independent over $K$, the ring $K[X_1, \ldots, X_h]$ is integral over $K[Z_1, \ldots, Z_h]$, and

$$J \cap K[Z_1, \ldots, Z_h] = \sum_{i=n+1}^h K[Z_1, \ldots, Z_h]Z_i = (Z_{n+1}, \ldots, Z_h).$$

We note that, by 14.9 (and 13.20 and 13.21), the family $(\phi(Z_i))_{i=1}^n$ is algebraically independent over $K$ and such that $A$ is integral over its subring $K[\phi(Z_1), \ldots, \phi(Z_n)]$.

Since $(Z_i)_{i=1}^n$ is algebraically independent over $K$, it follows from 1.16 that $K[Z_1, \ldots, Z_n]$ is essentially just a polynomial ring over $K$ in $n$ indeterminates $Z_1, \ldots, Z_n$. We can therefore apply 14.13 to this ring and its proper ideal $I' \cap K[Z_1, \ldots, Z_n]$: we deduce that there exist $W_1, \ldots, W_n \in K[Z_1, \ldots, Z_n]$ and $d \in \mathbb{N}_0$ with $0 \leq d \leq n$ such that $(W_i)_{i=1}^n$ is algebraically independent over $K$, the ring $K[Z_1, \ldots, Z_n]$ is integral over $K[W_1, \ldots, W_n]$, and

$$I' \cap K[W_1, \ldots, W_n] = \sum_{i=d+1}^n K[W_1, \ldots, W_n]W_i = (W_{d+1}, \ldots, W_n).$$

Set $Y_i := \phi(W_i)$ for all $i = 1, \ldots, n$: we show that $Y_1, \ldots, Y_n$ fulfil the requirements of the theorem.

Since the ring $K[Z_1, \ldots, Z_n]$ is integral over $K[W_1, \ldots, W_n]$, we quickly deduce from 13.20 and 13.21 (or directly) that $K[\phi(Z_1), \ldots, \phi(Z_n)]$ is integral over $K[Y_1, \ldots, Y_n]$; hence, by 13.23, $A$ is integral over $K[Y_1, \ldots, Y_n]$. Also, since $(\phi(Z_i))_{i=1}^n$ is algebraically independent over $K$, it follows that the ring $K[\phi(Z_1), \ldots, \phi(Z_n)]$ is an integral domain; by 12.29, its quotient field $K(\phi(Z_1), \ldots, \phi(Z_n))$ is algebraic over $K(Y_1, \ldots, Y_n)$, and it therefore follows from 12.55(vi) that $(Y_i)_{i=1}^n$ is algebraically independent over $K$. It remains only for us show that

$$I \cap K[Y_1, \ldots, Y_n] = \sum_{i=d+1}^n K[Y_1, \ldots, Y_n]Y_i = (Y_{d+1}, \ldots, Y_n),$$

and this is an easy consequence of the facts that $I' = \phi^{-1}(I)$, that

$$I' \cap K[W_1, \ldots, W_n] = \sum_{i=d+1}^n K[W_1, \ldots, W_n]W_i = (W_{d+1}, \ldots, W_n),$$

and that $\phi(W_i) = Y_i$ for all $i = 1, \ldots, n$. $\square$

**14.15** EXERCISE.   Show that, in the special case of 14.14 in which $K$ is infinite, and the affine $K$-algebra $A$ is generated (as a $K$-algebra) by $a_1, \ldots, a_h$, one can use the result of Exercise 14.11 to arrange that the $Y_1, \ldots, Y_d$ in the conclusion of Noether's Normalization Theorem are all $K$-linear combinations of $a_1, \ldots, a_h$. (This result has significance for algebraic geometry, although we shall not need it in this book.)

Our plan for the remainder of this chapter is to use Noether's Normalization Theorem to obtain important results about the dimensions of affine algebras over a field. It is therefore necessary for the fundamental definitions about dimensions of commutative rings to be available first, but before we discuss chains of prime ideals in commutative rings in detail, we introduce some terminology which will be useful later in the chapter.

**14.16** DEFINITION.   Let $A$ be a non-trivial affine algebra over the field $K$. A *Noether normalizing family for A* is a family $(Y_i)_{i=1}^n$ of elements of $A$ such that

(i) $(Y_i)_{i=1}^n$ is algebraically independent over $K$, and

(ii) $A$ is integral over $K[Y_1, \ldots, Y_n]$ (or, equivalently (in view of 13.20 and 13.21), $A$ is a finitely generated $K[Y_1, \ldots, Y_n]$-module.)

Thus Noether's Normalization Theorem 14.14 provides the existence of Noether normalizing families for non-trivial affine algebras over fields.

**14.17** DEFINITIONS.   Let $R$ be a non-trivial commutative ring.

(i) An expression

$$P_0 \subset P_1 \subset \ldots \subset P_n$$

(note the strict inclusions) in which $P_0, \ldots, P_n$ are prime ideals of $R$, is called a *chain of prime ideals of R*; the *length* of such a chain is the number of 'links', that is, 1 less than the number of prime ideals present. Thus the above displayed chain has length $n$.

Note that, for $P \in \operatorname{Spec}(R)$, we consider

$$P$$

to be a chain of prime ideals of $R$ of length 0. Since $R$ is non-trivial, there certainly exists at least one chain of prime ideals of $R$ of length 0.

(ii) A chain

$$P_0 \subset P_1 \subset \ldots \subset P_n$$

of prime ideals of $R$ is said to be *saturated* precisely when, for every $i \in \mathbb{N}$ with $1 \leq i \leq n$, there does not exist $Q \in \operatorname{Spec}(R)$ such that $P_{i-1} \subset Q \subset P_i$, that is, if and only if we cannot make a chain of length $n+1$ by the insertion of an additional prime ideal of $R$ strictly between two adjacent terms in the original chain.

(iii) A chain

$$P_0 \subset P_1 \subset \ldots \subset P_n$$

of prime ideals of $R$ is said to be *maximal* precisely when it is saturated, $P_n$ is a maximal ideal of $R$ and $P_0$ is a minimal prime ideal of the zero ideal $0$ of $R$ (see 3.52), that is, if and only if we cannot make a chain of length $n+1$ by the insertion of an additional prime ideal of $R$ at the beginning of, at the end of, or strictly between two adjacent terms in, the original chain.

(iv) The *dimension of $R$*, denoted by $\dim R$, is defined to be

$$\sup \{n \in \mathbb{N}_0 : \text{ there exists a chain of prime ideals of } R \text{ of length } n\}$$

if this supremum exists, and $\infty$ otherwise.

(v) Let $P \in \operatorname{Spec}(R)$. Then the *height of $P$*, denoted by $\operatorname{ht} P$ (or $\operatorname{ht}_R P$ if it is desired to emphasize the underlying ring) is defined to be the supremum of lengths of chains

$$P_0 \subset P_1 \subset \ldots \subset P_n$$

of prime ideals of $R$ for which $P_n = P$ if this supremum exists, and $\infty$ otherwise.

**14.18** REMARKS. Let $R$ be a non-trivial commutative ring.

(i) Note that $\dim R$ is either a non-negative integer or $\infty$; we do not define the dimension of the trivial commutative ring.

(ii) By 3.10, every prime ideal of $R$ is contained in a maximal ideal of $R$ (and, of course, maximal ideals are prime); also, by 3.53, every prime ideal of $R$ contains a minimal prime ideal of $0$. It follows that $\dim R$ is equal to the supremum of lengths of chains

$$P_0 \subset P_1 \subset \ldots \subset P_n$$

of prime ideals of $R$ with $P_n$ maximal and $P_0$ a minimal prime ideal of $0$. This is because the length $h$ of an arbitrary chain

$$P_0' \subset P_1' \subset \ldots \subset P_h'$$

of prime ideals of $R$ is bounded above by the length of a chain of this special type: if $P_0'$ is not a minimal prime ideal of $0$, then one can be inserted 'below' it; if $P_h'$ is not a maximal ideal of $R$, then one can be inserted 'above' it.

(iii) Thus, if $\dim R$ is finite, then

$$\dim R = \sup \{\operatorname{ht} M : M \text{ is a maximal ideal of } R\}$$
$$= \sup \{\operatorname{ht} P : P \in \operatorname{Spec}(R)\} .$$

(iv) It also follows from part (ii) above that, if $R$ is quasi-local with maximal ideal $M$ (see 3.12), then $\dim R = \operatorname{ht} M$.

(v) Let $S$ be a multiplicatively closed subset of $R$ and $P \in \operatorname{Spec}(R)$ be such that $P \cap S = \emptyset$, so that, by 5.32, $S^{-1}P \in \operatorname{Spec}(S^{-1}R)$. It is an easy consequence of the bijective, inclusion-preserving correspondence between $\{P \in \operatorname{Spec}(R) : P \cap S = \emptyset\}$ and $\operatorname{Spec}(S^{-1}R)$ established in 5.33 that $\operatorname{ht}_{S^{-1}R} S^{-1}P = \operatorname{ht}_R P$. To see this, let

$$P_0 \subset P_1 \subset \ldots \subset P_n$$

be a chain of prime ideals of $R$ with $P_n = P$, and let the extension and contraction notation of 2.41 refer to the natural ring homomorphism $R \to S^{-1}R$. By 5.33,

$$P_0^e \subset P_1^e \subset \ldots \subset P_n^e$$

is a chain of prime ideals of $S^{-1}R$ with $P_n^e = P^e = S^{-1}P$, and so $\operatorname{ht}_R P \le \operatorname{ht}_{S^{-1}R} S^{-1}P$. On the other hand, if

$$\mathcal{P}_0 \subset \mathcal{P}_1 \subset \ldots \subset \mathcal{P}_n$$

is a chain of prime ideals of $S^{-1}R$ with $\mathcal{P}_n = P^e$, then

$$\mathcal{P}_0^c \subset \mathcal{P}_1^c \subset \ldots \subset \mathcal{P}_n^c$$

is a chain of prime ideals of $R$ (also by 5.33) with $\mathcal{P}_n^c = P^{ec} = P$, and so we have $\operatorname{ht}_R P \ge \operatorname{ht}_{S^{-1}R} S^{-1}P$ as well.

(vi) It follows from parts (iv) and (v) above that, for $P \in \operatorname{Spec}(R)$,

$$\operatorname{ht} P = \operatorname{ht}_{R_P} PR_P = \dim R_P.$$

(vii) In part (v) above, we made considerable use of 5.33 to relate chains of prime ideals in a ring of fractions of $R$ to chains of prime ideals of $R$. In a similar way, we can make use of 2.39 and 3.28 to relate chains of prime ideals of a residue class ring $R/I$, where $I$ is a proper ideal of $R$, to chains of prime ideals of $R$ which contain $I$. For example, it follows from 2.39 and 3.28 that a chain of prime ideals of $R/I$ will have the form

$$P_0/I \subset P_1/I \subset \ldots \subset P_n/I,$$

where

$$P_0 \subset P_1 \subset \ldots \subset P_n$$

is a chain of prime ideals of $R$ for which $P_0 \supseteq I$. Using such ideas, we can see that $\dim R/I$ is equal to the supremum of lengths of chains

$$P_0 \subset P_1 \subset \ldots \subset P_n$$

of prime ideals of $R$ all of which contain $I$ if this supremum exists, and $\infty$ otherwise.

(viii) Let $P \in \operatorname{Spec}(R)$. Then, provided we adopt the natural conventions that $\infty + \infty = \infty$, that $n + \infty = \infty$ for all $n \in \mathbf{N}_0$ and that $\infty \leq \infty$, we have

$$\operatorname{ht} P + \dim R/P \leq \dim R.$$

This follows from part (vii) above and the observation that if

$$P_0 \subset P_1 \subset \ldots \subset P_n$$

is a chain of prime ideals of $R$ with $P_n = P$ and

$$P_0' \subset P_1' \subset \ldots \subset P_h'$$

is a chain of prime ideals of $R$ with $P_0' = P$, then

$$P_0 \subset P_1 \subset \ldots \subset P_n \subset P_1' \subset \ldots \subset P_h'$$

is a chain of prime ideals of $R$ of length $n + h$.

It is time we had some examples, even easy ones!

**14.19 EXAMPLES.** (i) By 8.39, every non-trivial commutative Artinian ring has dimension 0, because in such a ring every prime ideal is maximal.

(ii) In particular, a field has dimension 0.

(iii) In $\mathbf{Z}$, we have $0\mathbf{Z} \subset 2\mathbf{Z}$, a chain of prime ideals of length 1; by 3.34, every non-zero prime ideal of $\mathbf{Z}$ is maximal, and so it follows that there does not exist a chain of prime ideals of $\mathbf{Z}$ of length 2 since the middle term of such a chain would have to be maximal. Hence $\dim \mathbf{Z} = 1$.

**14.20 EXERCISE.** Let $R$ be a PID which is not a field, and let $P \in \operatorname{Spec}(R)$. State and prove a necessary and sufficient condition for $P$ to have height 1.

**14.21 EXERCISE.** Let $R$ be a UFD, and let $P \in \operatorname{Spec}(R)$. Show that $\operatorname{ht} P = 1$ if and only if $P = Rp$ for some irreducible element $p$ of $R$.

Less trivial examples of the use of the concepts of height and dimension will be given after our exploitation of Noether's Normalization Theorem later in the chapter. First of all, however, we show that some of the theorems of Chapter 13, especially the Incomparability Theorem 13.33, the Going-up Theorem 13.38 and the Going-down Theorem 13.41, yield useful relationships between various heights and dimensions when we have an 'integral extension' of commutative rings. The next three results are of this type; in view of Noether's Normalization Theorem, they will themselves be useful tools in the study of the dimension theory of affine algebras.

**14.22** PROPOSITION. *Let $R$ be a subring of the non-trivial commutative ring $S$, and suppose that $S$ is integral over $R$. Then* $\dim R = \dim S$.

*Proof.* Use the contraction notation of 2.41 in conjunction with the inclusion homomorphism $R \to S$. Let

$$Q_0 \subset Q_1 \subset \ldots \subset Q_n$$

be a chain of prime ideals of $S$. Then it follows from the Incomparability Theorem 13.33 and 3.27(ii) that

$$Q_0^c \subset Q_1^c \subset \ldots \subset Q_n^c$$

is a chain of prime ideals of $R$. Hence $\dim S \leq \dim R$.

Now suppose that

$$P_0 \subset P_1 \subset \ldots \subset P_n$$

is a chain of prime ideals of $R$. By the Lying-over Theorem 13.34, there exists $Q_0 \in \operatorname{Spec}(S)$ such that $Q_0^c = P_0$. It now follows from the Going-up Theorem 13.38 that there exists a chain

$$Q_0 \subset Q_1 \subset \ldots \subset Q_n$$

of prime ideals of $S$, and so $\dim R \leq \dim S$. $\square$

**14.23** COROLLARY. *Let $R$ be a subring of the commutative ring $S$, and suppose that $S$ is integral over $R$. Let $I$ be a proper ideal of $S$. Then* $\dim R/(I \cap R) = \dim S/I$.

*Proof.* This is immediate from 14.22 and 13.26(i), since the latter result shows that we may regard $R/(I \cap R)$ as a subring of $S/I$ in a natural way such that $S/I$ is integral over $R/(I \cap R)$. $\square$

**14.24** ♯EXERCISE. Let $R$ be a subring of the commutative ring $S$, and suppose that $S$ is integral over $R$. Let $Q \in \operatorname{Spec}(S)$. Show that $\operatorname{ht}_S Q \leq \operatorname{ht}_R(Q \cap R)$. (Use the Incomparability Theorem 13.33.)

Actually, we can do better than the result of 14.24 in circumstances where the Going-down Theorem 13.41 can be used.

**14.25** PROPOSITION. *Let $R$ be a subring of the integral domain $S$, and suppose that $R$ is integrally closed and that $S$ is integral over $R$. Let $Q \in \operatorname{Spec}(S)$. Then* $\operatorname{ht}_S Q = \operatorname{ht}_R(Q \cap R)$.

*Proof.* By 14.24, we have $\mathrm{ht}_S Q \leq \mathrm{ht}_R(Q \cap R)$. Let

$$P_0 \subset P_1 \subset \ldots \subset P_n$$

be a chain of prime ideals of $R$ such that $P_n = Q \cap R$. Then, by the Going-down Theorem 13.41, there exists a chain

$$Q_0 \subset Q_1 \subset \ldots \subset Q_n$$

of prime ideals of $S$ for which $Q_n = Q$. It follows that $\mathrm{ht}_S Q \geq \mathrm{ht}_R(Q \cap R)$, and so the proof is complete. $\square$

**14.26** EXERCISE. Let $R$ be a subring of the integral domain $S$, and suppose that $R$ is integrally closed and that $S$ is integral over $R$. Let $Q \in \mathrm{Spec}(S)$. Prove that $\mathrm{ht}_S Q + \dim S/Q = \dim S$ if and only if

$$\mathrm{ht}_R(Q \cap R) + \dim R/(Q \cap R) = \dim R.$$

**14.27** REMARKS. Let $A := K[X_1, \ldots, X_n]$.
  (i) Let $a_1, \ldots, a_n \in K$. It follows from Exercise 3.66 that

$$\mathrm{ht}_A(X_1 - a_1, \ldots, X_n - a_n) \geq n.$$

  (ii) We can therefore deduce from 14.6(i) that, in the special case in which $K$ is algebraically closed, every maximal ideal of $K[X_1, \ldots, X_n]$ has height at least $n$. We are going, in the course of the next few results, to use Noether's Normalization Theorem 14.14 to produce a substantial sharpening of this comment.

**14.28** THEOREM. *Let $A$ be a non-trivial affine algebra over a field $K$. Then all Noether normalizing families for $A$ have the same number of elements, $n$ say, and $n = \dim A$.*

*Proof.* Use Noether's Normalization Theorem 14.14 to see that there exists a Noether normalizing family $(Y_i)_{i=1}^n$ for $A$; thus this family is algebraically independent over $K$, and $A$ is integral over $K[Y_1, \ldots, Y_n]$. It is enough to prove that $n = \dim A$. We argue by induction on $n$. When $n = 0$, we see from 14.22 and 14.19(ii) that $A$ has dimension 0.

We therefore assume that $n > 0$ and the result has been proved for (non-trivial) affine $K$-algebras which have Noether normalizing families with fewer than $n$ elements. By 14.22 and 14.27(i),

$$\dim A = \dim K[Y_1, \ldots, Y_n] \geq n.$$

Let

$$Q_0 \subset Q_1 \subset \ldots \subset Q_m$$

be a chain of prime ideals of $A$: we must show that $m \leq n$. Let $P_i = Q_i \cap K[Y_1, \ldots, Y_n]$ for all $i = 0, \ldots, m$. By the Incomparability Theorem 13.33,

$$P_0 \subset P_1 \subset \ldots \subset P_m$$

is a chain of prime ideals of $K[Y_1, \ldots, Y_n]$. By 14.13, there exist a Noether normalizing family $(Z_i)_{i=1}^n$ for $K[Y_1, \ldots, Y_n]$ and $d \in \mathbb{N}_0$ with $0 \leq d \leq n$ such that

$$P_1 \cap K[Z_1, \ldots, Z_n] = \sum_{i=d+1}^n K[Z_1, \ldots, Z_n]Z_i = (Z_{d+1}, \ldots, Z_n).$$

Since $P_0 \subset P_1$ and $K[Y_1, \ldots, Y_n]$ is integral over $K[Z_1, \ldots, Z_n]$, it follows from the Incomparability Theorem 13.33 that $P_1 \cap K[Z_1, \ldots, Z_n] \neq 0$ and $d < n$.

Let $\bar{\phantom{x}} : K[Y_1, \ldots, Y_n] \to K[Y_1, \ldots, Y_n]/P_1$ denote the natural surjective ring homomorphism. Now it follows from 14.9 that $(\overline{Z_i})_{i=1}^d$ is a Noether normalizing family for $K[Y_1, \ldots, Y_n]/P_1$, while we see from 2.39 and 3.28 that

$$P_1/P_1 \subset P_2/P_1 \subset \ldots \subset P_m/P_1$$

is a chain of prime ideals of $K[Y_1, \ldots, Y_n]/P_1$. Hence, we can apply the inductive hypothesis to see that $m - 1 \leq d < n$, so that $m \leq n$. It therefore follows that $\dim A = n$, and so the inductive step is complete.

Thus the theorem is proved. $\square$

We have thus shown that every (non-trivial) affine algebra over a field $K$ has finite dimension. We have also shown that the polynomial ring $K[X_1, \ldots, X_n]$ has dimension $n$, because $(X_i)_{i=1}^n$ is a Noether normalizing family for $K[X_1, \ldots, X_n]$. This was hinted at in our discussion of transcendence degrees of field extensions in Chapter 12. In fact, we are now in a position to link together the ideas of transcendence degree and dimension for affine $K$-algebras which are domains.

**14.29** Corollary. *Suppose that the integral domain $A$ is an affine algebra over a field $K$, and let $L$ denote the field of fractions of $A$ (so that $L$ can be viewed as a field extension of $K$ in an obvious natural way). Then $\dim A = \text{tr.deg}_K L$.*

*Proof.* By Noether's Normalization Theorem 14.14 again, there exists a Noether normalizing family $(Y_i)_{i=1}^n$ for $A$; thus this family is algebraically independent over $K$, and $A$ is integral over $K[Y_1, \ldots, Y_n]$. Suppose that $b_1, \ldots, b_h \in A$ are such that $A = K[b_1, \ldots, b_h]$. Then $L =$

$K(b_1, \ldots, b_h)$ by 12.19 and 12.7(ii), and since $b_1, \ldots, b_h$ are all integral over $K[Y_1, \ldots, Y_n]$, it follows from 12.29 and 12.28 that $L$ is algebraic over its subfield $K(Y_1, \ldots, Y_n)$; hence $(Y_i)_{i=1}^n$ is a transcendence basis for $L$ over $K$. But $n = \dim A$ by 14.28, and therefore $\dim A = \text{tr.deg}_K L$. $\square$

Now that we know, for a field $K$, that $\dim K[X_1, \ldots, X_n] = n$, it follows from 14.27(ii) that, in the special case in which $K$ is algebraically closed, every maximal ideal of $K[X_1, \ldots, X_n]$ has height exactly $n$. We are now going to show, among other things, that this result remains true even if we drop the assumption that $K$ is algebraically closed. Actually, we shall do even better, because we shall show that every maximal chain of prime ideals (see 14.17(iii)) of $K[X_1, \ldots, X_n]$ (where $K$ is an arbitrary field) has length exactly $n$: this fact is a consequence of the next theorem, for which the following Lemma will be helpful.

**14.30** LEMMA. *Let $A$ be a non-trivial affine algebra over a field $K$, and let $(Y_i)_{i=1}^n$ be a Noether normalizing family for $A$. Let $Q_0 \subset Q_1$ be a saturated chain of prime ideals of $A$. Then*

$$Q_0 \cap K[Y_1, \ldots, Y_n] \subset Q_1 \cap K[Y_1, \ldots, Y_n]$$

*is a saturated chain of prime ideals of $K[Y_1, \ldots, Y_n]$.*

*Proof.* Set $Q_i \cap K[Y_1, \ldots, Y_n] =: P_i$ for $i = 0, 1$. By the Incomparability Theorem 13.33, $P_0 \subset P_1$ is a chain of prime ideals of $K[Y_1, \ldots, Y_n]$. Suppose that this chain is not saturated, and look for a contradiction. Then there exists $P \in \text{Spec}(K[Y_1, \ldots, Y_n])$ such that $P_0 \subset P \subset P_1$.

We can use 14.14 again to see that there exist a Noether normalizing family $(Z_i)_{i=1}^n$ for $K[Y_1, \ldots, Y_n]$ and $d \in \mathbb{N}_0$ with $0 \leq d \leq n$ such that

$$P_0 \cap K[Z_1, \ldots, Z_n] = \sum_{i=d+1}^n K[Z_1, \ldots, Z_n]Z_i = (Z_{d+1}, \ldots, Z_n).$$

Set $P_i \cap K[Z_1, \ldots, Z_n] =: P_i'$ for $i = 0, 1$ and $P \cap K[Z_1, \ldots, Z_n] =: P'$. Then

$$P_0' \subset P' \subset P_1'$$

is a chain of prime ideals of $K[Z_1, \ldots, Z_n]$. By 13.23, the family $(Z_i)_{i=1}^n$ is a Noether normalizing family for $A$. Let $\phi : A \to A/Q_0$ denote the natural surjective ring homomorphism. By 14.9, the family $(\phi(Z_i))_{i=1}^d$ is a Noether normalizing family for $A/Q_0$, and it is easy to deduce from the fact that

$$P_0' = Q_0 \cap K[Z_1, \ldots, Z_n] = \sum_{i=d+1}^n K[Z_1, \ldots, Z_n]Z_i = (Z_{d+1}, \ldots, Z_n)$$

that $(\phi(P_0') = 0$ and)

$$\phi(P_0') \subset \phi(P') \subset \phi(P_1')$$

is a chain of prime ideals of $K[\phi(Z_1), \ldots, \phi(Z_d)]$. But the latter integral domain is integrally closed (by 12.43 and 13.17), and $A/Q_0$ is integral over it. Also, by 3.28, $Q_1/Q_0 \in \mathrm{Spec}(A/Q_0)$, and it is easy to check that

$$(Q_1/Q_0) \cap K[\phi(Z_1), \ldots, \phi(Z_d)] = \phi(P_1').$$

We can therefore deduce from the Going-down Theorem 13.41 that there exists $Q \in \mathrm{Spec}(A)$ such that $Q_0 \subset Q \subset Q_1$, and this is a contradiction. The lemma is therefore proved. $\square$

**14.31** THEOREM. *Suppose that the integral domain $A$ is an affine algebra over a field $K$, and let $\dim A = n$. Then every maximal chain of prime ideals of $A$ has length exactly $n$.*

*Proof.* Of course, the length of a maximal chain of prime ideals of $A$ cannot exceed $n$: our task is to show that it cannot have length strictly less than $n$.

We argue by induction on $n$. When $n = 0$, the result is trivial, because an integral domain of dimension 0 must be a field. So suppose, inductively, that $n > 0$ and the result has been proved for integral domains which are affine $K$-algebras of dimension less than $n$.

By 14.28 and Noether's Normalization Theorem 14.14, there exists a Noether normalizing family for $A$ having $n$ elements, $(Y_i)_{i=1}^n$ say. Let

$$Q_0 \subset Q_1 \subset \ldots \subset Q_m$$

be a maximal chain of prime ideals of $A$, and set $P_i = Q_i \cap K[Y_1, \ldots, Y_n]$ for all $i = 0, \ldots, m$. Note that, since $A$ is a domain, $Q_0 = 0$; also, $Q_m$ must be a maximal ideal of $A$.

By the Incomparability Theorem 13.33, and 14.30,

$$P_0 \subset P_1 \subset \ldots \subset P_m$$

is a saturated chain of prime ideals of $K[Y_1, \ldots, Y_n]$. Note that $P_0 = 0 \cap K[Y_1, \ldots, Y_n] = 0$, and that $P_m$ is a maximal ideal of $K[Y_1, \ldots, Y_n]$, by 13.31. Thus

$$P_0 \subset P_1 \subset \ldots \subset P_m$$

is a maximal chain of prime ideals of $K[Y_1, \ldots, Y_n]$.

It follows that $\mathrm{ht}\, P_1 = 1$. Note that, in view of 1.16 and 1.42, the ring $K[Y_1, \ldots, Y_n]$ is a UFD. Let $0 \neq f \in P_1$: since $P_1$ is prime, one of the irreducible factors of $f$, say $p$, must lie in $P_1$. By 3.42, the principal ideal $(p)$ of $K[Y_1, \ldots, Y_n]$ is prime; hence $P_1 = (p)$, since $0 \subset (p) \subseteq P_1$ and $\mathrm{ht}\, P_1 = 1$.

Now we apply 14.12 to deduce that there exists a Noether normalizing family $(Z_i)_{i=1}^n$ for $K[Y_1, \ldots, Y_n]$ with $Z_n = p$ such that

$$P_1 \cap K[Z_1, \ldots, Z_n] = K[Z_1, \ldots, Z_n]Z_n = K[Z_1, \ldots, Z_n]p = (p).$$

Let $^- : K[Y_1, \ldots, Y_n] \to K[Y_1, \ldots, Y_n]/P_1$ denote the natural surjective ring homomorphism. It follows from 14.9 that $(\overline{Z_i})_{i=1}^{n-1}$ is a Noether normalizing family for $K[Y_1, \ldots, Y_n]/P_1$, while we see from 2.39 and 3.28 that

$$P_1/P_1 \subset P_2/P_1 \subset \ldots \subset P_m/P_1$$

is a maximal chain of prime ideals of the integral domain $K[Y_1, \ldots, Y_n]/P_1$. By 14.28, $\dim(K[Y_1, \ldots, Y_n]/P_1) = n - 1$ and so we can apply the inductive hypothesis to see that $m - 1 = n - 1$. This completes the inductive step. $\square$

**14.32** COROLLARY. *Suppose that the integral domain $A$ is an affine algebra over the field $K$. Then*

$$\mathrm{ht}\, P + \dim A/P = \dim A \qquad \text{for all } P \in \mathrm{Spec}(A).$$

*Proof.* Let $\dim A/P = h$ and $\mathrm{ht}\, P = m$. Thus there exists a saturated chain

$$P_0 \subset P_1 \subset \ldots \subset P_m$$

of prime ideals of $A$ with $P_m = P$ and $P_0 = 0$. Also, in view of 2.39 and 3.28, there exists a saturated chain

$$Q_0 \subset Q_1 \subset \ldots \subset Q_h$$

of prime ideals of $A$ with $Q_0 = P$ and $Q_h$ maximal. Now the chain

$$P_0 \subset \ldots \subset P_{m-1} \subset P \subset Q_1 \subset \ldots \subset Q_h$$

of prime ideals of $A$ must be maximal, and so $m + h = \dim A$ by 14.31. $\square$

**14.33** COROLLARY. *Suppose that the integral domain $A$ is an affine algebra over the field $K$. Then we have $\mathrm{ht}\, M = \dim A$ for every maximal ideal $M$ of $A$.*

*In particular, every maximal ideal of the polynomial ring $K[X_1, \ldots, X_n]$ has height exactly $n$.* $\square$

The result in the second paragraph of the above corollary was promised just before the statement of Lemma 14.30.

**14.34** EXERCISE. (i) Give an example of a Noetherian integral domain which has maximal ideals of different heights.

(ii) Give an example of an affine algebra $A$ (over a field) with a prime ideal $P$ for which $\operatorname{ht} P + \dim A/P < \dim A$.

**14.35** COROLLARY. *Let $A$ be a non-trivial affine algebra over the field $K$, and let $P, Q \in \operatorname{Spec}(A)$ with $P \subset Q$. Then all saturated chains of prime ideals from $P$ to $Q$ (that is, saturated chains of prime ideals of $A$ which have $P$ as smallest term and $Q$ as largest) have the same length, and this length is equal to $\dim A/P - \dim A/Q$.*

*Proof.* In view of 2.39 and 3.28, it is enough for us to show that all saturated chains of prime ideals of $A/P$ from $0 (= P/P)$ to $Q/P$ have length equal to $\dim A/P - \dim A/Q$. Now, by 2.40, $A/Q \cong (A/P)/(Q/P)$, and so it is enough for us to prove the claim in the special case in which $A$ is a domain and $P = 0$. However, in this case, the claim is an easy consequence of 14.31, 2.39 and 3.28, as we now show. Let

$$0 \subset Q_1 \subset \ldots \subset Q_{h-1} \subset Q$$

be a saturated chain of prime ideals of $A$, and note that, by the cited results, a saturated chain of prime ideals of $A$ from $Q$ to a maximal ideal must have length $\dim A/Q$. Put the second chain 'on top' of the first, so to speak, to obtain a (necessarily maximal) chain of prime ideals of length $h + \dim A/Q$; it now follows from 14.31 that $h + \dim A/Q = \dim A$. □

**14.36** EXERCISE. Let $A$ be a non-trivial affine algebra over the field $K$. Since $A$ is Noetherian, it follows from 8.17 that the zero ideal of $A$ has only finitely many minimal prime ideals: let these be $P_1, \ldots, P_n$. For each $i = 1, \ldots, n$, let $L_i$ denote the field of fractions of the integral domain $A/P_i$, so that $L_i$ is, in an obvious natural way, an extension field of $K$. Prove that
    (i) $\dim A = \max \{\operatorname{tr.deg}_K L_i : 1 \leq i \leq n\}$;
    (ii) if $\dim A = \operatorname{tr.deg}_K L_i$ for all $i = 1, \ldots, n$, then

$$\operatorname{ht} P + \dim A/P = \dim A \qquad \text{for all } P \in \operatorname{Spec}(A).$$

**14.37** EXERCISE. Let $A, B$ be non-trivial affine algebras over the field $K$, and suppose that $B$ is a $K$-subalgebra (see 8.9) of $A$. Prove that $\dim B \leq \dim A$. (Here are some hints: as in 14.36, let $P_1, \ldots, P_n$ be the minimal prime ideals of the zero ideal of $A$; remember 8.20; and note that if $(Z_i)_{i=1}^m$ is an algebraically independent family of elements of $A$, then $K[Z_1, \ldots, Z_m]$ is a domain.)

**14.38** FURTHER STEPS. Once again, we leave the interested reader to explore the geometrical significance of Noether's Normalization Theorem from texts such as Kunz [6] and Reid [13].

In this chapter, we have seen that the class of (non-trivial) affine algebras over a field $K$ provides a class of commutative Noetherian rings for which there is a highly satisfactory dimension theory. In the next chapter, we shall show that there is a good theory of dimension for general (non-trivial) commutative Noetherian rings, but the reader should be warned that some of the good properties of prime ideals in affine $K$-algebras established above, such as those in 14.31, 14.32, 14.33 and 14.35, do not apply in all commutative Noetherian rings. A detailed discussion of this is beyond the scope of this book; however, the reader might like to know that there are examples of commutative Noetherian rings exhibiting 'bad' behaviour in [9, Appendix].

# Chapter 15

# Dimension theory

In Chapter 14, we studied the highly satisfactory dimension theory for finitely generated commutative algebras over fields. Of course, finitely generated commutative algebras over fields form a subclass of the class of commutative Noetherian rings: in this chapter, we are going to study heights of prime ideals in a general commutative Noetherian ring $R$, and the dimension theory of such a ring.

The starting point will be Krull's Principal Ideal Theorem: this states that, if $a \in R$ is a non-unit of $R$ and $P \in \operatorname{Spec}(R)$ is a minimal prime ideal of the principal ideal $aR$ (see 8.17), then $\operatorname{ht} P \leq 1$. From this, we are able to go on to prove the Generalized Principal Ideal Theorem, which shows that, if $I$ is a proper ideal of $R$ which can be generated by $n$ elements, then $\operatorname{ht} P \leq n$ for every minimal prime ideal $P$ of $I$. A consequence is that each $Q \in \operatorname{Spec}(R)$ has finite height, because $Q$ is, of course, a minimal prime ideal of itself and every ideal of $R$ is finitely generated!

There are consequences for local rings: if $(R, M)$ is a local ring (recall from 8.26 that, in our terminology, a local ring is a commutative Noetherian ring which has exactly one maximal ideal), then $\dim R = \operatorname{ht} M$ by 14.18(iv), and so $R$ has finite dimension. In fact, we shall see that $\dim R$ is the least integer $i \in \mathsf{N}_0$ for which there exists an $M$-primary ideal that can be generated by $i$ elements.

We shall end the chapter, and the text of the book, by developing some of the properties of regular local rings; these form a very pleasing class of local rings.

**15.1 Lemma.** *Let $R$ be a commutative Noetherian ring and let $P$ be a minimal prime ideal of the proper ideal $I$ of $R$. Let $S$ be a multiplicatively closed subset of $R$ such that $P \cap S = \emptyset$. Then $S^{-1}P$ is a minimal prime* *ideal of the ideal $S^{-1}I$ of $S^{-1}R$.*

*Proof.* Use the extension and contraction notation of 2.41 in conjunction with the natural ring homomorphism $R \to S^{-1}R$. By 5.33, we have $P^e \in \text{Spec}(S^{-1}R)$, and, of course, $I^e \subseteq P^e$. Suppose that $P^e$ is not a minimal prime ideal of $I^e$, and look for a contradiction. Then, by 3.53, there exists a prime ideal $Q$ of $S^{-1}R$ such that $I^e \subseteq Q \subset P^e$. By 5.33, there exists $Q \in \text{Spec}(R)$ such that $Q \cap S = \emptyset$ and $\overline{Q}^e = Q$. Now contract back to $R$ and use 2.44(i) and 5.33 to see that

$$I \subseteq I^{ec} \subseteq Q^c = Q^{ec} = Q \subset P^{ec} = P,$$

contrary to the fact that $P$ is a minimal prime ideal of $I$. $\square$

The notion of the height of a prime ideal in a commutative ring was defined in 14.17(v).

**15.2** KRULL'S PRINCIPAL IDEAL THEOREM. *Let $R$ be a commutative Noetherian ring and let $a \in R$ be a non-unit. Let $P$ be a minimal prime ideal of the principal ideal $aR$ of $R$. Then $\text{ht } P \leq 1$.*

*Proof.* By 15.1, in the local ring $R_P$, the maximal ideal $PR_P$ is a minimal prime ideal of $(aR)R_P = (a/1)R_P$, and also $\text{ht}_{R_P} PR_P = \text{ht}_R P$ by 14.18(vi). It is therefore enough for us to prove this result under the additional hypotheses that $(R, M)$ is a local ring and $P = M$. We therefore assume that these hypotheses are in force during the remainder of the proof.

Suppose that $\text{ht } M > 1$ and look for a contradiction. Then there exists a chain

$$Q' \subset Q \subset M$$

of prime ideals of $R$ of length 2. Note that, since $M$ is a minimal prime ideal of $aR$ and is also the unique maximal ideal of $R$, it follows from 3.28 that $\text{Spec}(R/aR) = \{M/aR\}$. Hence, by 8.45, the ring $R/aR$ is an Artinian local ring.

We are now going to use the concept of symbolic prime power, introduced in 5.46: use the extension and contraction notation of 2.41 in conjunction with the natural ring homomorphism $R \to R_Q$, and recall that, for every $n \in \mathbb{N}$, the $n$-th symbolic power $Q^{(n)}$ of $Q$ is given by $Q^{(n)} = (Q^n)^{ec}$, and is a $Q$-primary ideal of $R$. Note that $Q^{(n)} \supseteq Q^{(n+1)}$ for each $n \in \mathbb{N}$. Hence

$$(Q^{(1)} + aR)/aR \supseteq (Q^{(2)} + aR)/aR \supseteq \ldots \supseteq (Q^{(n)} + aR)/aR \supseteq \ldots$$

is a descending chain of ideals in the Artinian ring $R/aR$, and so there exists $m \in \mathbb{N}$ such that $Q^{(m)} + aR = Q^{(m+1)} + aR$.

Now let $r \in Q^{(m)}$. Then $r = s + ac$ for some $s \in Q^{(m+1)}$, $c \in R$. But then $ac = r - s \in Q^{(m)}$, a $Q$-primary ideal of $R$, while $a \notin Q$ since $M$ is a minimal prime ideal of $aR$. Therefore $c \in Q^{(m)}$. It follows that

$$Q^{(m)} = Q^{(m+1)} + aQ^{(m)}.$$

But $a \in M$, and so

$$Q^{(m)}/Q^{(m+1)} = M(Q^{(m)}/Q^{(m+1)}).$$

Hence, by Nakayama's Lemma 8.24, we have $Q^{(m)} = Q^{(m+1)}$. Now extend back to $R_Q$ and use 2.43(ii) and 2.44(iii): we obtain

$$(Q^e)^m = (Q^m)^e = (Q^m)^{ece} = (Q^{(m)})^e = (Q^{(m+1)})^e = (Q^e)^{m+1}.$$

Another use of Nakayama's Lemma, this time on the finitely generated $R_Q$-module $(Q^e)^m$, shows that $(Q^e)^m = 0$.

Thus, in the local ring $R_Q$, the maximal ideal $Q^e$ is nilpotent, and so, by 3.47, is contained in every prime ideal of $R_Q$. But this contradicts the fact that $Q'^e \subset Q^e$ is a chain of prime ideals of $R_Q$. The proof is therefore complete. $\square$

**15.3** EXERCISE. Let $P, Q$ be prime ideals of the commutative Noetherian ring $R$ such that $P \subset Q$. Show that, if there exists one prime ideal of $R$ strictly between $P$ and $Q$ (that is, if the chain $P \subset Q$ is not saturated), then there are infinitely many. (If you find this difficult, try passing to $R/P$ and using the Prime Avoidance Theorem 3.61.)

The Principal Ideal Theorem leads straightaway to a far-reaching generalization.

**15.4** KRULL'S GENERALIZED PRINCIPAL IDEAL THEOREM. *Let $R$ be a commutative Noetherian ring and let $I$ be a proper ideal of $R$ which can be generated by $n$ elements. Then* ht $P \leq n$ *for each minimal prime ideal $P$ of $I$.*

*Proof.* We use induction on $n$. In the case when $n = 0$, we have $I = 0$, so that $P$ is a minimal prime ideal of the zero ideal of $R$ and therefore ht $P = 0$. In the case when $n = 1$, the claim follows from the Principal Ideal Theorem 15.2. We therefore assume, inductively, that $n > 1$ and the result has been proved for smaller values of $n$.

In view of 15.1, since $IR_P$ is an ideal of $R_P$ which can be generated by $n$ elements, it is enough, in order for us to complete the inductive step, to show that ht $P \leq n$ under the additional hypotheses that $(R, M)$ is a local ring and $P = M$, and so we make these assumptions.

By the maximal condition, given a non-maximal prime ideal $P'$ of $R$, there exists a non-maximal $P'' \in \operatorname{Spec}(R)$ such that $P' \subseteq P''$ and the chain $P'' \subset M$ of prime ideals is saturated. It is therefore sufficient for us to show that, for a non-maximal $Q \in \operatorname{Spec}(R)$ such that the chain $Q \subset M$ of prime ideals is saturated, we must have ht $Q \leq n - 1$.

We have $I \not\subseteq Q$, so that there exist $c_1, \ldots, c_n \in I$ with $c_n \notin Q$ and $I = \sum_{i=1}^{n} c_i R$. Now $M$ is the only prime ideal of $R$ which contains $Q + c_n R$, and so, by 8.45, the ring $R/(Q + c_n R)$ is an Artinian local ring. Now, by 8.41, 8.39 and 3.49, the maximal ideal in an Artinian local ring is nilpotent, and so there exists $h \in \mathbb{N}$ such that $c_i^h \in Q + c_n R$ for all $i = 1, \ldots, n - 1$. Hence there exist $d_1, \ldots, d_{n-1} \in Q$ and $r_1, \ldots, r_{n-1} \in R$ such that

$$c_i^h = d_i + r_i c_n \qquad \text{for all } i = 1, \ldots, n - 1.$$

Note that $\sum_{i=1}^{n-1} d_i R \subseteq Q$: our strategy is to show that $Q$ is a minimal prime ideal of $\sum_{i=1}^{n-1} d_i R$, and then to appeal to the inductive hypothesis.

Let $\overline{R} = R / \sum_{i=1}^{n-1} d_i R$ and let $^- : R \to \overline{R}$ denote the natural ring homomorphism. The above displayed equations show that any prime ideal $P'$ of $R$ which contains all of $d_1, \ldots, d_{n-1}, c_n$ must contain $c_1, \ldots, c_n$; hence $M$ is the one and only prime ideal of $R$ which contains all of $d_1, \ldots, d_{n-1}, c_n$. Therefore, in view of 3.28, the maximal ideal $M / \sum_{i=1}^{n-1} d_i R$ of $\overline{R}$ is a minimal prime ideal of the principal ideal $\overline{c_n} \overline{R}$. We can now use the Principal Ideal Theorem 15.2 to see that

$$\operatorname{ht}_{\overline{R}} \left( M / \sum_{i=1}^{n-1} d_i R \right) \leq 1;$$

hence $Q$ must be a minimal prime ideal of $\sum_{i=1}^{n-1} d_i R$, or else the chain

$$Q / \sum_{i=1}^{n-1} d_i R \subset M / \sum_{i=1}^{n-1} d_i R$$

of prime ideals of $\overline{R}$ could be extended 'downwards'. It therefore follows from the inductive hypothesis that ht $Q \leq n - 1$, and so the inductive step is complete.

The theorem is therefore proved. $\square$

**15.5 COROLLARY.** *Let $R$ be a commutative Noetherian ring.*

  (i) *Each prime ideal of $R$ has finite height. In particular, a local ring has finite dimension.*

(ii) *Let* $P, Q \in \operatorname{Spec}(R)$ *with* $P \subseteq Q$. *Then* $\operatorname{ht} P \leq \operatorname{ht} Q$, *and* $\operatorname{ht} P = \operatorname{ht} Q$ *if and only if* $P = Q$.

(iii) *The ring* $R$ *satisfies the descending chain condition on* prime *ideals.*

*Proof.* (i) Let $P \in \operatorname{Spec}(R)$. Since $R$ is Noetherian, $P$ is finitely generated, and can be generated by $n$ elements, say. Also $P$ is the unique minimal prime ideal of itself, and so it follows from the Generalized Principal Ideal Theorem 15.4 that $\operatorname{ht} P \leq n$.

The second statement follows from the first because the dimension of a local ring is equal to the height of its unique maximal ideal: see 14.18(iv).

(ii) Let $\operatorname{ht} P = n$, and let

$$P_0 \subset P_1 \subset \ldots \subset P_n$$

be a chain of prime ideals of $R$ with $P_n = P$. Then, if $P \neq Q$, the chain

$$P_0 \subset P_1 \subset \ldots \subset P_n \subset Q$$

of prime ideals of $R$ shows that $\operatorname{ht} Q \geq n + 1$. All the claims follow quickly from this.

(iii) For this, we just note that a strictly descending chain

$$P_1 \supset P_2 \supset \ldots \supset P_n$$

of prime ideals of $R$ must satisfy $n \leq \operatorname{ht} P_1 + 1$. $\square$

Our next major aim is the establishment of a sort of converse of the Generalized Principal Ideal Theorem 15.4. To prepare the ground for this, we extend the notion of height so that it applies to all ideals of a commutative Noetherian ring.

**15.6 DEFINITION and REMARKS.** Let $R$ be a commutative Noetherian ring and let $I$ be a proper ideal of $R$. Of course, there exist prime ideals of $R$ which contain $I$: we define the *height of* $I$, denoted by $\operatorname{ht} I$, by

$$\operatorname{ht} I = \min \left\{ \operatorname{ht} P : P \in \operatorname{Spec}(R) \text{ and } P \supseteq I \right\}.$$

Note that, when $I$ is prime, this new interpretation of '$\operatorname{ht} I$' coincides with our earlier one. Since, by 3.53, every prime ideal in $\operatorname{Var}(I)$ (see 3.48) contains a minimal prime ideal of $I$, and since, by 8.17, every prime ideal in $\operatorname{ass} I$ contains a minimal prime ideal of $I$, it follows from 15.5(ii) that

$$\operatorname{ht} I = \min \left\{ \operatorname{ht} P : P \text{ is a minimal prime ideal of } I \right\}$$
$$= \min \left\{ \operatorname{ht} P : P \in \operatorname{ass} I \right\}.$$

Note also that if $J$ is a second proper ideal of $R$ and $I \subseteq J$, then ht $I \leq$ ht $J$.

It is sometimes convenient to adopt the convention whereby the improper ideal $R$ of $R$ is regarded as having height $\infty$.

**15.7** REMARK. If $I$ is a proper ideal of the commutative Noetherian ring $R$ and $I$ can be generated by $n$ elements, then it follows from the Generalized Principal Ideal Theorem 15.4 that ht $I \leq n$.

**15.8** EXERCISE. Let $I, J$ be ideals in the commutative Noetherian ring $R$ such that $I \subset J$. Must it always be the case that ht $I <$ ht $J$? Justify your response.

**15.9** EXERCISE. Determine the heights of the proper ideals of $\mathbb{Z}$. Determine the heights of the proper ideals of a PID $R$.

**15.10** EXERCISE. Let $K$ be a field, and let $R := K[X, Y]$, the ring of polynomials over $K$ in two indeterminates $X$ and $Y$. Let $I := (X^2, XY)$. Find $\text{ht}_R I$. Can $I$ be generated by 1 element? Justify your response.

**15.11** EXERCISE. Let $K$ be a field, and let $R := K[X_1, X_2, X_3, X_4, X_5]$, the ring of polynomials over $K$ in five indeterminates $X_1, \ldots, X_5$. Determine the heights of each of the following ideals of $R$:
   (i) $(X_1, X_2, X_3, X_4)$;
   (ii) $(X_1X_5, X_2X_5, X_3X_5, X_4X_5)$;
   (iii) $(X_1, X_2) \cap (X_3, X_4)$;
   (iv) $(X_1X_3, X_2X_3, X_1X_4, X_2X_4)$;
   (v) $(X_1, X_2) \cap (X_3X_5, X_4X_5)$.

**15.12** LEMMA. *Let $I, P$ be ideals of the commutative Noetherian ring $R$ with $I \subseteq P$ and $P$ prime. Suppose that $\text{ht}\, I = \text{ht}\, P$. Then $P$ is a minimal prime ideal of $I$.*

*Proof.* Suppose that $P$ is not a minimal prime ideal of $I$. Then, by 3.53, there exists a minimal prime ideal $Q$ of $I$ such that $I \subseteq Q \subset P$. In view of 15.5(ii) and 15.6, we then have

$$\text{ht}\, I \leq \text{ht}\, Q < \text{ht}\, P,$$

contrary to hypothesis. $\square$

We are now in a position to prove the promised converse of the Generalized Principal Ideal Theorem.

**15.13** THEOREM. *Let $R$ be a commutative Noetherian ring and let $P \in$ Spec($R$); suppose that ht $P = n$. Then there exists an ideal $I$ of $R$ which can be generated by $n$ elements, has ht $I = n$, and is such that $I \subseteq P$.*

*Note.* It follows from 15.12 that the $I$ whose existence is asserted by the theorem will have $P$ as a minimal prime ideal.

*Proof.* We use induction on $n$. When $n = 0$, we just take $I = 0$ to find an ideal with the stated properties.

So suppose, inductively, that $n > 0$ and the claim has been proved for smaller values of $n$. Now there exists a chain

$$P_0 \subset \ldots \subset P_{n-1} \subset P_n$$

of prime ideals of $R$ with $P_n = P$. Note that ht $P_{n-1} = n - 1$: this is because ht $P_{n-1} <$ ht $P$ by 15.5(ii), while ht $P_{n-1} \geq n - 1$ by virtue of the above chain. We can therefore apply the inductive hypothesis to $P_{n-1}$: the conclusion is that there exists a proper ideal $J$ of $R$ which can be generated by $n - 1$ elements, $a_1, \ldots, a_{n-1}$ say, and which is such that $J \subseteq P_{n-1}$ and ht $J = n - 1$.

Note that, by 15.12, $P_{n-1}$ is actually a minimal prime ideal of $J$. Recall from 8.17 that $J$ has only finitely many minimal prime ideals; note also that, in view of the Generalized Principal Ideal Theorem 15.4 and the fact that ht $J = n-1$, all the minimal prime ideals of $J$ must have height exactly $n - 1$. Let the other minimal prime ideals of $J$, in addition to $P_{n-1}$, be $Q_1, \ldots, Q_t$. (In fact, $t$ could be 0, but this does not affect the argument significantly.)

We now use the Prime Avoidance Theorem 3.61 to see that

$$P \nsubseteq P_{n-1} \cup Q_1 \cup \ldots \cup Q_t:$$

if this were not the case, then it would follow from 3.61 that either $P \subseteq P_{n-1}$ or $P \subseteq Q_i$ for some $i$ with $1 \leq i \leq t$, and none of these possibilities can occur because ht $P = n$ while

$$\text{ht } P_{n-1} = \text{ht } Q_1 = \ldots = \text{ht } Q_t = n - 1.$$

Therefore, there exists

$$a_n \in P \setminus (P_{n-1} \cup Q_1 \cup \ldots \cup Q_t).$$

Define $I := \sum_{i=1}^{n} Ra_i = J + Ra_n$. We show that $I$ has all the desired properties. It is clear from its definition that $I$ can be generated by $n$ elements and that $I = J + Ra_n \subseteq P_{n-1} + P = P$. Thus, in order to complete the inductive step, it remains only to show that ht $I = n$.

Since $J \subseteq I \subseteq P$ and $\operatorname{ht} J = n - 1$, $\operatorname{ht} P = n$, we must have $\operatorname{ht} I = n - 1$ or $n$. Let us suppose that $\operatorname{ht} I = n - 1$ and look for a contradiction. Then there exists a minimal prime ideal $P'$ of $I$ such that $\operatorname{ht} P' = n - 1$. Now $J \subseteq I \subseteq P'$ and $\operatorname{ht} J = \operatorname{ht} P' = n - 1$. It therefore follows from 15.12 that $P'$ is one of the minimal prime ideals of $J$, that is, $P'$ is one of $P_{n-1}, Q_1, \ldots, Q_t$. But this is not possible because $a_n \in I \subseteq P'$ whereas $a_n$ belongs to none of $P_{n-1}, Q_1, \ldots, Q_t$. This contradiction shows that $\operatorname{ht} I = n$; thus the inductive step is complete. $\square$

**15.14 EXERCISE.** Let $R$ be a commutative Noetherian ring, and let ($n \in$ **N** and) $a_1, \ldots, a_n, b_1, \ldots, b_n \in R$ be such that

$$\operatorname{ht} \left( \sum_{j=1}^{i} Ra_j \right) = \operatorname{ht} \left( \sum_{j=1}^{i} Rb_j \right) = i \qquad \text{for all } i = 1, \ldots, n.$$

Show that there exist $c_1, \ldots, c_n \in R$ such that, for all $i = 1, \ldots, n$,

$$c_i \in \left( \sum_{j=1}^{i} Ra_j \right) \bigcap \left( \sum_{j=1}^{i} Rb_j \right) \qquad \text{and} \qquad \operatorname{ht} \left( \sum_{j=1}^{i} Rc_j \right) = i.$$

**15.15 COROLLARY** (of 15.13). *Let $R$ be a commutative Noetherian ring, and let $I$ be a proper ideal of $R$ which can be generated by $n$ elements. Let $P \in \operatorname{Spec}(R)$ be such that $I \subseteq P$. Then*

$$\operatorname{ht}_{R/I} P/I \leq \operatorname{ht}_R P \leq \operatorname{ht}_{R/I} P/I + n.$$

*Note.* Of course, in view of 3.28, the ideal $P/I$ of $R/I$ is prime.

*Proof.* First of all, it is an easy consequence of 14.18(vii) that

$$\operatorname{ht}_{R/I} P/I \leq \operatorname{ht}_R P.$$

Let $b_1, \ldots, b_n$ generate $I$. Set $\overline{R} = R/I$ and let $\overline{\phantom{x}} : R \to \overline{R}$ denote the natural ring homomorphism. Let $\operatorname{ht}_{\overline{R}} P/I = t$. By 15.12 and 15.13, there exist $a_1, \ldots, a_t \in R$ such that, in the ring $\overline{R}$, the prime ideal $P/I$ is a minimal prime ideal of $\sum_{i=1}^{t} \overline{Ra_i}$. Now

$$\sum_{i=1}^{t} \overline{Ra_i} = \left( \sum_{i=1}^{t} Ra_i + I \right) / I,$$

and it therefore follows from the considerations in 14.18(vii), 2.39 and 3.28 that $P$ must be a minimal prime ideal of

$$\sum_{i=1}^{t} Ra_i + I = \sum_{i=1}^{t} Ra_i + \sum_{i=1}^{n} Rb_i,$$

a proper ideal of $R$ which can be generated by $t + n$ elements. We can therefore deduce from the Generalized Principal Ideal Theorem 15.4 that $\operatorname{ht} P \leq t + n$, as required. $\square$

The above result 15.15 will be used in our discussion of regular local rings later in the chapter.

**15.16 EXERCISE.** Let $R$ be a commutative Noetherian ring, and let $a \in R$ be a non-unit and a non-zerodivisor. Let $P \in \operatorname{Spec}(R)$ be such that $a \in P$. Prove that

$$\operatorname{ht}_{R/Ra} P/Ra = \operatorname{ht}_R P - 1.$$

We are now going to apply our converse 15.13 of the Generalized Principal Ideal Theorem to local rings in order to produce another, very important, description of the dimension of such a ring. The following exercise is essentially revision: it is intended to bring back to the reader's mind important facts about local rings that were covered earlier in the book.

**15.17 ♯EXERCISE.** Let $(R, M)$ be a local ring, and let $Q$ be a proper ideal of $R$. Prove that the following statements are equivalent:
  (i) the $R$-module $R/Q$ has finite length;
  (ii) $\operatorname{Var}(Q) = \{M\}$;
  (iii) $\operatorname{ass}(Q) = \{M\}$;
  (iv) $Q$ is $M$-primary;
  (v) there exists $h \in \mathbf{N}$ such that $Q \supseteq M^h$;
  (vi) $\sqrt{Q} = M$.

**15.18 COROLLARY** (of 15.13). *Let $(R, M)$ be a local ring. Then $\dim R$ is equal to the least number of elements of $R$ that are needed to generate an $M$-primary ideal; in other words (and mathematical symbols!),*

$$\dim R = \min \left\{ i \in \mathbf{N}_0 : \exists\, a_1, \ldots, a_i \in R \text{ with } \sum_{j=1}^{i} Ra_j \ M\text{-primary} \right\}.$$

*Proof.* Set

$$d = \min \left\{ i \in \mathbf{N}_0 : \text{ there exist } a_1, \ldots, a_i \in R \text{ with } \sum_{j=1}^{i} Ra_j \ M\text{-primary} \right\}.$$

Note first of all that it follows from 14.18(iv) and the Generalized Principal Ideal Theorem 15.4 that $\dim R = \operatorname{ht} M \leq d$, simply because an $M$-primary ideal must have $M$ as its only minimal prime ideal. On the other hand, we can use 15.12 and 15.13 to see that there exists an ideal $Q$ of $R$ which has $M$ as a minimal prime ideal and which can be generated by $\dim R = \operatorname{ht} M$ elements. But every prime ideal of $R$ is contained in $M$, and so $M$ must be the one and only associated prime ideal of $Q$; hence $Q$ is $M$-primary. Thus there exists an $M$-primary ideal of $R$ which can be generated by $\dim R$ elements, and so $d \leq \dim R$. This completes the proof. $\square$

This alternative description of the dimension of a local ring leads immediately to the important concept of 'system of parameters'.

**15.19** DEFINITION. Let $(R, M)$ be a local ring of dimension $d$. By a *system of parameters for $R$* we mean a set of $d$ elements of $R$ which generate an $M$-primary ideal. It follows from 15.18 that each local ring does indeed possess a system of parameters. We say that $a_1, \ldots, a_d \in R$ *form a system of parameters for $R$* precisely when $\{a_1, \ldots, a_d\}$ is a system of parameters for $R$.

**15.20** EXERCISE. Let $(R, M)$ be a local ring of dimension $d$, and let $a_1, \ldots, a_d$ form a system of parameters for $R$. Let $n_1, \ldots, n_d \in \mathbb{N}$. Prove that $a_1^{n_1}, \ldots, a_d^{n_d}$ form a system of parameters for $R$.

**15.21** FURTHER STEPS. Let $(R, M)$ be a local ring. It follows from 15.17 that the $R$-module $R/M^n$ has finite length for all $n \in \mathbb{N}$. It can be shown that there exists a (necessarily uniquely determined) polynomial $f \in \mathbb{Q}[X]$ such that $\ell(R/M^n) = f(n)$ for all large values of the integer $n$, and it turns out that $\deg f$ is exactly $\dim R$. This result has not been covered in this book because a thorough approach to it would involve the theory of graded rings and graded modules, and there was not enough space to do justice to that topic. However, this result does provide another powerful and useful characterization of the dimension of a local ring, and any serious student of commutative algebra should be aware of it. The interested reader can find details in several of the books listed in the Bibliography, including [1, Chapter 11], [7, Chapter 5] and [8, Chapter 5].

**15.22** PROPOSITION. *Let $(R, M)$ be a local ring, and let $a_1, \ldots, a_t \in M$. Then*
$$\dim R - t \leq \dim R/(a_1, \ldots, a_t) \leq \dim R.$$
*Moreover, $\dim R/(a_1, \ldots, a_t) = \dim R - t$ if and only if $a_1, \ldots, a_t$ are all different and form a subset of a system of parameters for $R$.*

*Proof.* By 14.18(iv), $\dim R = \operatorname{ht} M$ and
$$\dim R/(a_1, \ldots, a_t) = \operatorname{ht}_{R/(a_1, \ldots, a_t)} M/(a_1, \ldots, a_t).$$

It is therefore immediate from 15.15 that

$$\dim R - t \leq \dim R/(a_1, \ldots, a_t) \leq \dim R.$$

Now set $\overline{R} := R/(a_1, \ldots, a_t)$, let $^- : R \to \overline{R}$ denote the natural ring homomorphism, let $\overline{M} := M/(a_1, \ldots, a_t)$, and let $d = \dim R$.

($\Rightarrow$) Suppose that $\dim \overline{R} = d - t$. Then $t \leq d$, and, by 15.18, there exist $a_{t+1}, \ldots, a_d \in M$ such that $\{\overline{a_{t+1}}, \ldots, \overline{a_d}\}$ is a system of parameters for $\overline{R}$. This means that $(a_1, \ldots, a_t, a_{t+1}, \ldots, a_d)/(a_1, \ldots, a_t)$ is an $\overline{M}$-primary ideal of $\overline{R}$. Hence, by 4.22 and 4.23, $(a_1, \ldots, a_d)$ is an $M$-primary ideal of $R$. It now follows from 15.18 that $(a_1, \ldots, a_t$ are all different and) $\{a_1, \ldots, a_d\}$ is a system of parameters for $R$.

($\Leftarrow$) Now suppose that $t \leq d$ and there exist $a_{t+1}, \ldots, a_d \in M$ such that $a_1, \ldots, a_t, a_{t+1}, \ldots, a_d$ form a system of parameters for $R$. This means that $(a_1, \ldots, a_d)$ is an $M$-primary ideal of $R$, so that, by 4.22 and 4.23, $(\overline{a_{t+1}}, \ldots, \overline{a_d})$ is an $\overline{M}$-primary ideal of $\overline{R}$. Hence, by 15.18, we have $d - t \geq \dim \overline{R}$. But it follows from the first part that $d - t \leq \dim \overline{R}$, and so the proof is complete. $\square$

**15.23 EXERCISE.** Let $(R, M)$ be a local ring of dimension $d \geq 1$, and let $\{a_1, \ldots, a_d\}$ and $\{b_1, \ldots, b_d\}$ be two systems of parameters for $R$. Show that there exists a system of parameters $\{c_1, \ldots, c_d\}$ for $R$ such that, for all $i = 1, \ldots, n$,

$$c_i \in \left(\sum_{j=1}^{i} Ra_j\right) \bigcap \left(\sum_{j=1}^{i} Rb_j\right).$$

**15.24 EXERCISE.** Let $(R, M)$ be a local ring, and let $G$ be a non-zero finitely generated $R$-module. We define the *dimension of* $G$, denoted by $\dim G$ (or $\dim_R G$), to be the dimension of the ring $R/\operatorname{Ann}(G)$.

Prove that $\dim G$ is equal to the least integer $i \in \mathbf{N}_0$ such that there exist $i$ elements $a_1, \ldots, a_i \in M$ for which $G/(a_1, \ldots, a_i)G$ has finite length.

**15.25 COROLLARY** (of 15.18). *Let $(R, M)$ be a local ring. Then* $\dim R \leq \operatorname{vdim}_{R/M} M/M^2$.

*Note.* The $R$-module $M/M^2$ is annihilated by $M$, and so, by 6.19, has a natural structure as a vector space over the residue field $R/M$ of $R$. By 9.3, the vector-space dimension $\operatorname{vdim}_{R/M} M/M^2$ is equal to the number of elements in each minimal generating set for $M$.

*Proof.* Of course, $M$ is itself an $M$-primary ideal of $R$, and so, in view of the comments in the above Note, the claim is immediate from 15.18, which shows that $\dim R$ is the least number of elements required to generate an $M$-primary ideal of $R$. $\square$

The above result leads immediately to the idea of regular local ring.

**15.26** DEFINITION. Let $(R, M)$ be a local ring. Then $R$ is said to be *regular* precisely when $\dim R = \text{vdim}_{R/M} M/M^2$.

**15.27** REMARKS. Let $(R, M)$ be a local ring of dimension $d$.

(i) As remarked in the note immediately following the statment of 15.25, it follows from 9.3 that $\text{vdim}_{R/M} M/M^2$ is the number of elements in each minimal generating set for $M$. In general, by 15.18, at least $d = \dim R$ (and perhaps more) elements are needed to generate $M$, and $R$ is regular precisely when $M$ can be generated by $d = \dim R$ elements.

(ii) Suppose that $R$ is regular and that $a_1, \ldots, a_d \in M$. It also follows from 9.3 that $a_1, \ldots, a_d$ generate $M$ if and only if their natural images $a_1 + M^2, \ldots, a_d + M^2$ in $M/M^2$ form a basis for this $R/M$-space, and that this is the case if and only if $a_1 + M^2, \ldots, a_d + M^2$ form a linearly independent family (over $R/M$).

Before we develop the theory of regular local rings, let us show that such things really do exist!

**15.28** EXAMPLES. (i) Let $R$ be a commutative Noetherian ring, and suppose that there exists a $P \in \text{Spec}(R)$ which has height $n$ and can be generated by $n = \text{ht}\, P$ elements, $a_1, \ldots, a_n$ say. Then the localization $R_P$ is a regular local ring of dimension $n$, because it is a local ring (by 8.3 and 5.20), it has dimension $n$ (by 14.18(vi)), and its maximal ideal

$$PR_P = \left(\sum_{i=1}^{n} Ra_i\right) R_P = \sum_{i=1}^{n} R_P \frac{a_i}{1}$$

can be generated by $n$ elements. This leads to a substantial supply of examples of regular local rings.

(ii) Let $p$ be a prime number. Then, in the ring $\mathbf{Z}$, we have $\text{ht}\, p\mathbf{Z} = 1$ by 14.19, and since $p\mathbf{Z}$ is a prime ideal of $\mathbf{Z}$ which can be generated by 1 element, it follows from (i) above that $\mathbf{Z}_{p\mathbf{Z}}$ is a regular local ring of dimension 1.

(iii) More generally, let $R$ be a PID which is not a field, and let $M$ be a maximal ideal of $R$. Then it follows from 3.34 that $M$ is a principal prime ideal of $R$ of height 1, and so it follows from (i) above that $R_M$ is a regular local ring of dimension 1. Regular local rings of this type play a significant rôle in algebraic number theory.

(iv) Let $K$ be a field, and let $R := K[X_1, \ldots, X_n]$, the ring of polynomials over $K$ in $n$ indeterminates $X_1, \ldots, X_n$. Let $a_1, \ldots, a_n \in K$. Then it follows from 3.15, 3.66 and 15.4 that the ideal $(X_1 - a_1, \ldots, X_n - a_n)$ of $R$ is a prime ideal of height $n$ which can clearly be generated by $n$ elements, and so it follows from (i) above that

$$K[X_1, \ldots, X_n]_{(X_1 - a_1, \ldots, X_n - a_n)}$$

is a regular local ring of dimension $n$. Regular local rings of this type play a significant rôle in algebraic geometry. (See also the comments in 14.8 (and 14.33).)

(v) It follows from (iv) above and the Nullstellensatz 14.6 that, when the field $K$ is algebraically closed, $K[X_1, \ldots, X_n]_M$ is a regular local ring of dimension $n$ for every maximal ideal $M$ of $R$.

(vi) Finally, it should not be overlooked that each field $K$ is a regular local ring of dimension 0, since the unique maximal ideal 0 of $K$ can be generated by 0 elements and, of course, $K$ has dimension 0.

**15.29** EXERCISE. Let $(R, M)$ be a local ring. Show that $R[[X]]$, the ring of formal power series over $R$ in the indeterminate $X$, is again a local ring, and $\dim R[[X]] = \dim R + 1$.

**15.30** EXERCISE. Let $(R, M)$ be a regular local ring. Show that $R[[X]]$, the ring of formal power series over $R$ in the indeterminate $X$, is again a regular local ring.

Deduce that, if $K$ is a field, then the ring $K[[X_1, \ldots, X_n]]$ of formal power series over $K$ in the $n$ indeterminates $X_1, \ldots, X_n$ is a regular local ring of dimension $n$.

The theory of regular local rings is very beautiful, but, unfortunately, some of it, especially that part which involves homological algebra, is beyond the scope of this book. However, we can develop some of the theory here, and we progress now towards the result that every regular local ring is an integral domain. Quite a few preliminary results are given first, and one interesting aspect of the eventual proof presented below is that we shall use the Prime Avoidance Theorem 3.61 in a situation where one of the ideals 'being avoided' (so to speak) is not prime.

**15.31** LEMMA. *Let $(R, M)$ be a local ring, and let $c \in M \setminus M^2$. Set $\overline{R} := R/Rc$ and $\overline{M} := M/Rc$, the maximal ideal of the local ring $\overline{R}$. Also, let $^{-}: R \to \overline{R}$ denote the natural surjective ring homomorphism. Then*

$$\mathrm{vdim}_{R/M}\, M/M^2 = \mathrm{vdim}_{\overline{R}/\overline{M}}\, \overline{M}/\overline{M}^2 + 1.$$

*Proof.* Let $n := \mathrm{vdim}_{\overline{R}/\overline{M}} \overline{M}/\overline{M}^2$, and let $a_1, \ldots, a_n \in M$ be such that their natural images in $\overline{M}/\overline{M}^2$ form a basis for this $\overline{R}/\overline{M}$-space. By 9.3, this means that $\overline{a_1}, \ldots, \overline{a_n}$ generate the ideal $\overline{M}$ of $\overline{R}$. Hence

$$M/Rc = \overline{M} = \sum_{i=1}^{n} \overline{Ra_i} = \left( \sum_{i=1}^{n} Ra_i + Rc \right)/Rc,$$

and so it follows from 2.37 that $M = \sum_{i=1}^{n} Ra_i + Rc$. Hence the $R/M$-space $M/M^2$ is spanned by $a_1 + M^2, \ldots, a_n + M^2, c + M^2$. In order to complete the proof, it is now enough for us to show that these $n+1$ elements form a linearly independent family over $R/M$. This we do.

So suppose $r_1, \ldots, r_n, s \in R$ are such that, in $M/M^2$,

$$\sum_{i=1}^{n} (r_i + M)(a_i + M^2) + (s + M)(c + M^2) = 0.$$

This means that $\sum_{i=1}^{n} r_i a_i + sc \in M^2$, so that, in $\overline{R}$, we have $\sum_{i=1}^{n} \overline{r_i a_i} \in \overline{M}^2$. Hence, in the $\overline{R}/\overline{M}$-space $\overline{M}/\overline{M}^2$, we have

$$\sum_{i=1}^{n} (\overline{r_i} + \overline{M})(\overline{a_i} + \overline{M}^2) = 0.$$

But $\overline{a_1} + \overline{M}^2, \ldots, \overline{a_n} + \overline{M}^2$ form a linearly independent family over $\overline{R}/\overline{M}$, and so $\overline{r_1}, \ldots, \overline{r_n} \in \overline{M}$. It follows from 2.37(i) that $r_1, \ldots, r_n \in M$.

We can therefore deduce from the relation $\sum_{i=1}^{n} r_i a_i + sc \in M^2$ that $sc \in M^2$. If we had $s \notin M$, then it would follow from 3.14 that $s$ would be a unit of $R$, so that $c = s^{-1}sc \in M^2$, a contradiction. Hence $s \in M$ too, so that $a_1 + M^2, \ldots, a_n + M^2, c + M^2$ do form a linearly independent family over $R/M$, as required. $\square$

**15.32** COROLLARY. *Let $(R, M)$ be a regular local ring, and let $c \in M \backslash M^2$. Then $R/Rc$ is a regular local ring and*

$$\dim R/Rc = \dim R - 1.$$

*Proof.* Note that the hypotheses imply that

$$\dim R = \mathrm{vdim}_{R/M} M/M^2 \geq 1.$$

Set $\overline{R} := R/Rc$ and $\overline{M} := M/Rc$, the maximal ideal of the local ring $\overline{R}$. By 15.15, we have $\mathrm{ht}_{\overline{R}} \overline{M} \geq \mathrm{ht}_R M - 1$. Hence, in view of 14.18(iv) and 15.25,

$$\mathrm{vdim}_{\overline{R}/\overline{M}} \overline{M}/\overline{M}^2 \geq \dim \overline{R} = \mathrm{ht}_{\overline{R}} \overline{M} \geq \mathrm{ht}_R M - 1 = \dim R - 1.$$

But in view of 15.31 and the fact that $R$ is regular,

$$\dim R - 1 = \text{vdim}_{R/M} M/M^2 - 1 = \text{vdim}_{\overline{R}/\overline{M}} \overline{M}/\overline{M}^2.$$

We have thus shown that

$$\text{vdim}_{\overline{R}/\overline{M}} \overline{M}/\overline{M}^2 \geq \dim \overline{R} \geq \dim R - 1 = \text{vdim}_{\overline{R}/\overline{M}} \overline{M}/\overline{M}^2,$$

from which it is immediate that $\overline{R}$ is a regular local ring with dimension 1 less than that of $R$. $\square$

**15.33** LEMMA. *Let $(R, M)$ be a local ring which is not a domain, and suppose that $P$ is a prime ideal of $R$ which is principal. Then $\text{ht}\, P = 0$, that is, $P$ is a minimal prime ideal of $0$.*

*Proof.* Suppose that $\text{ht}\, P > 0$ and look for a contradiction. Thus there exists $Q \in \text{Spec}(R)$ such that $Q \subset P$. Now, since $P$ is principal, there exists $p \in P$ such that $P = Rp$. Note that $p \notin Q$, or else $P = Rp \subseteq Q$, which is not possible. The strategy of the proof is to show that $Q \subseteq P^n$ for all $n \in \mathbb{N}$, and then to appeal to Krull's Intersection Theorem 8.25.

Let $a \in Q$; of course $a \in P$. Suppose, inductively, that $n \in \mathbb{N}$ and we have shown that $a \in P^n$. Now $P^n = Rp^n$, and so there exists $b \in R$ such that $a = bp^n$. Now $a \in Q$, a prime ideal of $R$, and $p \notin Q$; hence $b \in Q \subset P$, so that $a = bp^n \in P^{n+1}$. This completes the inductive step; we have proved that $Q \subseteq \bigcap_{n=1}^{\infty} P^n$. But the latter intersection is $0$, by Krull's Intersection Theorem 8.25, and so $Q = 0$. This contradicts the fact that $R$ is not a domain. $\square$

**15.34** THEOREM. *A regular local ring is an integral domain.*

*Proof.* Let $(R, M)$ be a regular local ring of dimension $d$. We are going to argue by induction on $d$.

In the case where $d = 0$, we see from 15.27(i) that $M$ can be generated by 0 elements, and so $M = 0$. This means that $R$ is a field, and so certainly an integral domain.

Now suppose, inductively, that $d > 0$, and that the result has been proved for all regular local rings of dimension less than $d$. Let us suppose that $R$ is not a domain, and look for a contradiction. Since

$$\text{vdim}_{R/M} M/M^2 = \dim R = d > 0,$$

we have $M \supset M^2$. Let $c \in M \setminus M^2$. Note that, by 15.32, the local ring $R/Rc$ is regular of dimension $d - 1$. Hence, by the inductive assumption,

$Rc \in \operatorname{Spec}(R)$. It now follows from 15.33, since we are assuming that $R$ is not a domain, that ht $Rc = 0$, that is, $Rc$ is a minimal prime ideal of 0.

By 8.17, there are only finitely many minimal prime ideals of 0: let these be $P_1, \ldots, P_s$. We have thus shown that $M \setminus M^2 \subseteq \bigcup_{i=1}^{s} P_i$, so that

$$M \subseteq M^2 \cup P_1 \cup P_2 \cup \ldots \cup P_s.$$

We now apply the Prime Avoidance Theorem 3.61: we deduce that either $M \subseteq M^2$ or $M \subseteq P_i$ for some $i$ with $1 \leq i \leq s$. However, neither of these is possible: $M \subseteq M^2$ would be in contradiction to the fact, observed above, that $M^2 \subset M$, while $M \subseteq P_i$ for some $i$ between 1 and $s$ would mean that

$$d = \dim R = \operatorname{ht} M \leq \operatorname{ht} P_i = 0,$$

which is also a contradiction.

Thus $R$ is a domain, and the inductive step is complete. □

**15.35** EXERCISE. Let $A = \mathbf{R}[X_1, \ldots, X_{n+1}]$ (where $n \in \mathbf{N}$), the ring of polynomials over the real field $\mathbf{R}$ in the $n+1$ indeterminates $X_1, \ldots, X_{n+1}$. Let $P := (X_1, \ldots, X_{n+1})$, a prime ideal of $A$ by 3.15. Let $Q := (X_n^2 + X_{n+1}^2)$. Show that $A_P/QA_P$ is a local integral domain of dimension $n$ which is not regular.

**15.36** EXERCISE. Construct an example of a regular local ring whose field of fractions and residue field have different characteristics.

Recall from 15.19 that a system of parameters for a $d$-dimensional local ring $(R, M)$ is a set of $d$ elements of $M$ which generates an $M$-primary ideal. In the case where $R$ is regular, $M$ itself can be generated by $d$ elements, and so there is at least one system of parameters for $R$ which generates $M$. Accordingly, we make the following definition.

**15.37** DEFINITION. Let $(R, M)$ be a regular local ring of dimension $d$. A *regular system of parameters for $R$* is a set of $d$ elements of $R$ which generate $M$, that is, a system of parameters for $R$ which actually generates $M$.

**15.38** THEOREM. *Let $(R, M)$ be a regular local ring of dimension $d > 0$, and let $\{u_1, \ldots, u_d\}$ be a regular system of parameters for $R$. Then, for each $i = 1, \ldots, d$, the local ring $R/(u_1, \ldots, u_i)$ is regular of dimension $d - i$. Furthermore,*

$$0 \subset (u_1) \subset (u_1, u_2) \subset \ldots \subset (u_1, \ldots, u_i) \subset \ldots \subset (u_1, \ldots, u_d)$$

*is a saturated chain of prime ideals of $R$ (of length $d$).*

*Proof.* Let $i \in \mathbb{N}$ with $1 \leq i \leq d$. By 15.22, the local ring $\overline{R} := R/(u_1, \ldots, u_i)$ has dimension $d - i$; further, if we let $^- : R \to \overline{R}$ denote the natural ring homomorphism, then the maximal ideal $M/(u_1, \ldots, u_i)$ of $\overline{R}$ can be generated by the $d - i$ elements $\overline{u_{i+1}}, \ldots, \overline{u_d}$ (make an obvious interpretation in the case where $i = d$). Thus $\overline{R}$ is a regular local ring, and so, by 15.34, is an integral domain. Therefore, by 3.23, the ideal $(u_1, \ldots, u_i)$ of $R$ is prime.

Next, note that all the inclusion relations in

$$0 \subseteq (u_1) \subseteq (u_1, u_2) \subseteq \ldots \subseteq (u_1, \ldots, u_i) \subseteq \ldots \subseteq (u_1, \ldots, u_d)$$

must be strict, since $\dim R/(u_1, \ldots, u_i) = d - i$ for all $i = 1, \ldots, d$. Thus we do indeed have a chain of prime ideals of $R$ as displayed in the statement of the theorem, and it must be saturated simply because its length is $d = \dim R$. $\square$

**15.39 Exercise.** Let $(R, M)$ be a regular local ring of dimension $d > 0$, and let $\{u_1, \ldots, u_d\}$ be a regular system of parameters for $R$. Show that

$$((u_1, \ldots, u_{i-1}) : u_i) = (u_1, \ldots, u_{i-1}) \quad \text{for all } i = 1, \ldots, d.$$

(Of course, in the case in which $i = 1$, the above condition is to be interpreted as $(0 : u_1) = 0$.)

**15.40 Further Steps.** This exercise shows, in fact, that $(u_i)_{i=1}^d$ is a particular example of what is called an '$R$-sequence' or a 'regular sequence on $R$'. The theory of such sequences, which is beyond the scope of this book, leads on to the important ideas of 'grade' and 'Cohen-Macaulay ring', and the interested reader will find much on these topics in some of the references cited in the Bibliography, such as, for example, [8, Chapter 6]. In addition, 'Gorenstein rings' form an important subclass of the class of Cohen-Macaulay rings: Gorenstein rings have interesting connections with irreducible ideals.

The following lemma is in preparation for a discussion of regular local rings of dimension 1.

**15.41 Lemma.** *Let $(R, M)$ be an Artinian local ring such that $M$ is principal. Then every ideal of $R$ is a power of $M$, and so is principal.*

*Proof.* If $M = 0$, then $R$ is a field, and the claim is clear in this case.

Therefore we suppose that $M \neq 0$: let $b$ be a generator for $M$. Let $I$ be a non-zero proper ideal of $R$. By 8.39 and 8.41, there exists $t \in \mathbb{N}$ such that $M^t = 0$. Hence, there exists $h \in \mathbb{N}$ such that $I \subseteq M^h = Rb^h$ but

$I \not\subseteq M^{h+1} = Rb^{h+1}$. There exists $a \in I \setminus M^{h+1}$: we have $a = rb^h$ for some $r \in R$, and, moreover, $r \notin M$. It now follows from 3.11 that $r$ is a unit of $R$, and so $b^h = r^{-1}a \in I$. Hence $M^h = Rb^h = I$. It follows that every ideal of $R$ (including the zero one!) is a power of $M$. $\square$

**15.42** THEOREM. *Let $(R, M)$ be a local integral domain of dimension 1. Then the following statments are equivalent:*
  (i) *$R$ is regular;*
  (ii) *every non-zero ideal of $R$ is a power of $M$;*
  (iii) *there exists $a \in R$ such that each non-zero ideal of $R$ has the form $Ra^h$ for some $h \in \mathbf{N}_0$;*
  (iv) *$R$ is a PID;*
  (v) *$R$ is integrally closed.*

*Proof.* (i) $\Rightarrow$ (ii) By definition, $M$ is principal. Let $I$ be an ideal of $R$ with $0 \subset I \subset R$. The only prime ideals of $R$ are $M$ and 0; hence $I$ is $M$-primary (by 15.17), and there exists $t \in \mathbf{N}$ such that $M^t \subseteq I$. By 8.45, the ring $R/M^t$ is an Artinian local ring, and its maximal ideal is principal. Hence, by 15.41, $I/M^t$ is a power of $M/M^t$, and so $I$ is a power of $M$.

(ii) $\Rightarrow$ (iii) Since $\mathrm{vdim}_{R/M} M/M^2 \geq 1$ by 15.25, we have $M \supset M^2$: let $a \in M \setminus M^2$. By assumption, $Ra = M^n$ for some $n \in \mathbf{N}$. Since $a \notin M^2$, we must have $n = 1$, and so $M = Ra$. Since every non-zero ideal of $R$ is a power of $M$, it now follows that each non-zero ideal of $R$ has the form $Ra^h$ for some $h \in \mathbf{N}_0$.

(iii) $\Rightarrow$ (iv) This is clear.

(iv) $\Rightarrow$ (v) Use 13.17 in conjunction with the fact (3.39) that a PID is a UFD.

(v) $\Rightarrow$ (i) Let $a \in M \setminus \{0\}$. By 15.17 again, $Ra$ is $M$-primary and contains a power of $M$: let $t$ be the least $i \in \mathbf{N}$ such that $M^i \subseteq Ra$. Then $M^{t-1} \not\subseteq Ra$: let $b \in M^{t-1} \setminus Ra$.

Let $K$ denote the field of fractions of $R$. Set $c := a/b \in K$, and note that $c^{-1} = b/a \in K \setminus R$ (since $b \notin Ra$). Thus, by assumption, $c^{-1}$ is not integral over $R$. Note that

$$c^{-1}M := \left\{ c^{-1}r : r \in M \right\}$$

is an $R$-submodule of $K$ and, moreover, is actually contained in $R$ since $bM \subseteq M^t \subseteq Ra$. Thus $c^{-1}M$ is, in fact, an ideal of $R$. Our strategy is to show that this ideal is actually $R$.

Suppose that this is not the case, so that $c^{-1}M \subseteq M$. This means that the finitely generated $R$-module $M$ is closed under multiplication by elements of the subring $R[c^{-1}]$ of $K$, and so has a natural structure as $R[c^{-1}]$-module. Since $R$ is a domain, $M$ is a faithful $R[c^{-1}]$-module, and so it follows from 13.20 that $c^{-1}$ is integral over $R$. This contradiction shows that $c^{-1}M = R$.

It follows that $M = Rc$ is principal, so that $R$ is regular. $\square$

**15.43 EXERCISE.** Let $R$ be a Noetherian integral domain of dimension 1. Prove that the following statements are equivalent:

(i) $R$ is integrally closed;

(ii) each non-zero proper ideal of $R$ can be uniquely (apart from the order of the factors) expressed as a product of prime ideals of $R$;

(iii) for every $P \in \mathrm{Spec}(R)$, the localization $R_P$ is a regular local ring.

A Noetherian integral domain of dimension 1 satisfying the above equivalent conditions is called a *Dedekind domain*.

**15.44 EXERCISE.** Let $(R, M)$ be a regular local ring of dimension 1. Note that $M^n \supset M^{n+1}$ for all $n \in \mathbf{N}_0$, by Nakayama's Lemma 8.24. For each $a \in R \setminus \{0\}$, there is, by 15.42, a unique $t \in \mathbf{N}_0$ such that $Ra = M^t$: define $v(a) = t$. Also, set $v(0) = \infty$.

Let $K$ denote the field of fractions of $R$. Show that $v$ can be uniquely extended to a function $\tilde{v} : K \to \mathbf{Z} \cup \{\infty\}$ such that

(i) $\tilde{v}(bc) = \tilde{v}(b) + \tilde{v}(c)$ for all $b, c \in K$, and

(ii) $\tilde{v}(b + c) \geq \min\{\tilde{v}(b), \tilde{v}(c)\}$ for all $b, c \in K$.

(Here, the natural conventions that $\infty + \infty = \infty$, that $\infty + n = \infty$ for all $n \in \mathbf{Z}$, and that $\infty > n$ for all $n \in \mathbf{Z}$ and $\infty \geq \infty$, are to be employed.)

Show also that $\{b \in K : \tilde{v}(b) \geq 0\} = R$.

**15.45 FURTHER STEPS.** In 15.44, and with the notation of that exercise, the mapping $\tilde{v} : K \to \mathbf{Z} \cup \{\infty\}$ is an example of a 'discrete valuation' (on $K$); in fact, another name for a regular local ring of dimension 1 is 'discrete valuation ring'. These rings form a subclass of the class of valuation rings, which were alluded to in 13.45. Once again, we shall have to leave the interested reader to explore the details from other texts, such as [8, Chapter 4] or [1, Chapter 9].

Our last theorem in this book is Hilbert's Syzygy Theorem, which serves to give a small hint about the value of homological algebra as a tool in commutative algebra. We first provide one preliminary lemma.

**15.46 LEMMA.** *Let $G$ be a non-zero, finitely generated module over the local ring $(R, M)$, and let $s := \mathrm{vdim}_{R/M} G/MG$. By 9.3, $G$ can be generated by $s$ elements: suppose that it is generated by $g_1, \ldots, g_s$. Let $F$ be the free $R$-module of rank $s$ given by $F = \bigoplus_{i=1}^{s} R_i$, where $R_i = R$ for all $i = 1, \ldots, s$.*

*Let $\psi : F \to G$ be the $R$-epimorphism for which*

$$\psi((r_1, \ldots, r_s)) = \sum_{i=1}^{s} r_i g_i \quad \text{for all } (r_1, \ldots, r_s) \in F.$$

*Then* $\operatorname{Ker} \psi$ *is finitely generated and* $\operatorname{Ker} \psi \subseteq MF.$

*Proof.* By 7.21, the $R$-module $F$ is Noetherian, and so $\operatorname{Ker} \psi$ is finitely generated by 7.13.

Let $(r_1, \ldots, r_s) \in \operatorname{Ker} \psi$, so that $\sum_{i=1}^{s} r_i g_i = 0$. By 9.3, the $R/M$-space $G/MG$ is generated by $g_1 + MG, \ldots, g_s + MG$, and so these $s$ elements must form a linearly independent family over $R/M$. It follows that $r_i \in M$ for all $i = 1, \ldots, s$. Hence

$$(r_1, \ldots, r_s) = r_1(1, 0, \ldots, 0) + \cdots + r_s(0, \ldots, 0, 1) \in MF.$$

This completes the proof. $\square$

**15.47** HILBERT'S SYZYGY THEOREM. *Let $(R, M)$ be a regular local ring of dimension $d$, and let $G$ be a finitely generated $R$-module. Then there exists an exact sequence*

$$0 \longrightarrow F_d \xrightarrow{f_d} F_{d-1} \longrightarrow \cdots \longrightarrow F_i \xrightarrow{f_i} \cdots \longrightarrow F_1 \xrightarrow{f_1} F_0 \xrightarrow{f_0} G \longrightarrow 0$$

*of $R$-modules and $R$-homomorphisms in which $F_0, F_1, \ldots, F_d$ are all finitely generated free $R$-modules.*

*Note.* The sequence

$$0 \longrightarrow F_d \xrightarrow{f_d} F_{d-1} \longrightarrow \cdots \longrightarrow F_i \xrightarrow{f_i} \cdots \longrightarrow F_1 \xrightarrow{f_1} F_0 \xrightarrow{f_0} G \longrightarrow 0$$

is called a *finite free resolution of $G$ of length $d$*, and the theorem shows that $G$ has 'finite homological dimension'. Note that it is not claimed that all the $F_i$ are non-zero, and, indeed, some finitely generated $R$-modules will have finite free resolutions of shorter length if $d > 0$.

*Proof.* First note that, when $d = 0$, the regular local ring $R$ is a field, and the claim is an easy consequence of the standard theory of finite-dimensional vector spaces. Hence we can, and do, assume that $d > 0$.

Let $s_0 := \operatorname{vdim}_{R/M} G/MG$. Use 15.46 to construct a free $R$-module $F_0$ of rank $s_0$ and an $R$-epimorphism $f_0 : F_0 \to G$ for which $K_0 := \operatorname{Ker} f_0 \subseteq MF_0$. If $K_0 = 0$, take $F_1 = 0$ and $f_1 : F_1 \to F_0$ to be the zero homomorphism. If $K_0 \neq 0$, then apply 15.46 to the (finitely generated) $R$-module $K_0$ to obtain a free $R$-module $F_1$ of finite rank and an $R$-epimorphism

$\psi : F_1 \to K_0$ such that Ker $\psi \subseteq MF_1$; in this case, let $f_1 : F_1 \to F_0$ be the composition of $\psi$ and the inclusion homomorphism from $K_0$ into $F_0$. Note that $K_1 := \text{Ker } f_1 = \text{Ker } \psi \subseteq MF_1$.

Continue in this way: after $d+1$ steps, we arrive at an exact sequence

$$F_d \xrightarrow{f_d} F_{d-1} \longrightarrow \cdots \longrightarrow F_i \xrightarrow{f_i} \cdots \longrightarrow F_1 \xrightarrow{f_1} F_0 \xrightarrow{f_0} G \longrightarrow 0$$

of $R$-modules and $R$-homomorphisms in which $F_0, F_1, \ldots, F_d$ are all finitely generated free $R$-modules and $K_i := \text{Ker } f_i \subseteq MF_i$ for all $i = 0, \ldots, d$. Note that, in view of our construction and 15.46, each $F_i$ is actually a direct sum of finitely many copies of $R$.

It is enough for us to prove that $K_d = 0$, and this is what we shall do.

Let $\{u_1, \ldots, u_d\}$ be a regular system of parameters for $R$, and set $P_i = (u_1, \ldots, u_i)$ for all $i = 1, \ldots, d$ and $P_0 = 0$. By 15.38, $P_i \in \text{Spec}(R)$ for all $i = 0, \ldots, d$. We aim next to show that

$$K_i \cap P_j F_i = P_j K_i \quad \text{for all } i, j \in \mathbb{N}_0 \text{ with } 0 \leq j \leq i \leq d.$$

Of course, $K_i \cap P_j F_i \supseteq P_j K_i$ for such $i$ and $j$, and it is the opposite inclusion which needs some work. We prove that by induction on $j$: there is nothing to prove when $j = 0$.

Thus we suppose, inductively, that $0 < j \leq d$ and we have proved that $K_i \cap P_{j-1} F_i \subseteq P_{j-1} K_i$ for all $i \in \mathbb{N}_0$ with $j - 1 \leq i \leq d$. Now consider an integer $i$ with $j \leq i \leq d$, and let $g \in K_i \cap P_j F_i$. If $F_i = 0$, then $g \in P_j K_i$; thus we assume $F_i \neq 0$. Thus $g = \sum_{k=1}^{j} u_k m_k$ for some $m_1, \ldots, m_j \in F_i$, and $f_i(g) = 0$. Hence $0 = f_i(g) = \sum_{k=1}^{j} u_k f_i(m_k)$, so that, in $F_{i-1}$,

$$u_j f_i(m_j) = - \sum_{k=1}^{j-1} u_k f_i(m_k).$$

Thus, bearing in mind that $F_{i-1}$ is a direct sum of finitely many copies of $R$, we see that all the components of $u_j f_i(m_j)$ belong to $P_{j-1}$. But $P_{j-1}$ is prime and $u_j \notin P_{j-1}$ by 15.38; thus, all the components of $f_i(m_j)$ belong to $P_{j-1}$. Hence, by the inductive hypothesis and the fact that $K_{i-1} = \text{Im } f_i$, we deduce that

$$f_i(m_j) \in P_{j-1} F_{i-1} \cap K_{i-1} \subseteq P_{j-1} K_{i-1} = P_{j-1} \text{Im } f_i.$$

Thus there exist $z_1, \ldots, z_{j-1} \in F_i$ such that

$$f_i(m_j) = \sum_{k=1}^{j-1} u_k f_i(z_k).$$

Now let $z_j = m_j - \sum_{k=1}^{j-1} u_k z_k$ $(\in F_i)$. Then $f_i(z_j) = 0$, and so $z_j \in K_i$. Therefore $g - u_j z_j \in K_i$. But

$$g - u_j z_j = \sum_{k=1}^{j} u_k m_k - u_j \left( m_j - \sum_{k=1}^{j-1} u_k z_k \right) = \sum_{k=1}^{j-1} u_k (m_k + u_j z_k)$$
$$\in K_i \cap P_{j-1} F_i \subseteq P_{j-1} K_i$$

in view of the inductive hypothesis. Hence $g \in P_j K_i$, as required. We have thus shown that $K_i \cap P_j F_i \subseteq P_j K_i$, and so the inductive step is complete.

We have thus proved that

$$K_i \cap P_j F_i = P_j K_i \qquad \text{for all } i, j \in \mathbf{N}_0 \text{ with } 0 \le j \le i \le d.$$

In particular, $K_d \cap P_d F_d = P_d K_d$. But $P_d = M$ and $K_d \subseteq M F_d$. Hence $K_d = M K_d$, and so $K_d = 0$ by Nakayama's Lemma 8.24. Thus the proof is complete. $\square$

**15.48** EXERCISE. Let $M$ be a finitely generated module over the principal ideal domain $R$. Show that there exists an exact sequence

$$0 \longrightarrow F_1 \longrightarrow F_0 \longrightarrow M \longrightarrow 0$$

of $R$-modules and $R$-homomorphisms in which $F_1$ and $F_0$ are finitely generated free $R$-modules, that is, there exists a finite free resolution of $M$ of length 1.

**15.49** EXERCISE. Let $p, q$ be irreducible elements in a UFD $R$, and suppose that $p$ and $q$ are not associates. Show that the sequence

$$0 \longrightarrow R \xrightarrow{f_2} R \oplus R \xrightarrow{f_1} R \xrightarrow{f_0} R/(p, q) \longrightarrow 0,$$

in which $f_0$ is the natural epimorphism and the homomorphisms $f_1$ and $f_2$ are defined by $f_1((r, s)) = rp + sq$ for all $r, s \in R$ and $f_2(r) = (-rq, rp)$ for all $r \in R$, is exact.

**15.50** EXERCISE. Let $R$ be a Dedekind domain and let $I, J, K$ be ideals of $R$. Show that

(i) $I \cap (J + K) = (I \cap J) + (I \cap K)$;

(i) $I + (J \cap K) = (I + J) \cap (I + K)$.

**15.51** EXERCISE. Let $(n \in \mathbf{N}$ and) $R_1, \ldots, R_n$ be Artinian local rings; let $R$ denote the direct product ring $R_1 \times \cdots \times R_n$. Show that

(i) for $i = 1, \ldots, n$, the ring $R_i$ is isomorphic to a localization of $R$;

(ii) if, for each $i = 1, \ldots, n$, every ideal of $R_i$ is principal, then every ideal of $R$ is principal.

**15.52** EXERCISE. Let $R$ be a Dedekind domain.

(i) Let $I$ be a non-zero, proper ideal of $R$. Show that every ideal of $R/I$ is principal. (You might find 15.42 and 15.51 helpful.)

(ii) Show that each ideal of $R$ can be generated by 2 (or fewer) elements.

**15.53** FURTHER STEPS. It is hoped that Hilbert's Syzygy Theorem will tempt the reader to take the advice, offered several times in this book, to study homological algebra for use as a tool in commutative algebra. It can, in fact, be shown that the converse of 15.47 is true (that is, that if $(R, M)$ is a local ring with the property that every non-zero finitely generated $R$-module has a finite free resolution, then $R$ is regular), so that one arrives at a 'homological' characterization of regular local rings. This leads to an elegant proof of the fact that a localization of a regular local ring is again regular, and this is regarded as a spectacular achievement of homological algebra. The reader can find accounts of these ideas in, for example, [8, Chapter 7] and [11, Chapter 9] (but he or she will need to know some homological algebra to understand the proofs!).

Although we have now reached the end of our road as far as this book is concerned, it is hoped that some of the comments under 'Further Steps' in the various chapters will have tempted the reader to take his or her studies of commutative algebra beyond this point: there are many avenues leading onwards, there is much to enjoy, and there are several good books to enjoy it from. Have fun!

# Bibliography

[1] M. F. ATIYAH and I. G. MACDONALD, *Introduction to commutative algebra* (Addison-Wesley, Reading, Massachusetts, 1969).

[2] E. T. COPSON, *Metric spaces,* Cambridge Tracts in Mathematics and Mathematical Physics 57 (Cambridge University Press, 1968).

[3] P. R. HALMOS, *Naive set theory* (Van Nostrand Reinhold, New York, 1960).

[4] B. HARTLEY and T. O. HAWKES, *Rings, modules and linear algebra* (Chapman and Hall, London, 1970).

[5] I. KAPLANSKY, *Commutative rings* (Allyn and Bacon, Boston, 1970).

[6] E. KUNZ, *Introduction to commutative algebra and algebraic geometry* (Birkhäuser, Basel, 1985).

[7] H. MATSUMURA, *Commutative algebra* (Benjamin/Cummings, Reading, Massachusetts, 1980).

[8] H. MATSUMURA, *Commutative ring theory* (Cambridge University Press, 1986).

[9] M. NAGATA, *Local rings* (Interscience, New York, 1962).

[10] D. G. NORTHCOTT, *Ideal theory,* Cambridge Tracts in Mathematics and Mathematical Physics 42 (Cambridge University Press, 1953).

[11] D. G. NORTHCOTT, *An introduction to homological algebra* (Cambridge University Press, 1960).

[12] D. G. NORTHCOTT, *Lessons on rings, modules and multiplicities* (Cambridge University Press, 1968).

[13] M. REID, *Undergraduate algebraic geometry*, London Mathematical Society Student Texts 12 (Cambridge University Press, 1988).

[14] J. J. ROTMAN, *Notes on homological algebra* (Van Nostrand Reinhold, New York, 1970).

[15] D. SHARPE, *Rings and factorization* (Cambridge University Press, 1987).

[16] I. STEWART, *Galois theory* (Chapman and Hall, London, 1973).

[17] I. STEWART and D. TALL, *Complex analysis* (Cambridge University Press, 1983).

[18] O. ZARISKI and P. SAMUEL, *Commutative Algebra*, Vol. I, Graduate Texts in Mathematics 28 (Springer, Berlin, 1975).

[19] O. ZARISKI and P. SAMUEL, *Commutative Algebra*, Vol. II, Graduate Texts in Mathematics 29 (Springer, Berlin, 1975).

# Index